I0055018

Introduction to Digital Signal Processing

Introduction to Digital Signal Processing

Edited by
Leon Beach

Larsen & Keller
www.larsen-keller.com

Introduction to Digital Signal Processing
Edited by Leon Beach
ISBN: 978-1-63549-087-9 (Hardback)

© 2017 Larsen & Keller

Larsen & Keller
Published by Larsen and Keller Education,
5 Penn Plaza,
19th Floor,
New York, NY 10001, USA

Cataloging-in-Publication Data

Introduction to digital signal processing / edited by Leon Beach.
p. cm.
Includes bibliographical references and index.
ISBN 978-1-63549-087-9
1. Signal processing--Digital techniques. 2. Digital communications. I. Beach, Leon.
TK5102.9 .I58 2017
621.382 2--dc23

This book contains information obtained from authentic and highly regarded sources. All chapters are published with permission under the Creative Commons Attribution Share Alike License or equivalent. A wide variety of references are listed. Permissions and sources are indicated; for detailed attributions, please refer to the permissions page. Reasonable efforts have been made to publish reliable data and information, but the authors, editors and publisher cannot assume any responsibility for the vailidity of all materials or the consequences of their use.

Trademark Notice: All trademarks used herein are the property of their respective owners. The use of any trademark in this text does not vest in the author or publisher any trademark ownership rights in such trademarks, nor does the use of such trademarks imply any affiliation with or endorsement of this book by such owners.

The publisher's policy is to use permanent paper from mills that operate a sustainable forestry policy. Furthermore, the publisher ensures that the text paper and cover boards used have met acceptable environmental accreditation standards.

Printed and bound in the United States of America.

For more information regarding Larsen and Keller Education and its products, please visit the publisher's website www.larsen-keller.com

Table of Contents

Preface

Digital signal processing refers to the science and practice of using digital processes to perform operations related to signal processing. It is closely related to analog signal processing. The fundamental applications of this field are in biomedical engineering, digital image processing, sensor array processing, statistical signal processing, speech signal processing, radar, seismic data processing, etc. This book elucidates new techniques and their applications in a multidisciplinary approach. Such selected concepts that redefine this field have been presented in it. Coherent flow of topics, student-friendly language and extensive use of examples make this textbook an invaluable source of knowledge for the students.

A detailed account of the significant topics covered in this book is provided below:

Chapter 1- Digital signal processing is the usage of digital processing. The applications of digital signal processing include audio and speech signal processing, sonar, radar, digital image processing, among others. This chapter will provide an integrated understanding of digital signal processing.

Chapter 2- The section on digital data offers an insightful focus, keeping in mind the complex subject matter. Some of the topics related to digital data are digital media, digital video, digital audio, digital photography etc. The following section will not only provide an overview, it will also delve deep into the topics related to it.

Chapter 3- Linear time-invariant theory is derived from applied mathematics. Some of the applications of linear time-invariant theory are circuits, signal processing, control theory and other technical areas. The other theories and concepts that have been explained in this text are delta modulation, sample rate conversion, quantization and discretization. This text helps the reader in developing a better understanding of the main concepts related to digital signal processing.

Chapter 4- Digital signal processing has a number of techniques and methods; some of these are Z-transform, advanced Z-transform, matched Z- transform method, Zak transform, etc. The topics elaborated in this chapter will help in gaining a better perspective about the techniques and methods of digital signal processing.

Chapter 5- A digital signal processor is a microprocessor that is used in the operational needs of digital signal processing. The aim of digital signal processors is to measure and compress analog signals. Image processors, media processors and video scalers are also explained in the section.

Chapter 6- Sampling is the reduction of continuous signal to a discrete signal. Sampling can be classified into undersampling, oversampling, upsampling and decimation. This chapter has been carefully written to provide an easy understanding of the varied facets of sampling.

Chapter 7- A filter is a device that is used in the process of removing unwanted components from a signal. Filters can be linear, casual, analog or digital, discrete time or passive or active type etc. The topics discussed in this text are adaptive filters, digital filters, recursive least squares filters, finite impulsive responses, infinite impulse responses etc. The major components of filters are discussed in this chapter.

Chapter 8- Devices related to digital signal processing are analog-to-digital converter, digital-to-analog converter, time-to-digital converter and reconstruction filter. Analog-to-digital converters are systems that help in transforming an analog signal to a digital signal whereas a digital-to-analog convertor is a device that converts a digital signal into an analog signal. This chapter helps discusses in detail the devices that are related to digital signal processing.

Chapter 9- This text provides a plethora of the allied fields of digital signal processing for better comprehension. Audio signal processing, digital image processing and statistical signal processing are discussed in this chapter. Audio signal processing is the modification of audio signals with the help of audio effect whereas digital image processing is the use of computer algorithms to implement image processing on digital images. This section is a compilation of various fields of digital signal processing of the broader subject matter.

It gives me an immense pleasure to thank our entire team for their efforts. Finally in the end, I would like to thank my family and colleagues who have been a great source of inspiration and support.

Editor

1

Introduction to Digital Signal Processing

Digital signal processing is the usage of digital processing. The applications of digital signal processing include audio and speech signal processing, sonar, radar, digital image processing, among others. This chapter will provide an integrated understanding of digital signal processing.

Digital Signal Processing

Digital signal processing (DSP) is the use of digital processing, such as by computers, to perform a wide variety of signal processing operations. The signals processed in this manner are a sequence of numbers that represent samples of a continuous variable in a domain such as time, space, or frequency.

Digital signal processing and analog signal processing are subfields of signal processing. DSP applications include audio and speech signal processing, sonar, radar and other sensor array processing, spectral estimation, statistical signal processing, digital image processing, signal processing for telecommunications, control of systems, biomedical engineering, seismic data processing, among others.

Digital signal processing can involve linear or nonlinear operations. Nonlinear signal processing is closely related to nonlinear system identification and can be implemented in the time, frequency, and spatio-temporal domains.

The application of digital computation to signal processing allows for many advantages over analog processing in many applications, such as error detection and correction in transmission as well as data compression. DSP is applicable to both streaming data and static (stored) data.

Signal Sampling

The increasing use of computers has resulted in the increased use of, and need for, digital signal processing. To digitally analyze and manipulate an analog signal, it must be digitized with an analog-to-digital converter. Sampling is usually carried out in two stages, discretization and quantization. Discretization means that the signal is divided into equal intervals of time, and each interval is represented by a single measurement of amplitude. Quantization means each amplitude measurement is approximated by a value from a finite set. Rounding real numbers to integers is an example.

The Nyquist–Shannon sampling theorem states that a signal can be exactly reconstructed from its samples if the sampling frequency is greater than twice the highest frequency of the signal, but this requires an infinite number of samples. In practice, the sampling frequency is often significantly higher than twice that required by the signal's limited bandwidth.

Theoretical DSP analyses and derivations are typically performed on discrete-time signal models with no amplitude inaccuracies (quantization error), "created" by the abstract process of sampling. Numerical methods require a quantized signal, such as those produced by an analog-to-digital converter (ADC). The processed result might be a frequency spectrum or a set of statistics. But often it is another quantized signal that is converted back to analog form by a digital-to-analog converter (DAC).

Domains

In DSP, engineers usually study digital signals in one of the following domains: time domain (one-dimensional signals), spatial domain (multidimensional signals), frequency domain, and wavelet domains. They choose the domain in which to process a signal by making an informed assumption (or by trying different possibilities) as to which domain best represents the essential characteristics of the signal. A sequence of samples from a measuring device produces a temporal or spatial domain representation, whereas a discrete Fourier transform produces the frequency domain information, that is, the frequency spectrum.

Time and Space Domains

The most common processing approach in the time or space domain is enhancement of the input signal through a method called filtering. Digital filtering generally consists of some linear transformation of a number of surrounding samples around the current sample of the input or output signal. There are various ways to characterize filters; for example:

- A "linear" filter is a linear transformation of input samples; other filters are "non-linear". Linear filters satisfy the superposition condition, i.e. if an input is a weighted linear combination of different signals, the output is a similarly weighted linear combination of the corresponding output signals.
- A "causal" filter uses only previous samples of the input or output signals; while a "non-causal" filter uses future input samples. A non-causal filter can usually be changed into a causal filter by adding a delay to it.

- A "time-invariant" filter has constant properties over time; other filters such as adaptive filters change in time.
- A "stable" filter produces an output that converges to a constant value with time, or remains bounded within a finite interval. An "unstable" filter can produce an output that grows without bounds, with bounded or even zero input.
- A "finite impulse response" (FIR) filter uses only the input signals, while an "infinite impulse response" filter (IIR) uses both the input signal and previous samples of the output signal. FIR filters are always stable, while IIR filters may be unstable.

A filter can be represented by a block diagram, which can then be used to derive a sample processing algorithm to implement the filter with hardware instructions. A filter may also be described as a difference equation, a collection of zeroes and poles or, if it is an FIR filter, an impulse response or step response.

The output of a linear digital filter to any given input may be calculated by convolving the input signal with the impulse response.

Frequency Domain

Signals are converted from time or space domain to the frequency domain usually through the Fourier transform. The Fourier transform converts the signal information to a magnitude and phase component of each frequency. Often the Fourier transform is converted to the power spectrum, which is the magnitude of each frequency component squared.

The most common purpose for analysis of signals in the frequency domain is analysis of signal properties. The engineer can study the spectrum to determine which frequencies are present in the input signal and which are missing.

In addition to frequency information, phase information is often needed. This can be obtained from the Fourier transform. With some applications, how the phase varies with frequency can be a significant consideration.

Filtering, particularly in non-realtime work can also be achieved by converting to the frequency domain, applying the filter and then converting back to the time domain. This is a fast, O(n log n) operation, and can give essentially any filter shape including excellent approximations to brickwall filters.

There are some commonly used frequency domain transformations. For example, the cepstrum converts a signal to the frequency domain through Fourier transform, takes the logarithm, then applies another Fourier transform. This emphasizes the harmonic structure of the original spectrum.

Frequency domain analysis is also called *spectrum-* or *spectral analysis*.

Z-plane Analysis

Filters for analog processing have what's called an *infinite impulse response* (IIR). Digital filters come in both IIR and FIR types. FIR filters have many advantages, but are computationally more demanding. Whereas FIR filters are always stable, IIR filters have feedback loops that may resonate when stimulated with certain input signals. The Z-transform provides a tool for analyzing potential stability issues of digital IIR filters. It is analogous to the Laplace transform, which is used to design analog IIR filters.

Wavelet

An example of the 2D discrete wavelet transform that is used in JPEG2000. The original image is high-pass filtered, yielding the three large images, each describing local changes in brightness (details) in the original image. It is then low-pass filtered and downscaled, yielding an approximation image; this image is high-pass filtered to produce the three smaller detail images, and low-pass filtered to produce the final approximation image in the upper-left.

In numerical analysis and functional analysis, a discrete wavelet transform (DWT) is any wavelet transform for which the wavelets are discretely sampled. As with other wavelet transforms, a key advantage it has over Fourier transforms is temporal resolution: it captures both frequency *and* location information (location in time).

Applications

The main applications of DSP are audio signal processing, audio compression, digital image processing, video compression, speech processing, speech recognition, digital communications, digital synthesizers, radar, sonar, financial signal processing, seismology and biomedicine. Specific examples are speech compression and transmission in digital mobile phones, room correction of sound in hi-fi and sound reinforcement applications, weather forecasting, economic forecasting, seismic data processing, analysis and control of industrial processes, medical imaging such as CAT scans and MRI, MP3 compression, computer graphics, image manipulation, hi-fi loudspeaker crossovers and equalization, and audio effects for use with electric guitar amplifiers.

Implementation

DSP algorithms have long been run on general-purpose computers and digital signal processors. DSP algorithms are also implemented on purpose-built hardware such as application-specific integrated circuit (ASICs). Additional technologies for digital signal processing include more powerful general purpose microprocessors, field-programmable gate arrays (FPGAs), digital signal controllers (mostly for industrial applications such as motor control), and stream processors.

Depending on the requirements of the application, digital signal processing tasks can be implemented on general purpose computers.

Often when the processing requirement is not real-time, processing is economically done with an existing general-purpose computer and the signal data (either input or output) exists in data files. This is essentially no different from any other data processing, except DSP mathematical techniques (such as the FFT) are used, and the sampled data is usually assumed to be uniformly sampled in time or space. For example: processing digital photographs with software such as *Photoshop*.

However, when the application requirement is real-time, DSP is often implemented using specialized microprocessors such as the DSP56000, the TMS320, or the SHARC. These often process data using fixed-point arithmetic, though some more powerful versions use floating point. For faster applications FPGAs might be used. Beginning in 2007, multicore implementations of DSPs have started to emerge from companies including Freescale and Stream Processors, Inc. For faster applications with vast usage, ASICs might be designed specifically. For slow applications, a traditional slower processor such as a microcontroller may be adequate. Also a growing number of DSP applications are now being implemented on embedded systems using powerful PCs with multi-core processors.

Digital Signal (Signal Processing)

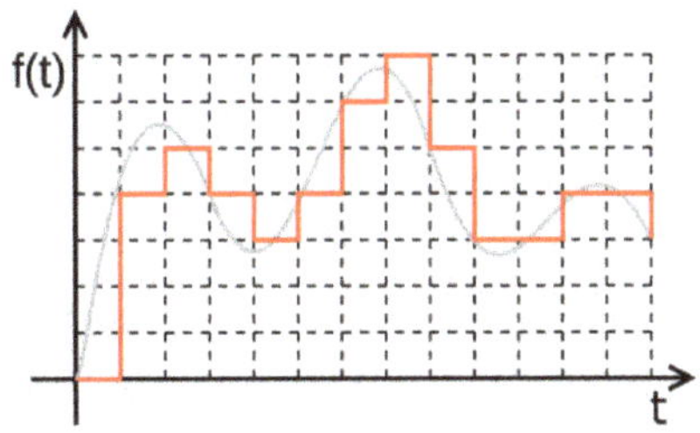

Digital signal (red) is the sampled and rounded representation of the grey analog signal

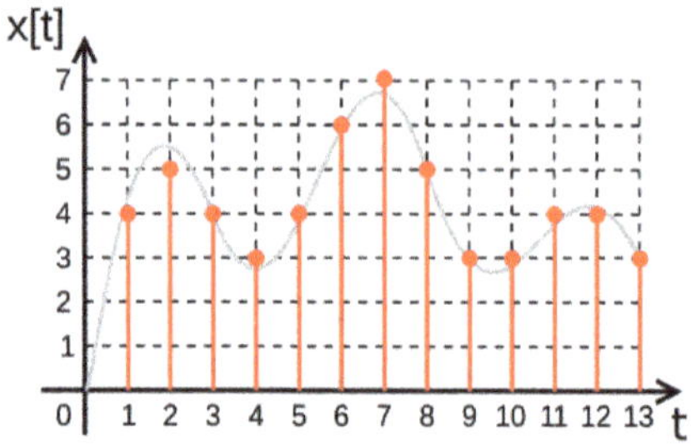

A digital signal (red) that is produced by sampling may be considered discrete in time as well as by value, and is equivalent to a series of numbers, 4,5,4,3,4,6, etc.

In the context of digital signal processing (DSP), a digital signal is a discrete-time signal for which not only the time but also the amplitude has discrete values; in other words, its samples take on only values from a discrete set (a countable set that can be mapped one-to-one to a subset of integers). If that discrete set is finite, the discrete values can be represented with digital words of a finite width. Most commonly, these discrete values are represented as fixed-point words (either proportional to the waveform values or companded) or floating-point words.

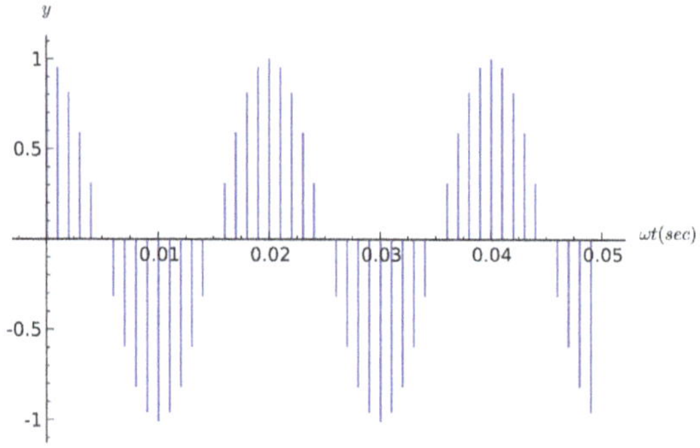

Discrete cosine waveform with frequency of 50 Hz and a sampling rate of 1000 samples/sec, easily satisfying the sampling theorem for reconstruction of the original cosine function from samples.

The process of analog-to-digital conversion produces a digital signal. The conversion process can be thought of as occurring in two steps:

1. sampling, which produces a continuous-valued discrete-time signal, and
2. quantization, which replaces each sample value by an approximation selected from a given discrete set (for example by truncating or rounding).

It can be shown that for signal frequencies strictly below the Nyquist limit that the original continuous-valued continuous-time signal can be almost perfectly reconstructed, down to the (often very low) limit set by the quantisation.

Common practical digital signals are represented as 8-bit (256 levels), 16-bit (65,536 levels), 24-bit (16.8 million levels) and 32-bit (4.3 billion levels). But the number of quantization levels is not necessarily limited to powers of two.

References

- Billings, Stephen A. (Sep 2013). Nonlinear System Identification: NARMAX Methods in the Time, Frequency, and Spatio-Temporal Domains. UK: Wiley. ISBN 978-1-119-94359-4.
- Broesch, James D.; Stranneby, Dag; Walker, William (2008-10-20). Digital Signal Processing: Instant access (1 ed.). Butterworth-Heinemann - Newnes. p. 3. ISBN 9780750689762.
- Stranneby, Dag; Walker, William (2004). Digital Signal Processing and Applications (2nd ed.). Elsevier. ISBN 0-7506-6344-8.
- Smith, Steven W. (2002-11-06). “3”. Digital Signal Processing: A Practical Guide for Engineers and Scientists. Demystifying Technology. 1 (1 ed.). Newnes. pp. 35–39. ISBN 075067444X.
- Harris, Frederic J. (2004-05-24). “1.1”. Multirate Signal Processing for Communication Systems. Upper Saddle River, NJ: Prentice Hall PTR. p. 2. ISBN 0131465112.
- Vaseghi, Saeed V. (2009-03-02). “1.4”. Advanced Digital Signal Processing and Noise Reduction (4 ed.). Chichester, West Suffix, United Kingdom: John Wiley & Sons. p. 23. ISBN 0470754060.
- Diniz, Paulo S. R.; Eduardo A. B. da Silva; Sergio L. Netto (2010-09-13). “1.1”. Digital Signal Processing: System Analysis and Design (2 ed.). New York & UK: Cambridge University Press. p. 5. ISBN 0521887755.
- Manolakis, Dimitris G.; Vinay K. Ingle (2011-11-21). “1.1.1”. Applied Digital Signal Processing: Theory and Practice. Cambridge, UK: Cambridge University Press. p. 5. ISBN 0521110025.
- Ingle, Vinay K.; John G. Proakis (2011-01-01). “1.1”. Digital Signal Processing Using MATLAB (3 ed.). Stamford, CT: CL Engineering. p. 3. ISBN 1111427372.

2

Digital Data: An Overview

The section on digital data offers an insightful focus, keeping in mind the complex subject matter. Some of the topics related to digital data are digital media, digital video, digital audio, digital photography etc. The following section will not only provide an overview, it will also delve deep into the topics related to it.

Digital Data

Digital data, in information theory and information systems, are discrete, discontinuous representations of information or works, as contrasted with continuous, or analog signals which behave in a continuous manner, or represent information using a continuous function.

Although digital representations are the subject matter of discrete mathematics, the information represented can be either discrete, such as numbers and letters, or it can be continuous, such as sounds, images, and other measurements.

The term is most commonly used in computing and electronics, especially where real-world information is converted to binary numeric form as in digital audio and digital photography.

Symbol to Digital Conversion

Since symbols (for example, alphanumeric characters) are not continuous, representing symbols digitally is rather simpler than conversion of continuous or analog information to digital. Instead of sampling and quantization as in analog-to-digital conversion, such techniques as polling and encoding are used.

A symbol input device usually consists of a group of switches that are polled at regular intervals to see which switches are switched. Data will be lost if, within a single polling interval, two switches are pressed, or a switch is pressed, released, and pressed again. This polling can be done by a specialized processor in the device to prevent burdening the main CPU. When a new symbol has been entered, the device typically sends an interrupt, in a specialized format, so that the CPU can read it.

For devices with only a few switches (such as the buttons on a joystick), the status of each can be encoded as bits (usually 0 for released and 1 for pressed) in a single word.

This is useful when combinations of key presses are meaningful, and is sometimes used for passing the status of modifier keys on a keyboard (such as shift and control). But it does not scale to support more keys than the number of bits in a single byte or word.

Devices with many switches (such as a computer keyboard) usually arrange these switches in a scan matrix, with the individual switches on the intersections of x and y lines. When a switch is pressed, it connects the corresponding x and y lines together. Polling (often called scanning in this case) is done by activating each x line in sequence and detecting which y lines then have a signal, thus which keys are pressed. When the keyboard processor detects that a key has changed state, it sends a signal to the CPU indicating the scan code of the key and its new state. The symbol is then encoded, or converted into a number, based on the status of modifier keys and the desired character encoding.

A custom encoding can be used for a specific application with no loss of data. However, using a standard encoding such as ASCII is problematic if a symbol such as 'ß' needs to be converted but is not in the standard.

It is estimated that in the year 1986 less than 1% of the world's technological capacity to store information was digital and in 2007 it was already 94%. The year 2002 is assumed to be the year when human kind was able to store more information in digital than in analog format (the "beginning of the digital age").

Properties of Digital Information

All digital information possesses common properties that distinguish it from analog data with respect to communications:

- Synchronization: Since digital information is conveyed by the sequence in which symbols are ordered, all digital schemes have some method for determining the beginning of a sequence. In written or spoken human languages synchronization is typically provided by pauses (spaces), capitalization, and punctuation. Machine communications typically use special synchronization sequences.
- Language: All digital communications require a *formal language*, which in this context consists of all the information that the sender and receiver of the digital communication must both possess, in advance, in order for the communication to be successful. Languages are generally arbitrary and specify the meaning to be assigned to particular symbol sequences, the allowed range of values, methods to be used for synchronization, etc.
- Errors: Disturbances (noise) in analog communications invariably introduce some, generally small deviation or error between the intended and actual communication. Disturbances in a digital communication do not result in errors unless the disturbance is so large as to result in a symbol being misinterpret-

ed as another symbol or disturb the sequence of symbols. It is therefore generally possible to have an entirely error-free digital communication. Further, techniques such as check codes may be used to detect errors and guarantee error-free communications through redundancy or retransmission. Errors in digital communications can take the form of *substitution errors* in which a symbol is replaced by another symbol, or *insertion/deletion* errors in which an extra incorrect symbol is inserted into or deleted from a digital message. Uncorrected errors in digital communications have unpredictable and generally large impact on the information content of the communication.

- Copying: Because of the inevitable presence of noise, making many successive copies of an analog communication is infeasible because each generation increases the noise. Because digital communications are generally error-free, copies of copies can be made indefinitely.
- Granularity: The digital representation of a continuously variable analog value typically involves a selection of the number of symbols to be assigned to that value. The number of symbols determines the precision or resolution of the resulting datum. The difference between the actual analog value and the digital representation is known as *quantization error*. For example, if the actual temperature is 23.234456544453 degrees, but if only two digits (23) are assigned to this parameter in a particular digital representation, the quantizing error is: 0.234456544453. This property of digital communication is known as *granularity*.
- Compressible: According to Miller, "Uncompressed digital data is very large, and in its raw form would actually produce a larger signal (therefore be more difficult to transfer) than analog data. However, digital data can be compressed. Compression reduces the amount of bandwidth space needed to send information. Data can be compressed, sent and then decompressed at the site of consumption. This makes it possible to send much more information and result in, for example, digital television signals offering more room on the airwave spectrum for more television channels."

Historical Digital Systems

Even though digital signals are generally associated with the binary electronic digital systems used in modern electronics and computing, digital systems are actually ancient, and need not be binary or electronic.

- Written text (due to the limited character set and the use of discrete symbols - the alphabet in most cases)
- The *abacus* was created sometime between 1000 BC and 500 BC, it later became a form of calculation frequency. Nowadays it can be used as a very ad-

vanced, yet basic digital calculator that uses beads on rows to represent numbers. Beads only have meaning in discrete up and down states, not in analog in-between states.

- A *beacon* is perhaps the simplest non-electronic digital signal, with just two states (on and off). In particular, *smoke signals* are one of the oldest examples of a digital signal, where an analog "carrier" (smoke) is modulated with a blanket to generate a digital signal (puffs) that conveys information.
- Morse code uses six digital states—dot, dash, intra-character gap (between each dot or dash), short gap (between each letter), medium gap (between words), and long gap (between sentences)—to send messages via a variety of potential carriers such as electricity or light, for example using an electrical telegraph or a flashing light.
- The Braille system was the first binary format for character encoding, using a six-bit code rendered as dot patterns.
- Flag semaphore uses rods or flags held in particular positions to send messages to the receiver watching them some distance away.
- International maritime signal flags have distinctive markings that represent letters of the alphabet to allow ships to send messages to each other.
- More recently invented, a modem modulates an analog "carrier" signal (such as sound) to encode binary electrical digital information, as a series of binary digital sound pulses. A slightly earlier, surprisingly reliable version of the same concept was to bundle a sequence of audio digital "signal" and "no signal" information (i.e. "sound" and "silence") on magnetic cassette tape for use with early home computers.

Digital Media

Hard drives store information in binary form and so are considered a type of physical digital media.

Digital media are any media that are encoded in a machine-readable format. Digital media can be created, viewed, distributed, modified and preserved on digital electronics devices. Computer programs and software; digital imagery, digital video; video games; web pages and websites, including social media; data and databases; digital audio, such as mp3s; and e-books are examples of digital media. Digital media are frequently contrasted with print media, such as printed books, newspapers and magazines, and other traditional or analog media, such as pictures, film or audio tape.

Combined with the Internet and personal computing, digital media has caused disruption in publishing, journalism, entertainment, education, commerce and politics. Digital media has also posed new challenges to copyright and intellectual property laws, fostering an open content movement in which content creators voluntarily give up some or all of their legal rights to their work. The ubiquity of digital media and its effects on society suggest that we are at the start of a new era in industrial history, called the Information Age, perhaps leading to a paperless society in which all media are produced and consumed on computers. However, challenges to a digital transition remain, including outdated copyright laws, censorship, the digital divide, and the specter of a digital dark age, in which older media becomes inaccessible to new or upgraded information systems. Digital media has a significant, wide-ranging and complex impact on society and culture.

History

Before Electronics

Analog computers, such as Babbage's Difference Engine, use physical, i.e. tangible, parts and actions to control operations

Machine-readable media predates the Internet, modern computers and electronics. Machine-readable codes and information were first conceptualized by Charles Babbage in the early 1800s. Babbage imagined that these codes would provide instructions for his Difference Engine and Analytical Engine, machines he designed to solve the problem of error in calculations. Between 1822 and 1823, Ada Lovelace, a mathematician, wrote the first instructions for calculating numbers on Babbage's engines. Lovelace's instructions are now believed to be the first computer program.

Though the machines were designed to perform analytical tasks, Lovelace anticipated the potential social impact of computers and programming, writing, "For, in so distributing and combining the truths and the formulae of analysis, that they may become most easily and rapidly amenable to the mechanical combinations of the engine, the relations and the nature of many subjects in that science are necessarily thrown into new lights, and more profoundly investigated... there are in all extensions of human power, or additions to human knowledge, various collateral influences, besides the main and primary object attained." Other early machine-readable media include the instructions for player pianos and jacquard looms.

Digital Computers

01010111 01101001 01101011
01101001 01110000 01100101
01100100 01101001 01100001

Digital codes, like binary, can be changed without reconfiguring mechanical parts

Though they used machine-readable media, Babbage's engines, player pianos, jacquard looms and many other early calculating machines were themselves analog computers, with physical, mechanical parts. The first truly digital media came into existence with the rise of digital computers. Digital computers use binary code and Boolean logic to store and process information, allowing one machine in one configuration to perform many different tasks. The first modern, programmable, digital computers, the Manchester Mark 1 and the EDSAC, were independently invented between 1948 and 1949. Though different in many ways from modern computers, these machines had digital software controlling their logical operations. They were encoded in binary, a system of ones and zeroes that are combined to make hundreds of characters. The 1s and 0s of binary are the "digits" of digital media.

"As We May Think"

While digital media came into common use in the early 1950s, the *conceptual* foundation of digital media is traced to the work of scientist and engineer Vannevar Bush and his celebrated essay "As We May Think," published in the *Atlantic Monthly* in 1945. Bush envisioned a system of devices that could be used to help scientists, doctors, historians and others, store, analyze and communicate information. Calling this then-imaginary device a "memex", Bush wrote:

The owner of the memex, let us say, is interested in the origin and properties of the bow and arrow. Specifically he is studying why the short Turkish bow was apparently superior to the English long bow in the skirmishes of the Crusades. He has dozens of possibly pertinent books and articles in his memex. First he runs through an encyclopedia, finds an interesting but sketchy article, leaves it projected. Next, in a history, he finds another pertinent item, and ties the two together. Thus he goes, building a trail of many items. Occasionally he inserts a comment of his own, either linking it into the main trail or joining it by a side trail to a particular item. When it becomes evident that the elastic properties of available materials had a great deal to do with the bow, he branches off on a side trail which takes him through textbooks on elasticity and tables of physical constants. He inserts a page of longhand analysis of his own. Thus he builds a trail of his interest through the maze of materials available to him.

Bush hoped that the creation of this memex would be the work of scientists after World War II. Though the essay predated digital computers by several years, "As We May Think," anticipated the potential social and intellectual benefits of digital media and provided the conceptual framework for digital scholarship, the World Wide Web, wikis and even social media. It was recognized as a significant work even at the time of its publication.

Impact

The Digital Revolution

In the years since the invention of the first digital computers, computing power and storage capacity have increased exponentially. Personal computers and smartphones put the ability to access, modify, store and share digital media in the hands of billions of people. Many electronic devices, from digital cameras to drones have the ability to create, transmit and view digital media. Combined with the World Wide Web and the Internet, digital media has transformed 21st century society in a way that is frequently compared to the cultural, economic and social impact of the printing press. The change has been so rapid and so widespread that it has launched an economic transition from an industrial economy to an information-based economy, creating a new period in human history known as the Information Age or the digital revolution.

The transition has created some uncertainty about definitions. Digital media, new media, multimedia, and similar terms all have a relationship to both the engineering innovations and cultural impact of digital media. The blending of digital media with other media, and with cultural and social factors, is sometimes known as new media or "the new media." Similarly, digital media seems to demand a new set of communications skills, called transliteracy, media literacy, or digital literacy. These skills include not only the ability to read and write—traditional literacy—but the ability to navigate the Internet, evaluate sources , and create digital content. The idea that we are moving toward a fully digital, paperless society is accompanied by the fear that we may soon—or currently—be facing a digital dark age, in which older media are no longer accessible on modern devices or using modern methods of scholarship. Digital media has a significant, wide-ranging and complex effect on society and culture.

Disruption in Industry

Compared with print media, the mass media, and other analog technologies, digital media are easy to copy, store, share and modify. This quality of digital media has led to significant changes in many industries, especially journalism, publishing, education, entertainment, and the music business. The overall effect of these changes is so far-reaching that it is difficult to quantify. For example, in movie-making, the transition from analog film cameras to digital cameras is nearly complete. The transition has economic benefits to Hollywood, making distribution easier and making it possible to add high-quality digital effects to films. At the same time, it has affected the analog special effects, stunt, and animation industries in Hollywood. It has imposed painful costs on small movie theaters, some of which did not or will not survive the transition to digital. The effect of digital media on other media industries is similarly sweeping and complex.

In journalism, digital media and citizen journalism have led to the loss of thousands of jobs in print media and the bankruptcy of many major newspapers. But the rise of digital journalism has also created thousands of new jobs and specializations. E-books and self-publishing are changing the book industry, and digital textbooks and other media-inclusive curricula are changing primary and secondary education. In academia, digital media has led to a new form of scholarship, called digital scholarship, and new fields of study, such as digital humanities and digital history. It has changed the way libraries are used and their role in society. Every major media, communications and academic endeavor is facing a period of transition and uncertainty related to digital media.

Individual as Content Creator

Digital media has also allowed individuals to be much more active in content creation. Anyone with access to computers and the Internet can participate in social media and contribute their own writing, art, videos, photography and commentary to the Internet, as well as conduct business online. This has come to be known as citizen journalism. This spike in user created content is due to the development of the internet as well as the way in which users interact with media today. The release of technologies such mobile devices allow for easier and quicker access to all things media. Many media production tools that were once only available to a few are now free and easy to use. The cost of devices that can access the internet is dropping steadily, and now personal ownership of multiple digital devices is becoming standard. These elements have significantly affected political participation. Digital media is seen by many scholars as having a role in Arab Spring, and crackdowns on the use of digital and social media by embattled governments are increasingly common. Many governments restrict access to digital media in some way, either to prevent obscenity or in a broader form of political censorship.

User-generated content raises issues of privacy, credibility, civility and compensation for cultural, intellectual and artistic contributions. The spread of digital media, and the wide range of literacy and communications skills necessary to use it effectively, have deepened the digital divide between those who have access to digital media and those who don't.

The rising of digital media has made the consumer's audio collection more precise and personalized. It is no longer necessary to purchase an entire album if the consumer is ultimately interested in only a few audio files.

Web only News

As the internet becomes more and more prevalent, more companies are beginning to distribute content through internet only means. Prime time audiences have dropped 23% for News Corp, the worlds largest broadcasting channel. With the loss of viewers there is a loss of revenue but not as bad as what would be expected. While the dollar amount dropped roughly 2%, overall cable revenue was up about 5% which is slower growth than what was expected. Cisco Inc released its latest forecast and the numbers are all trending to internet news to continue to grow at a rate where it will be quadruple by 2018.

As of 2012, the worlds largest internet only media company, The Young Turks, are averaging 750,000 users per day, and are continuing to grow, currently having over 2 billion views across all The Young Turks controlled channels, which covers world news, sports, movie reviews, college focused content and a round table style discussion channel.

Copyright Challenges

Digital media pose many challenges to current copyright and intellectual property laws. The ease of creating, modifying and sharing digital media makes copyright enforcement a challenge, and copyright laws are widely seen as outdated. For example, under current copyright law, common Internet memes are probably illegal to share in many countries. Legal rights are at least unclear for many common Internet activities, such as posting a picture that belongs to someone else to a social media account, covering a popular song on a YouTube video, or writing fanfiction. Over the last decade the concept of fair use has been applied to many online medias.

To resolve some of these issues, content creators can voluntarily adopt open or copyleft licenses, giving up some of their legal rights, or they can release their work to the public domain. Among the most common open licenses are Creative Commons licenses and the GNU Free Documentation License, both of which are in use on Wikipedia. Open licenses are part of a broader open content movement that pushes for the reduction or removal of copyright restrictions from software, data and other digital media.

Additional software has been developed in order to protect digital media. digital rights management (DRM) is used to digitally copyright material and allows users to use that media for specific cases. For example, DRM allows a movie producer to rent a movie at a lower price than selling the movie, restricting the movie rental license length, rather than only selling the movie at full price. Additionally, DRM can prevent unauthorized sharing or modification of media.

Digital Media is numerical, networked and interactive system of links and databases that allows us to navigate from one bit of content or webpage to another.

One form of Digital media that is becoming a phenomenon is in the form of a digital magazine. What exactly is a digital magazine? Due to the economic importance of digital magazines, the Audit Bureau of Circulations integrated the definition of this medium in its latest report (March 2011): a digital magazine involves the distribution of a magazine content by electronic means; it may be a replica. This is the out dated definition of what a digital magazine is. A digital magazine should not be, in fact, a replica of the print magazine in PDF, as was common practice in recent years. It should, rather, be a magazine that is, in essence, interactive and created from scratch to a digital platform (Internet, mobile phones, private networks, iPad or other device). The barriers for digital magazine distribution are thus decreasing. At the same time digitizing platforms are broadening the scope of where digital magazines can be published, such as within websites and on smartphones. With the improvements of tablets and digital magazines are becoming visually enticing and readable magazines with it graphic arts.

Digital Video

Digital video is a representation of moving visual images in the form of encoded digital data. This is in contrast to analog video, which represents moving visual images with analog signals. Digital video comprises a series digital images displayed in rapid succession. In contrast, one of the key analog video methods, motion picture film, uses a series of photographs which are projected in rapid succession. Standard film stocks such as 16 mm and 35 mm record at 24 frames per second. For video, there are two frame rate standards: NTSC, at about 30 frames per second, and PAL at 25 frames per second.

Digital video was first introduced commercially in 1986 with the Sony D1 format, which recorded an uncompressed standard definition component video signal in digital form instead of the high-band analog forms that had been commonplace until then.

Digital video can be copied with no degradation in quality. In contrast, when analog sources are copied, they experience generation loss. Digital video can also be stored on hard disks or streamed over the Internet to end users who watch content on a desktop computer screen or a digital Smart TV. In everyday practice, digital video content such as TV shows and movies also includes a digital audio soundtrack.

History

Starting in the late 1970s to the early 1980s, several types of video production equipment that were digital in their internal workings were introduced, such as time base correctors (TBC) and digital video effects (DVE) units. They operated by taking a standard analog composite video input and digitizing it internally. This made it easier to either correct or enhance the video signal, as in the case of a TBC, or to manipulate and add effects to the video, in the case of a DVE unit. The digitized and processed video information was then converted back to standard analog video for output.

Later on in the 1970s, manufacturers of professional video broadcast equipment, such as Bosch (through their Fernseh division), RCA, and Ampex developed prototype digital videotape recorders (VTR) in their research and development labs. Bosch's machine used a modified 1" Type B transport, and recorded an early form of CCIR 601 digital video. Ampex's prototype digital video recorder used a modified 2" Quadruplex VTR (an Ampex AVR-3), but fitted with custom digital video electronics, and a special "octaplex" 8-head headwheel (regular analog 2" Quad machines only used 4 heads). The audio on Ampex's prototype digital machine, nicknamed by its developers as "Annie", still recorded the audio in analog as linear tracks on the tape, like 2" Quad. None of these machines from these manufacturers were ever marketed commercially, however.

Digital video was first introduced commercially in 1986 with the Sony D1 format, which recorded an uncompressed standard definition component video signal in digital form instead of the high-band analog forms that had been commonplace until then. Due to its expense, and the requirement of component video connections using 3 cables (such as YPbPr or RGB component video) to and from a D1 VTR that most television facilities were not wired for (composite NTSC or PAL video using one cable was the norm for most of them at that time), D1 was used primarily by large television networks and other component-video capable video studios.

In 1988, Sony and Ampex co-developed and released the D2 digital videocassette format, which recorded video digitally without compression in ITU-601 format, much like D1. But D2 had the major difference of encoding the video in composite form to the NTSC standard, thereby only requiring single-cable composite video connections to and from a D2 VCR, making it a perfect fit for the majority of television facilities at the time. This made D2 quite a successful format in the television broadcast industry throughout the late '80s and the '90s. D2 was also widely used in that era as the master tape format for mastering laserdiscs (prior to D2, most laserdiscs were mastered using analog 1" Type C videotape).

D1 & D2 would eventually be replaced by cheaper systems using video compression, most notably Sony's Digital Betacam (still heavily used as an electronic field production (EFP) recording format by professional television producers) that were introduced into the network's television studios. Other examples of digital video formats utilizing compression were Ampex's DCT (the first to employ such when introduced in 1992), the industry-standard DV and MiniDV (and its professional variations, Sony's DVCAM and Panasonic's DVCPRO), and Betacam SX, a lower-cost variant of Digital Betacam using MPEG-2 compression.

One of the first digital video products to run on personal computers was *PACo: The PICS Animation Compiler* from The Company of Science & Art in Providence, RI, which was developed starting in 1990 and first shipped in May 1991. PACo could stream unlimited-length video with synchronized sound from a single file (with the ".CAV" file extension) on CD-ROM. Creation required a Mac; playback was possible on Macs, PCs, and Sun Sparcstations. In 1992, Bernard Luskin, Philips Interactive Media, and Eric

Doctorow, Paramount Worldwide Video, successfully put the first fifty videos in digital MPEG 1 on CD, developed the packaging and launched movies on CD, leading to advancing versions of MPEG, and to DVD.

QuickTime, Apple Computer's architecture for time-based and streaming data formats appeared in June, 1991. Initial consumer-level content creation tools were crude, requiring an analog video source to be digitized to a computer-readable format. While low-quality at first, consumer digital video increased rapidly in quality, first with the introduction of playback standards such as MPEG-1 and MPEG-2 (adopted for use in television transmission and DVD media), and then the introduction of the DV tape format allowing recordings in the format to be transferred direct to digital video files (containing the same video data recorded on the transferred DV tape) on an editing computer and simplifying the editing process, allowing non-linear editing systems (NLE) to be deployed cheaply and widely on desktop computers with no external playback/recording equipment needed, save for the computer simply requiring a FireWire port to interface to the DV-format camera or VCR. The widespread adoption of digital video has also drastically reduced the bandwidth needed for a high-definition video signal (with HDV and AVCHD, as well as several commercial variants such as DVCPRO-HD, all using less bandwidth than a standard definition analog signal) and tapeless camcorders based on flash memory and often a variant of MPEG-4.

Overview

Digital video comprises a series of orthogonal bitmap digital images displayed in rapid succession at a constant rate. In the context of video these images are called frames. We measure the rate at which frames are displayed in frames per second (FPS). Since every frame is an orthogonal bitmap digital image it comprises a raster of pixels. If it has a width of W pixels and a height of H pixels we say that the frame size is W*x*H. Pixels have only one property, their color. The color of a pixel is represented by a fixed number of bits. The more bits the more subtle variations of colors can be reproduced. This is called the color depth (CD) of the video.

An example video can have a duration (T) of 1 hour (3600*sec*), a frame size of 640x480 *(WxH)* at a color depth of 24*bits* and a frame rate of 25*fps*. This example video has the following properties:

- pixels per frame = 640 * 480 = 307,200
- bits per frame = 307,200 * 24 = 7,372,800 = 7.37*Mbits*
- bit rate (BR) = 7.37 * 25 = 184.25*Mbits/sec*
- video size (VS) = 184*Mbits/sec* * 3600*sec* = 662,400*Mbits* = 82,800*Mbytes* = 82.8*Gbytes*

The most important properties are *bit rate* and *video size*. The formulas relating those two with all other properties are:

BR = W * H * CD * FPS

VS = BR * T = W * H * CD * FPS * T

(units are: BR in bit/s, W and H in pixels, CD in bits, VS in bits, T in seconds)

while some secondary formulas are:

pixels_per_frame = W * H

pixels_per_second = W * H * FPS

bits_per_frame = W * H * CD

Interlacing

In interlaced video each *frame* is composed of two *halves of an image*. The first half contains only the odd-numbered lines of a full frame. The second half contains only the even-numbered lines. Those halves are referred to individually as *fields*. Two consecutive fields compose a full frame. If an interlaced video has a frame rate of 15 frames per second the field rate is 30 fields per second. All the properties and formulas discussed here apply equally to interlaced video but one should be careful not to confuse the fields per second rate with the frames per second rate.

Properties of Compressed Video

The above are accurate for uncompressed video. Because of the relatively high bit rate of uncompressed video, video compression is extensively used. In the case of compressed video each frame requires a small percentage of the original bits. Assuming a compression algorithm that shrinks the input data by a factor of CF, the bit rate and video size would equal to:

BR = W * H * CD * FPS / CF

VS = BR * T / CF

Note that it is not necessary that all frames are equally compressed by a factor of CF. In practice they are not, so CF is the *average* factor of compression for *all* the frames taken together.

The above equation for the bit rate can be rewritten by combining the compression factor and the color depth like this:

BR = W * H * (CD / CF) * FPS

The value (CD / CF) represents the average bits per pixel (BPP). As an example, if we have a color depth of 12bits/pixel and an algorithm that compresses at 40x, then BPP equals 0.3 (12/40). So in the case of compressed video the formula for bit rate is:

$$BR = W * H * BPP * FPS$$

The same formula is valid for uncompressed video because in that case one can assume that the "compression" factor is 1 and that the average bits per pixel equal the color depth.

Bit Rate and BPP

As is obvious by its definition bit rate is a measure of the rate of information content of the digital video stream. In the case of uncompressed video, bit rate corresponds directly to the quality of the video (remember that bit rate is proportional to every property that affects the video quality). Bit rate is an important property when transmitting video because the transmission link must be capable of supporting that bit rate. Bit rate is also important when dealing with the storage of video because, as shown above, the video size is proportional to the bit rate and the duration. Bit rate of uncompressed video is too high for most practical applications. Video compression is used to greatly reduce the bit rate. BPP is a measure of the efficiency of compression. A true-color video with no compression at all may have a BPP of 24 bits/pixel. Chroma subsampling can reduce the BPP to 16 or 12 bits/pixel. Applying jpeg compression on every frame can reduce the BPP to 8 or even 1 bits/pixel. Applying video compression algorithms like MPEG1, MPEG2 or MPEG4 allows for fractional BPP values.

Constant Bit Rate Versus Variable Bit Rate

As noted above BPP represents the *average* bits per pixel. There are compression algorithms that keep the BPP almost constant throughout the entire duration of the video. In this case we also get video output with a constant bit rate (CBR). This CBR video is suitable for real-time, non-buffered, fixed bandwidth video streaming (e.g. in videoconferencing). Noting that not all frames can be compressed at the same level because quality is more severely impacted for scenes of high complexity some algorithms try to constantly adjust the BPP. They keep it high while compressing complex scenes and low for less demanding scenes. This way one gets the best quality at the smallest average bit rate (and the smallest file size accordingly). Of course when using this method the bit rate is variable because it tracks the variations of the BPP.

Technical Overview

Standard film stocks such as 16 mm and 35 mm record at 24 frames per second. For video, there are two frame rate standards: NTSC, which shoot at 30/1.001 (about 29.97) frames per second or 59.94 fields per second, and PAL, 25 frames per second or 50 fields per second. Digital video cameras come in two different image capture formats: interlaced and deinterlaced / progressive scan. Interlaced cameras record the image in alternating sets of lines: the odd-numbered lines are scanned, and then the even-numbered lines are scanned, then the odd-numbered lines are scanned again, and so on. One set of odd or even lines is referred to as a "field", and a consecutive pairing of two fields of opposite parity is called a *frame*. Deinterlaced cameras records each frame as distinct, with all scan lines being captured at the same moment in time. Thus, interlaced video captures samples the scene motion twice as often as progressive video does, for the same number of frames per second. Progressive-scan camcorders generally produce a slightly sharper image. However, motion may not be as smooth as interlaced video which uses 50 or 59.94 fields per second, particularly if they employ the 24 frames per second standard of film.

Digital video can be copied with no degradation in quality. No matter how many generations of a digital source is copied, it will still be as clear as the original first generation of digital footage. However a change in parameters like frame size as well as a change of the digital format can decrease the quality of the video due to new calculations that have to be made. Digital video can be manipulated and edited to follow an order or sequence on an NLE, or non-linear editing workstation, a computer-based device intended to edit video and audio. More and more, videos are edited on readily available, increasingly affordable consumer-grade computer hardware and software. However, such editing systems require ample disk space for video footage. The many video formats and parameters to be set make it quite impossible to come up with a specific number for how many minutes need how much time.

Digital video has a significantly lower cost than 35 mm film. In comparison to the high cost of film stock, the tape stock (or other electronic media used for digital video recording, such as flash memory or hard disk drive) used for recording digital video is very inexpensive. Digital video also allows footage to be viewed on location without the expensive chemical processing required by film. Also physical deliveries of tapes and broadcasts do not apply anymore. Digital television (including higher quality HDTV) started to spread in most developed countries in early 2000s. Digital video is also used in modern mobile phones and video conferencing systems. Digital video is also used for Internet distribution of media, including streaming video and peer-to-peer movie distribution. However even within Europe are lots of TV-Stations not broadcasting in HD, due to restricted budgets for new equipment for processing HD.

Many types of video compression exist for serving digital video over the internet and on optical disks. The file sizes of digital video used for professional editing are generally not practical for these purposes, and the video requires further compression with codecs such as Sorenson, H.264 and more recently Apple ProRes especially for HD. Probably the most widely used formats for delivering video over the internet are MPEG4, Quicktime, Flash and Windows Media, while MPEG2 is used almost exclusively for DVDs, providing an exceptional image in minimal size but resulting in a high level of CPU consumption to decompress.

As of 2011, the highest resolution demonstrated for digital video generation is 35 megapixels (8192 x 4320). The highest speed is attained in industrial and scientific high speed cameras that are capable of filming 1024x1024 video at up to 1 million frames per second for brief periods of recording.

Interfaces and Cables

Many interfaces have been designed specifically to handle the requirements of uncompressed digital video (from roughly 400 Mbit/s to 10 Gbit/s):

- High-Definition Multimedia Interface
- Digital Visual Interface
- Serial Digital Interface
- DisplayPort
- Digital component video
- Unified Display Interface
- FireWire
- USB

The following interface has been designed for carrying MPEG-Transport compressed video:

- DVB-ASI

Compressed video is also carried using UDP-IP over Ethernet. Two approaches exist for this:

- Using RTP as a wrapper for video packets
- 1-7 MPEG Transport Packets are placed directly in the UDP packet

Storage Formats

Encoding

All current formats, which are listed below, are PCM based.

- CCIR 601 used for broadcast stations
- MPEG-4 good for online distribution of large videos and video recorded to flash memory
- MPEG-2 used for DVDs, Super-VCDs, and many broadcast television formats
- MPEG-1 used for video CDs

- H.261
- H.263
- H.264 also known as *MPEG-4 Part 10*, or as *AVC*, used for Blu-ray Discs and some broadcast television formats

Tapes

- Betacam SX, Betacam IMX, Digital Betacam, or DigiBeta – Commercial video systems by Sony, based on original Betamax technology
- D-VHS – MPEG-2 format data recorded on a tape similar to S-VHS
- D1, D2, D3, D5, D9 (also known as Digital-S) – various SMPTE commercial digital video standards
- Digital8 – DV-format data recorded on Hi8-compatible cassettes; largely a consumer format
- DV, MiniDV – used in most of today's videotape-based consumer camcorders; designed for high quality and easy editing; can also record high-definition data (HDV) in MPEG-2 format
- DVCAM, DVCPRO – used in professional broadcast operations; similar to DV but generally considered more robust; though DV-compatible, these formats have better audio handling.
- DVCPRO50, DVCPROHD support higher bandwidths as compared to Panasonic's DVCPRO.
- HDCAM was introduced by Sony as a high-definition alternative to DigiBeta.
- MicroMV – MPEG-2-format data recorded on a very small, matchbook-sized cassette; obsolete
- ProHD – name used by JVC for its MPEG-2-based professional camcorders

Digital Audio

Audio levels display on a digital audio recorder (Zoom H4n)

Digital audio is technology that can be used to record, store, generate, manipulate, and reproduce sound using audio signals that have been encoded in digital form. Following significant advances in digital audio technology during the 1970s, it gradually replaced analog audio technology in many areas of sound production, sound recording (tape systems were replaced with digital recording systems), sound engineering and telecommunications in the 1990s and 2000s.

A microphone converts sound (a singer's voice or the sound of an instrument playing) to an analog electrical signal, then an analog-to-digital converter (ADC)—typically using pulse-code modulation—converts the analog signal into a digital signal. This digital signal can then be recorded, edited and modified using digital audio tools. When the sound engineer wishes to listen to the recording on headphones or loudspeakers (or when a consumer wishes to listen to a digital sound file of a song), a digital-to-analog converter performs the reverse process, converting a digital signal back into an analog signal, which analog circuits amplify and send to a loudspeaker.

Digital audio systems may include compression, storage, processing and transmission components. Conversion to a digital format allows convenient manipulation, storage, transmission and retrieval of an audio signal. Unlike analog audio, in which making copies of a recording leads to degradation of the signal quality, when using digital audio, an infinite number of copies can be made without any degradation of signal quality.

Overview

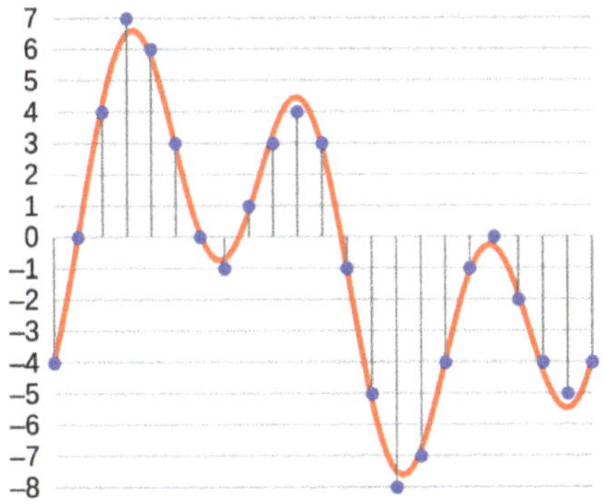

A sound wave, in red, represented digitally, in blue (after sampling and 4-bit quantization).

Digital audio technologies in the 2010s are used in the recording, manipulation, mass-production, and distribution of sound, including recordings of songs, instrumental pieces, podcasts, sound effects, and other sounds. Modern online music distribution depends on digital recording and data compression. The availability of music as data files, rather than as physical objects, has significantly reduced the costs of distribution. Before digital audio, the music industry distributed and sold music by selling physical copies of albums in the form of records, tapes, and then CDs. With digital audio and online distribution systems such as iTunes, companies sell digital sound files to consumers, which the consumer receives over the Internet. This digital audio/Internet distribution model is much less expensive than producing physical copies of recordings, packaging them and shipping them to stores.

An analog audio system captures sounds, and converts their physical waveforms into electrical representations of those waveforms by use of a transducer, such as a microphone. The sounds are then stored, as on tape, or transmitted. The process is reversed for playback: the audio signal is amplified and then converted back into physical waveforms via a loudspeaker. Analog audio retains its fundamental wave-like characteristics throughout its storage, transformation, duplication, and amplification.

Analog audio signals are susceptible to noise and distortion, due to the innate characteristics of electronic circuits and associated devices. Disturbances in a digital system do not result in error unless the disturbance is so large as to result in a symbol being misinterpreted as another symbol or disturb the sequence of symbols. It is therefore generally possible to have an entirely error-free digital audio system in which no noise or distortion is introduced between conversion to digital format, and conversion back to analog.

A digital audio signal may be encoded for correction of any errors that might occur in the storage or transmission of the signal, but this is not strictly part of the digital audio process. This technique, known as "channel coding", is essential for broadcast or recorded digital systems to maintain bit accuracy. The discrete time and level of the binary signal allow a decoder to recreate the analog signal upon replay. Eight to Fourteen Bit Modulation is a channel code used in the audio Compact Disc (CD).

Conversion Process

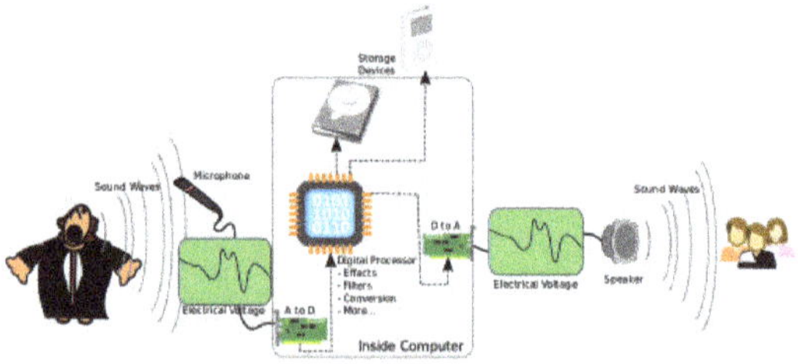

The lifecycle of sound from its source, through an ADC, digital processing, a DAC, and finally as sound again.

A digital audio system starts with an ADC that converts an analog signal to a digital signal. The ADC runs at a specified sampling rate and converts at a known bit resolution. CD audio, for example, has a sampling rate of 44.1 kHz (44,100 samples per second), and has 16-bit resolution for each stereo channel. Analog signals that have not already been bandlimited must be passed through an anti-aliasing filter before conversion, to prevent the distortion that is caused by audio signals with frequencies higher than the Nyquist frequency, which is half of the system's sampling rate.

A digital audio signal may be stored or transmitted. Digital audio can be stored on a CD, a digital audio player, a hard drive, a USB flash drive, or any other digital data storage device. The digital signal may then be altered through digital signal processing, where it may be filtered or have effects applied. Audio data compression techniques, such as MP3, Advanced Audio Coding, Ogg Vorbis, or FLAC, are commonly employed to reduce the file size. Digital audio can be streamed to other devices.

For playback, digital audio must be converted back to an analog signal with a DAC. DACs run at a specific sampling rate and bit resolution, but may use oversampling, upsampling or downsampling to convert signals that have been encoded with a different sampling rate.

History in Commercial Recording

Pulse-code modulation was invented by British scientist Alec Reeves in 1937 and was used in telecommunications applications long before its first use in commercial broadcast and recording. Commercial digital recording was pioneered in Japan by NHK and Nippon Columbia, also known as Denon, in the 1960s. The first commercial digital recordings were released in 1971.

The BBC also began to experiment with digital audio in the 1960s. By the early 1970s it had developed a 2-channel recorder, and in 1972 it deployed a digital audio transmission system that linked their broadcast center to their remote transmitters.

The first 16-bit PCM recording in the United States was made by Thomas Stockham at the Santa Fe Opera in 1976, on a Soundstream recorder. An improved version of the Soundstream system was used to produce several classical recordings by Telarc in 1978. The 3M digital multitrack recorder in development at the time was based on BBC technology. The first all-digital album recorded on this machine was Ry Cooder's *Bop till You Drop* in 1979. British record label Decca began development of its own 2-track digital audio recorders in 1978 and released the first European digital recording in 1979.

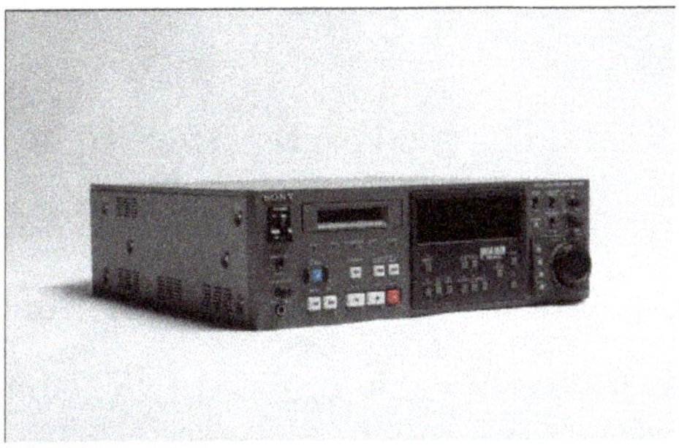

Sony digital audio recorder PCM-7030

Popular digital multitrack recorders produced by Sony and Mitsubishi in the early 1980s helped to bring about digital recording's acceptance by the major record companies. The 1982 introduction of the CD popularized digital audio with consumers.

Technologies

- Digital audio broadcasting
- Digital Audio Broadcasting (DAB)
- HD Radio
- Digital Radio Mondiale (DRM)
- In-band on-channel (IBOC)

Storage technologies

- Digital audio player
- Digital Audio Tape (DAT)
- Digital Compact Cassette (DCC)
- Compact Disc (CD)
- Hard disk recorder
- DVD-Audio
- MiniDisc
- Super Audio CD
- Blu-ray Disc (BD)
- Various audio file formats

Digital Audio Interfaces

Audio-specific interfaces include:

- AC'97 (Audio Codec 1997) interface between integrated circuits on PC motherboards
- Intel High Definition Audio - modern replacement for AC'97
- ADAT interface
- AES3 interface with XLR connectors, common in professional audio equipment
- S/PDIF - either over coaxial cable or TOSLINK, common in consumer audio equipment and derived from AES3
- AES47 - professional AES3-style digital audio over Asynchronous Transfer Mode networks
- I^2S (Inter-IC sound) interface between integrated circuits in consumer electronics

- MADI (Multichannel Audio Digital Interface)
- MIDI - low-bandwidth interconnect for carrying instrument data; cannot carry sound but can carry digital sample data in non-realtime
- TDIF, TASCAM proprietary format with D-sub cable
- A2DP via Bluetooth

Several interfaces are engineered to carry digital video and audio together, including HDMI and DisplayPort.

Any digital bus can carry digital audio. In professional architectural or installation applications, many digital audio Audio over Ethernet protocols and interfaces exist.

Digital Photography

Nikon D700 — a 12.1-megapixel full-frame DSLR

The Canon PowerShot A95

Digital photography is a form of photography that uses cameras containing arrays of electronic photodetectors to capture images focused by a lens, as opposed to an exposure on photographic film. The captured images are digitized and stored as a computer file ready for further digital processing, viewing, digital publishing or printing.

Until the advent of such technology, photographs were made by exposing light sensitive photographic film and paper, which were processed in liquid chemical solutions to develop and stabilize the image. Digital photographs are typically created solely by computer-based photoelectric and mechanical techniques, without wet bath chemical processing.

Digital photography is one of several forms of digital imaging. Digital images are also created by non-photographic equipment such as computer tomography scanners and radio telescopes. Digital images can also be made by scanning other printed photographic images or negatives.

The first consumer digital cameras were marketed in the late 1990s. Professionals gravitated to digital slowly, and were won over when their professional work required using digital files to fulfill the demands of employers and/or clients, for faster turn-around than conventional methods would allow. Starting around 2007, digital cameras were incorporated in cellphones and in the following years cellphone cameras became widespread, particularly due to their connectivity to social media websites and email. Since 2010, the digital point-and-shoot and DSLR formats have also seen competition from the mirrorless digital camera format, which typically provides better image quality than the point-and-shoot or cellphone formats but comes in a smaller size and shape than the typical DSLR. Many mirrorless cameras accept interchangeable lenses and have advanced features through an electronic viewfinder, which replaces the through-the-lens finder image of the SLR format.

The Digital Camera

History

The first flyby spacecraft image of Mars was taken from Mariner 4 on July 15, 1965 with a camera system designed by NASA/JPL. It used a video camera tube followed by a digitizer, rather than a mosaic of solid state sensor elements, so it was not what we usually define as a digital camera, but it produced a digital image that was stored on tape for later slow transmission back to earth.

The first recorded attempt at building a digital camera was in 1975 by Steven Sasson, an engineer at Eastman Kodak. It used the then-new solid-state CCD (charge-coupled device, a high-speed semiconductor) image sensor chips developed by Fairchild Semiconductor in 1973. The camera weighed 8 pounds (3.6 kg), recorded black and white images to a cassette tape, had a resolution of 0.01 megapixels (10,000 pixels), and took 23 seconds to capture its first image in December 1975. The prototype camera was a technical exercise, not intended for production.

The first true digital camera that recorded images as a computerized file was likely the Fuji DS-1P of 1988, which recorded to a 16 MB internal memory card that used a battery to keep the data in memory. This camera was never marketed internationally, and has not been confirmed to have shipped even in Japan.

The first commercially available digital camera was the 1990 Dycam Model 1; it also sold as the Logitech Fotoman. It used a CCD image sensor, stored pictures digitally, and connected directly to a computer for downloading images.

Sensors

Image sensors read the intensity of light, and digital memory devices store the digital image information as RGB color space or as raw data.

The two main types of sensors are charge-coupled devices (CCD), in which the photocharge is shifted to a central charge-to-voltage converter, and CMOS or active pixel sensors.

Multifunctionality and connectivity

Except for some linear array type of cameras at the highest-end and simple web cams at the lowest-end, a digital memory device (usually a memory card; floppy disks and CD-RWs are less common) is used for storing images, which may be transferred to a computer later.

Digital cameras can take pictures, and may also record sound and video. Some can be used as webcams, some can use the PictBridge standard to connect to a printer without using a computer, and some can display pictures directly on a television set. Similarly, many camcorders can take still photographs, and store them on videotape or on flash memorycards with the same functionality as digital cameras.

Digital photography is one of the most exceptional instances of the shift from converting conventional analog information to digital information. This shift is so tremendous because it was a chemical and mechanical process and became an all digital process with a built in computer in all digital cameras.

Performance Metrics

The quality of a digital image is a composite of various factors, many of which are similar to those of film cameras. Pixel count (typically listed in megapixels, millions of pixels) is only one of the major factors, though it is the most heavily marketed figure of merit. Digital camera manufacturers advertise this figure because consumers can use it to easily compare camera capabilities. It is not, however, the major factor in evaluating a digital camera for most applications. The processing system inside the camera that turns the raw data into a color-balanced and pleasing photograph is usually more critical, which is why some 4+ megapixel cameras perform better than higher-end cameras.

Image at left has a higher pixel count than the one to the right, but lower spatial resolution.

Resolution in pixels is not the only measure of image quality. A larger sensor with the same number of pixels generally produces a better image than a smaller one. One of the most important differences is an improvement in image noise. This is one of the advantages of digital SLR (single-lens reflex) cameras, which have larger sensors than simpler cameras (so-called point and shoot cameras) of the same resolution.

- Lens quality: resolution, distortion, dispersion.
- Capture medium: CMOS, CCD, negative film, reversal film etc.
- Capture format: pixel count, digital file type (RAW, TIFF, JPEG), film format (135 film, 120 film, 5x4, 10x8).
- Processing: digital and / or chemical processing of 'negative' and 'print'.

Pixel Counts

The number of pixels n for a given maximum resolution (w horizontal pixels by h vertical pixels) is the product $n = w \times h$. This yields e. g. 1.92 megapixels (1,920,000 pixels) for an image of 1600×1200. The majority of compact as well as some DSLR cameras have a 4:3 aspect ratio, i.e. $w/h = 4/3$. According to *Digital Photography Review*, the 4:3 ratio is because "computer monitors are 4:3 ratio, old CCDs always had a 4:3 ratio, and thus digital cameras inherited this aspect ratio."

The pixel count quoted by manufacturers can be misleading as it may not be the number of full-color pixels. For cameras using single-chip image sensors the number claimed is the total number of single-color-sensitive photosensors, whether they have different locations in the plane, as with the Bayer sensor, or in stacks of three co-located photosensors as in the Foveon X3 sensor. However, the images have different numbers of RGB pixels: Bayer-sensor cameras produce as many RGB pixels as photosensors via demosaicing (interpolation), while Foveon sensors produce uninterpolated image files with one-third as many RGB pixels as photosensors. Comparisons of megapixel ratings of these two types of sensors are sometimes a subject of dispute.

The relative increase in detail resulting from an increase in resolution is better compared by looking at the number of pixels across (or down) the picture, rather than the total number of pixels in the picture area. For example, a sensor of 2560 × 1600 sensor elements is described as "4 megapixels" (2560 × 1600 = 4,096,000). Increasing to 3200 × 2048 increases the pixels in the picture to 6,553,600 (6.5 megapixels), a factor of 1.6, but the pixels per cm in the picture (at the same image size) increases by only 1.25 times. A measure of the comparative increase in linear resolution is the square root of the increase in area resolution, i.e., megapixels in the entire image.

Dynamic Range

Practical imaging systems both digital and film, have a limited "dynamic range": the range of luminosity that can be reproduced accurately. Highlights of the subject that are too bright are rendered as white, with no detail; shadows that are too dark are rendered as black. The loss of detail is not abrupt with film, or in dark shadows with digital sensors: some detail is retained as brightness moves out of the dynamic range. "Highlight burn-out" of digital sensors, however, can be abrupt, and highlight detail may be lost. And as the sensor elements for different colors saturate in turn, there can be gross hue or saturation shift in burnt-out highlights.

Some digital cameras can show these blown highlights in the image review, allowing the photographer to re-shoot the picture with a modified exposure. Others compensate for the total contrast of a scene by selectively exposing darker pixels longer. A third technique is used by Fujifilm in its FinePix S3 Pro digital SLR. The image sensor contains additional photodiodes of lower sensitivity than the main ones; these retain detail in parts of the image too bright for the main sensor.

High dynamic range imaging (HDR) addresses this problem by increasing the dynamic range of images by either

- increasing the dynamic range of the image sensor or
- by using exposure bracketing and post-processing the separate images to create a single image with a higher dynamic range.

HDR images curtail burn-outs and black-outs.

Storage

Many camera phones and most digital cameras use memory cards having flash memory to store image data. The majority of cards for separate cameras are SD (Secure Digital) format; many are CompactFlash (CF) and the other formats are rare. XQD card format was the last new form of card, targeted at high-definition camcorders and high-resolution digital photo cameras. Most modern digital cameras also use internal memory for a limited capacity for pictures that can be transferred to or from the card or through the camera's connections; even without a memory card inserted into the camera.

Memory cards can hold vast numbers of photos, requiring attention only when the memory card is full. For most users, this means hundreds of quality photos stored on the same memory card. Images may be transferred to other media for archival or personal use. Cards with high speed and capacity are suited to video and burst mode (capture several photographs in a quick succession).

Because photographers rely on the integrity of image files, it is important to take proper care of memory cards. Common advocacy calls for formatting of the cards after transferring the images onto a computer. However, since all cameras only do quick formatting of cards, it is advisable to carry out a more thorough formatting using appropriate software on a PC once in a while. Effectively, this involves scanning of the cards to search for possible errors.

Market Impact

In late 2002, 2-megapixel cameras were available in the United States for less than $100, with some 1-megapixel cameras for under $60. At the same time, many discount stores with photo labs introduced a "digital front end", allowing consumers to obtain true chemical prints (as opposed to ink-jet prints) in an hour. These prices were similar to those of prints made from film negatives. However, because digital images have a different aspect ratio than 35 mm film images, people have started to realize that 4x6 inch prints crop some of the image off the print. Some photofinishers have started offering prints with the same aspect ratio as the digital cameras record.

In July 2003, digital cameras entered the disposable camera market with the release of the Ritz Dakota Digital, a 1.2-megapixel (1280 x 960) CMOS-based digital camera costing only $11 (USD). Following the familiar single-use concept long in use with film cameras, Ritz intended the Dakota Digital for single use. When the pre-programmed 25-picture limit is reached, the camera is returned to the store, and the consumer receives back prints and a CD-ROM with their photos. The camera is then refurbished and resold.

Since the introduction of the Dakota Digital, a number of similar single-use digital cameras have appeared. Most single-use digital cameras are nearly identical to the original Dakota Digital in specifications and function, though a few include superior specifications and more advanced functions (such as higher image resolutions and LCD screens). Most, if not all these single-use digital cameras cost less than $20 (USD), not including processing. However, the huge demand for complex digital cameras at competitive prices has often caused manufacturing shortcuts, evidenced by a large increase in customer complaints over camera malfunctions, high parts prices, and short service life. Some digital cameras offer only a 90-day warranty.

Since 2003, digital cameras have outsold film cameras. Prices of 35mm compact cameras have dropped with manufacturers further outsourcing to countries such as China. Kodak announced in January 2004 that they would no longer sell Kodak-branded film cameras in the developed world. In January 2006, Nikon followed suit and announced they would stop production of all but two models of their film cameras. They will continue to produce the low-end Nikon FM10, and the high-end Nikon F6. In the same month, Konica Minolta announced it was pulling out of the camera business altogether. The price of 35mm and APS (Advanced Photo System) compact cameras have dropped, probably due to direct competition from digital and the resulting growth of the offer of second-hand film cameras. Pentax have reduced production of film cameras but not halted it. The technology has improved so rapidly that one of Kodak's film cameras was discontinued before it was awarded a "camera of the year" award later in the year. The decline in film camera sales has also led to a decline in purchases of film for such cameras. In November 2004, a German division of Agfa-Gevaert, AgfaPhoto, split off. Within six months it filed for bankruptcy. Konica Minolta Photo Imaging, Inc. ended production of Color film and paper worldwide by March 31, 2007. In addition, by 2005, Kodak employed less than a third of the employees it had twenty years earlier. It is not known if these job losses in the film industry have been offset in the digital image industry. Digital cameras have decimated the film photography industry through declining use of the expensive film rolls and development chemicals previously required to develop the photos. This has had a dramatic effect on companies such as Fuji, Kodak, and Agfa. Many stores that formerly offered photofinishing services or sold film no longer do, or have seen a tremendous decline. In 2012, Kodak filed for bankruptcy after struggling to adapt to the changing industry.

In addition, digital photography has resulted in some positive market impacts as well. The increasing popularity of products such as digital photo frames and canvas prints is a direct result of the increasing popularity of digital photography.

Digital camera sales peaked in March 2012 averaging about 11 million units a month, but sales have declined significantly ever since. By March 2014, about 3 million were purchased each month, about 30 percent of the peak sales total. The decline may have bottomed out, with sales average hovering around 3 million a month. The main competitor is smartphones, most of which have built-in digital cameras, which routinely get better. They also offer the ability to record videos.

Social Impact

Until the advent of the digital camera, amateur photographers could either buy print or slide film for their cameras. Slides could be developed and shown to an audience using a slide projector. Digital photography revolutionized the industry by eliminating the delay and cost. The ease of viewing, transferring, editing and distributing allowed consumers to manage their digital photos with ordinary home computers rather than specialized equipment.

Camera phones, being the majority of cameras, have arguably the largest impact. The user can set their Smartphones to upload their products to the Internet, preserving them even if the camera is destroyed or the images deleted. Some high street photography shops have self-service kiosks that allow images to be printed directly from smartphones via Bluetooth technology.

Archivists and historians have noticed the transitory nature of digital media. Unlike film and print, which are tangible and immediately accessible to a person, digital image storage is ever-changing, with old media and decoding software becoming obsolete or inaccessible by new technologies. Historians are concerned that we are creating a historical void where information and details about an era would have been lost within either failed or inaccessible digital media. They recommend that professional and amateur users develop strategies for digital preservation by migrating stored digital images from old technologies to new. Scrapbookers who may have used film for creating artistic and personal memoirs may need to modify their approach to digital photo books to personalize them and retain the special qualities of traditional photo albums.

The web has been a popular medium for storing and sharing photos ever since the first photograph was published on the web by Tim Berners-Lee in 1992 (an image of the CERN house band Les Horribles Cernettes). Today photo sharing sites such as Flickr, Picasa and PhotoBucket, as well as social Web sites, are used by millions of people to share their pictures.

Recent Research and Innovation

Research and development continues to refine the lighting, optics, sensors, processing, storage, display, and software used in digital photography. Here are a few examples.

- 3D models can be created from collections of normal images. The resulting scene can be viewed from novel viewpoints, but creating the model is very computationally intensive. An example is Microsoft's Photosynth, which provides some models of famous places as examples.
- Panoramic photographs can be created directly in camera without the need for any external processing. Some cameras feature a 3D Panorama capability, combining shots taken with a single lens from different angles to create a sense of depth.
- High dynamic range cameras and displays are commercially available. Sensors with dynamic range in excess of 1,000,000:1 are in development, and software is also available to combine multiple non-HDR images (shot with different exposures) into an HDR image.
- Motion blur can be dramatically removed by a flutter shutter (a flickering shutter that adds a signature to the blur, which postprocessing recognizes). It is not yet commercially available.

- Advanced bokeh techniques use a hardware system of 2 sensors, one to take the photo as usual while the other records depth information. Bokeh effect and refocusing can then be applied to an image after the photo is taken.
- In advanced camera or camcorders, manipulating the sensitivity of the sensor not one, but 2 or more neutral density filters are available.
- An object's specular reflection can be captured using computer-controlled lights and sensors. This is needed to create attractive images of oil paintings, for instance. It is not yet commercially available, but some museums are starting to use it.
- Dust reduction systems help keep dust off of image sensors. Originally introduced only by a few cameras like Olympus DSLRs, have now become standard in most models and brands of detachable lens camera, except the low-end/ cheap ones.

Other areas of progress include improved sensors, more powerful software, advanced camera processors (sometimes using more than one processor, e.g., the Canon 7d camera has 2 Digic 4 processors), enlarged gamut displays, built in GPS & WiFi, and computer-controlled lighting.

Comparison with Film Photography

Advantages Already in Consumer Level Cameras

The primary advantage of consumer-level digital cameras is the low recurring cost, as users need not purchase photographic film. Processing costs may be reduced or even eliminated. Digicams tend also to be easier to carry and to use, than comparable film cameras. They more easily adapt to modern use of pictures. Some, particularly those that are smartphones, can send their pictures directly to e-mail or web pages or other electronic distribution.

Advantages of Professional Digital Cameras

The Golden Gate Bridge retouched for painterly light effects

- Immediate image review and deletion is possible; lighting and composition can be assessed immediately, which ultimately conserves storage space.
- High volume of images to medium ratio; allowing for extensive photography sessions without changing film rolls. To most users a single memory card is sufficient for the lifetime of the camera whereas film rolls are a re-incurring cost of film cameras.
- Faster workflow: Management (colour and file), manipulation and printing tools are more versatile than conventional film processes. However, batch processing of RAW files can be time consuming, even on a fast computer.
- Precision and reproducibility of processing: since processing in the digital domain is purely numerical, image processing using deterministic (non-random) algorithms is perfectly reproducible and eliminates variations common with photochemical processing that make many image processing techniques difficult if not impractical.
- Digital manipulation: A digital image can be modified and manipulated much easier and faster than with traditional negative and print methods. The digital image to the right was captured in Raw image format, processed and output in 3 different ways from the source RAW file, then merged and further processed for color saturation and other special effects to produce a more dramatic result than was originally captured with the RAW image.

Manufacturers such as Nikon and Canon have promoted the adoption of digital single-lens reflex cameras (DSLRs) by photojournalists. Images captured at 2+ megapixels are deemed of sufficient quality for small images in newspaper or magazine reproduction. Eight- to 24-megapixel images, found in modern digital SLRs, when combined with high-end lenses, can approximate the detail of film prints from 35 mm film based SLRs.

Disadvantages of Digital Cameras

- High ISO image noise may manifest as multicolored speckles in digital images, rather than the less-objectionable "grain" of high-ISO film. While this speckling can be removed by noise-reduction software, either in-camera or on a computer, this can have a detrimental effect on image quality as fine detail may be lost in the process.
- As with any sampled signal, the combination of regular (periodic) pixel structure of common electronic image sensors and regular (periodic) structure of (typically man-made) objects being photographed can cause objectionable aliasing artefacts, such as false colors when using cameras using a Bayer pattern sensor. Aliasing is also present in film, but typically manifests itself in less obvious ways (such as increased granularity) due to the stochastic grain structure (stochastic sampling) of film.

For many consumers, the advantages of digital cameras outweigh the disadvantages. Some professional photographers still prefer film. Concerns that have been raised by professional photographers include: editing and post-processing of RAW files can take longer than 35mm film, downloading a large number of images to a computer can be time-consuming, shooting in remote sites requires the photographer to carry a number of batteries, equipment failure—while all cameras may fail, some film camera problems (e.g., meter or rangefinder problems, failure of only some shutter speeds) can be worked around. As time passes, it is expected that more professional photographers will switch to digital.

Equivalent Features

Image noise / grain

Noise in a digital camera's image may sometimes be visually similar to film grain in a film camera.

Speed of use

Turn of the century digital cameras had a long start-up delay compared to film cameras, i.e., the delay from when they are turned on until they are ready to take the first shot, but this is no longer the case for modern digital cameras with start-up times under 1/4 seconds.

Frame rate

While some film cameras could reach up to 10 fps, like the Canon EOS-1V HS, professional digital SLR cameras can take still photographs at highest frame rates. While the Sony SLT technology allows rates of up to 12 fps, the Canon EOS-1Dx can take stills at a 14 fps rate. The Nikon F5 is limited to 36 continuous frames (the length of the film) while the Canon EOS-1D Mark III is able to take about 110 high definition JPEG images before its buffer must be cleared and the remaining space on the storage media can be used.

Image longevity

Depending on the materials and how they are stored, analog photographic film and prints may fade as they age. Similarly, the media on which digital images are stored or printed can decay or become corrupt, leading to a loss of image integrity.

Colour reproduction

Colour reproduction (gamut) is dependent on the type and quality of film or sensor used and the quality of the optical system and film processing. Different films and sensors have different color sensitivity; the photographer needs to understand his equipment, the light conditions, and the media used to ensure accurate colour reproduction. Many digital cameras offer RAW format (sensor data), which makes it possible to choose color space in the development stage regardless of camera settings.

Even in RAW format, however, the sensor and the camera's dynamics can only capture colors within the gamut supported by the hardware. When that image is transferred for reproduction on any device, the best possible gamut is the gamut that the end device supports. For a monitor, it is the gamut of the display device. For a photographic print, it is the gamut of the device that prints the image on a specific type of paper. Color gamut or Color space is an abstract term that describes an area where points of color fit in a three-dimensional space.

Professional photographers often use specially designed and calibrated monitors that help them to reproduce color accurately and consistently.

Frame Aspect Ratios

Most digital point & shoot cameras have an aspect ratio of 1.33 (4:3), the same as analog television or early movies. However, a 35 mm picture's aspect ratio is 1.5 (3:2). Several digital cameras take photos in either ratio, and nearly all digital SLRs take pictures in a 3:2 ratio, as most can use lenses designed for 35 mm film. Some photo labs print photos on 4:3 ratio paper, as well as the existing 3:2. In 2005 Panasonic launched the first consumer camera with a native aspect ratio of 16:9, matching HDTV. This is similar to a 7:4 aspect ratio, which was a common size for APS film. Different aspect ratios is one of the reasons consumers have issues when cropping photos. An aspect ratio of 4:3 translates to a size of 4.5"x6.0". This loses half an inch when printing on the "standard" size of 4"x6", an aspect ratio of 3:2. Similar cropping occurs when printing on other sizes, i.e., 5"x7", 8"x10", or 11"x14".

References

- Miller, Vincent (2011). Understanding digital culture. London: Sage Publications. sec. "Convergence and the contemporary media experience". ISBN 978-1-84787-497-9.
- Pavlik, John; McIntosh, Shawn. Converging Media (Fourth ed.). Oxford University Press. pp. 237–239. ISBN 978-0-19-934230-3.
- Pavlik, John; Mclntosh, Shawn. Converging Media (fourth ed.). Oxford University Press. p. 89. ISBN 978-0-19-934230-3.
- Dewar, James A. (1998). "The information age and the printing press: looking backward to see ahead". RAND Corporation. Retrieved 29 March 2014.
- O'Carroll, Eoin (10 December 2012). "Ada Lovelace: what did the first computer program do?". Christian Science Monitor. Retrieved 29 March 2014.
- Copeland, B. Jack (Fall 2008). "The modern history of computing". The Stanford Encyclopedia of Philosophy. Stanford University. Retrieved 31 March 2014.
- Simpson, Rosemary; Allen Renear; Elli Mylonas; Andries van Dam (March 1996). "50 years after "As We May Think": the Brown/MIT Vannevar Bush symposium" (PDF). Interactions. pp. 47–67.

Retrieved 29 March 2014.

- Mynatt, Elizabeth. "As we may think: the legacy of computing research and the power of human cognition". Computing Research Association. Retrieved 30 March 2014.
- Bazillion, Richard (2001). "Academic libraries in the digital revolution" (PDF). Educause Quarterly. Retrieved 31 March 2014.
- Cusumano, Catherine (18 March 2013). "Changeover in film technology spells end for age of analog". Brown Daily Herald. Retrieved 31 March 2014.
- McCracken, Erin (5 May 2013). "Last reel: Movie industry's switch to digital hits theaters -- especially small ones -- in the wallet". York Daily Record. Retrieved 29 March 2014.
- Kirchhoff, Suzanne M. (9 September 2010). "The U.S. newspaper industry in transition" (PDF). Congressional Research Service. Retrieved 29 March 2014.
- Zara, Christopher (2 October 2012). "Job growth in digital journalism is bigger than anyone knows". International Business Times. Retrieved 29 March 2014.

3

Theories and Concepts of Digital Signal Processing

Linear time-invariant theory is derived from applied mathematics. Some of the applications of linear time-invariant theory are circuits, signal processing, control theory and other technical areas. The other theories and concepts that have been explained in this text are delta modulation, sample rate conversion, quantization and discretization. This text helps the reader in developing a better understanding of the main concepts related to digital signal processing.

Linear Time-invariant Theory

Linear time-invariant theory, commonly known as LTI system theory, comes from applied mathematics and has direct applications in NMR spectroscopy, seismology, circuits, signal processing, control theory, and other technical areas. It investigates the response of a linear and time-invariant system to an arbitrary input signal. Trajectories of these systems are commonly measured and tracked as they move through time (e.g., an acoustic waveform), but in applications like image processing and field theory, the LTI systems also have trajectories in spatial dimensions. Thus, these systems are also called *linear translation-invariant* to give the theory the most general reach. In the case of generic discrete-time (i.e., sampled) systems, *linear shift-invariant* is the corresponding term. A good example of LTI systems are electrical circuits that can be made up of resistors, capacitors, and inductors.

Overview

The defining properties of any LTI system are *linearity* and *time invariance.*

- *Linearity* means that the relationship between the input and the output of the system is a linear map: If input $x_1(t)$ produces response $y_1(t)$, and input $x_2(t)$ produces response $y_2(t)$, then the *scaled* and *summed* input $a_1x_1(t)+a_2x_2(t)$ produces the scaled and summed response $a_1y_1(t)+a_2y_2(t)$ where a_1 and a_2 are real scalars. It follows that this can be extended to an arbitrary number of terms, and so for real numbers $c_1, c_2, \ldots, c_k$,,

Input $\sum_k c_k x_k(t)$ produces output $\sum_k c_k y_k(t)$.

In particular,

$$\text{Input } \int_{-\infty}^{\infty} c_\omega x_\omega(t)\mathrm{d}\omega \text{ produces output } \int_{-\infty}^{\infty} c_\omega y_\omega(t)\mathrm{d}\omega \quad \textbf{(Eq.1)}$$

where c_ω and x_ω are scalars and inputs that vary over a continuum indexed by ω. Thus if an input function can be represented by a continuum of input functions, combined "linearly", as shown, then the corresponding output function can be represented by the corresponding continuum of output functions, *scaled* and *summed* in the same way.

- *Time invariance* means that whether we apply an input to the system now or T seconds from now, the output will be identical except for a time delay of T seconds. That is, if the output due to input $x(t)$ is $y(t)$, then the output due to input $x(t-T)$ is $y(t-T)$. Hence, the system is time invariant because the output does not depend on the particular time the input is applied.

The fundamental result in LTI system theory is that any LTI system can be characterized entirely by a single function called the system's impulse response. The output of the system is simply the convolution of the input to the system with the system's impulse response. This method of analysis is often called the *time domain* point-of-view. The same result is true of discrete-time linear shift-invariant systems in which signals are discrete-time samples, and convolution is defined on sequences.

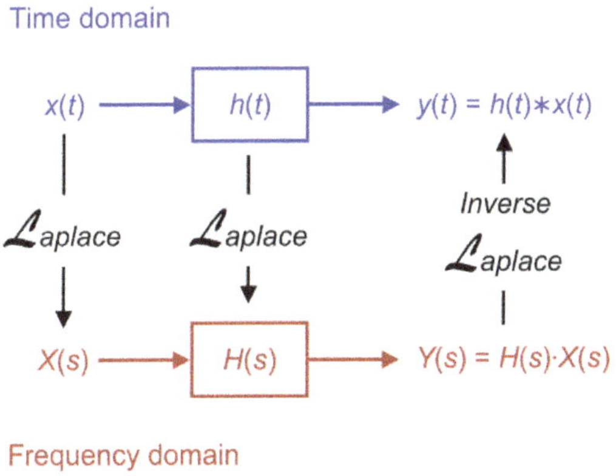

Relationship between the time domain and the frequency domain

Equivalently, any LTI system can be characterized in the *frequency domain* by the system's transfer function, which is the Laplace transform of the system's impulse response (or Z transform in the case of discrete-time systems). As a result of the properties of these transforms, the output of the system in the frequency domain is the product of the transfer function and the transform of the input. In other words, convolution in the time domain is equivalent to multiplication in the frequency domain.

For all LTI systems, the eigenfunctions, and the basis functions of the transforms, are complex exponentials. This is, if the input to a system is the complex waveform Ae^{st} for some complex amplitude A and complex frequency s, the output will be some complex constant times the input, say Be^{st} for some new complex amplitude B. The ratio B/A is the transfer function at frequency s.

Since sinusoids are a sum of complex exponentials with complex-conjugate frequencies, if the input to the system is a sinusoid, then the output of the system will also be a sinusoid, perhaps with a different amplitude and a different phase, but always with the same frequency upon reaching steady-state. LTI systems cannot produce frequency components that are not in the input.

LTI system theory is good at describing many important systems. Most LTI systems are considered "easy" to analyze, at least compared to the time-varying and/or nonlinear case. Any system that can be modeled as a linear homogeneous differential equation with constant coefficients is an LTI system. Examples of such systems are electrical circuits made up of resistors, inductors, and capacitors (RLC circuits). Ideal spring–mass–damper systems are also LTI systems, and are mathematically equivalent to RLC circuits.

Most LTI system concepts are similar between the continuous-time and discrete-time (linear shift-invariant) cases. In image processing, the time variable is replaced with two space variables, and the notion of time invariance is replaced by two-dimensional shift invariance. When analyzing filter banks and MIMO systems, it is often useful to consider vectors of signals.

A linear system that is not time-invariant can be solved using other approaches such as the Green function method. The same method must be used when the initial conditions of the problem are not null.

Continuous-time Systems

Impulse Response and Convolution

The behavior of a linear, continuous-time, time-invariant system with input signal x(t) and output signal y(t) is described by the convolution integral:

$$y(t) = x(t) * h(t) \overset{\text{def}}{=} \int_{-\infty}^{\infty} x(t-\tau)\cdot h(\tau)\,\mathrm{d}\tau$$

$$= \int_{-\infty}^{\infty} x(\tau)\cdot h(t-\tau)\mathrm{d}\tau, \qquad \text{(using commutativity)}$$

where $h(t)$ is the system's response to an impulse: $x(\tau)=\delta(\tau)$. $y(t)$ is therefore proportional to a weighted average of the input function $x(\tau)$. The weighting function is $h(-\tau)$, simply shifted by amount t. As t changes, the weighting function emphasizes different parts of the input function. When $h(\tau)$ is zero for all negative τ, $y(t)$ depends only on values of x prior to time t, and the system is said to be causal.

To understand why the convolution produces the output of an LTI system, let the notation $\{x(u-\tau);u\}$ represent the function $x(u-\tau)$ with variable u and constant τ. And let the shorter notation $\{x\}$ represent $\{x(u);u\}$. Then a continuous-time system transforms an input function, $\{x\}$, into an output function, $\{y\}$. And in general, every value of the output can depend on every value of the input. This concept is represented by:

$$y(t) \stackrel{\text{def}}{=} O_t\{x\},$$

where O_t is the transformation operator for time t. In a typical system, $y(t)$ pends most heavily on the values of x that occurred near time t. Unless the transform itself changes with the output function is just constant, and the system is uninteresting.

For a linear system, O must satisfy Eq.1 :

$$O_t\left\{\int_{-\infty}^{\infty} c_\tau\, x_\tau(u)\mathrm{d}\tau;\ u\right\} = \int_{-\infty}^{\infty} c_\tau\, \underbrace{y_\tau(t)}_{O_t\{x_\tau\}}\mathrm{d}\tau. \qquad \textbf{\textit{(Eq.2)}}$$

And the time-invariance requirement is:

$$\begin{aligned} O_t\{x(u-\tau);\ u\} &= y(t-\tau) \\ &\stackrel{\text{def}}{=} O_{t-\tau}\{x\}. \end{aligned} \qquad \textbf{\textit{(Eq.3)}}$$

In this notation, we can write the impulse response as $h(t) \stackrel{\text{def}}{=} O_t\{\delta(u);u\}$.

Similarly:

$$h(t-\tau) \stackrel{\text{def}}{=} O_{t-\tau}\{\delta(u);u\}$$

$$= O_t\{\delta(u-\tau);u\}. \qquad \text{(using **Eq.3**)}$$

Substituting this result into the convolution integral:

$$x(t)*h(t) = \int_{-\infty}^{\infty} x(\tau)\cdot h(t-\tau)\mathrm{d}\tau$$

$$= \int_{-\infty}^{\infty} x(\tau)\cdot O_t\{\delta(u-\tau);\ u\}\mathrm{d}\tau,$$

which has the form of the right side of Eq.2 for the case $c_\tau = x(\tau)$ and $x_\tau(u) = \delta(u-\tau)$. Eq.2 then allows this continuation:

$$x(t)*h(t) = O_t\left\{\int_{-\infty}^{\infty} x(\tau)\cdot\delta(u-\tau)\mathrm{d}\tau;\ u\right\}$$

$$= O_t\{x(u);\ u\}$$

$$\stackrel{\text{def}}{=} y(t).$$

In summary, the input function, $\{x\}$, can be represented by a continuum of time-shifted impulse functions, combined «linearly», as shown at Eq.1. The system's linearity property allows the system's response to be represented by the corresponding continuum of impulse responses, combined in the same way. And the time-invariance property allows that combination to be represented by the convolution integral.

The mathematical operations above have a simple graphical simulation.

Exponentials as Eigenfunctions

An eigenfunction is a function for which the output of the operator is a scaled version of the same function. That is,

$$\mathcal{H}f = \lambda f,$$

where f is the eigenfunction and λ is the eigenvalue, a constant.

The exponential functions Ae^{st}, where $A, s \in \mathbb{C}$, are eigenfunctions of a linear, time-invariant operator. A simple proof illustrates this concept. Suppose the input is $x(t) = Ae^{st}$. The output of the system with impulse response $\imath(t)$ is then

$$\int_{-\infty}^{\infty} h(t-\tau) Ae^{s\tau}\,\mathrm{d}\tau$$

which, by the commutative property of convolution, is equivalent to

$$\overbrace{\int_{-\infty}^{\infty} h(\tau) Ae^{s(t-\tau)}\,\mathrm{d}\tau}^{\mathcal{H}f} = \int_{-\infty}^{\infty} h(\tau) Ae^{st} e^{-s\tau}\,\mathrm{d}\tau = Ae^{st} \int_{-\infty}^{\infty} h(\tau) e^{-s\tau}\,\mathrm{d}\tau$$

$$= \underbrace{\overbrace{Ae^{st}}^{f}}_{\text{Input}} \underbrace{\overbrace{H(s)}^{\lambda}}_{\text{Scalar}},$$

where the scalar

$$H(s) \stackrel{\text{def}}{=} \int_{-\infty}^{\infty} h(t) e^{-st}\,\mathrm{d}t$$

is dependent only on the parameter s.

So the system's response is a scaled version of the input. In particular, for any $A, s \in \mathbb{C}$, the system output is the product of the input Ae^{st} and the constant $H(s)$. Hence, Ae^{st} is an eigenfunction of an LTI system, and the corresponding eigenvalue is $H(s)$.

Direct Proof

It is also possible to directly derive complex exponentials as eigenfunctions of LTI systems.

Let's set $v(t) = e^{i\omega t}$ some complex exponential and $v_a(t) = e^{i\omega(t+a)}$ a time-shifted version of it.

$H[v_a](t) = e^{i\omega a} H[v](t)$ by linearity with respect to the constant $e^{i\omega a}$.

$H[v_a](t) = H[v](t+a)$ by time invariance of H.

So $H[v](t+a) = e^{i\omega a} H[v](t)$. Setting $t = 0$ and renaming we get :

$$H[v](\tau) = e^{i\omega\tau} H[v](0)$$

i.e. that a complex exponential $e^{i\omega\tau}$ as input will give a complex exponential of same frequency as output.

Fourier and Laplace Transforms

The eigenfunction property of exponentials is very useful for both analysis and insight into LTI systems. The Laplace transform

$$H(s) \overset{\text{def}}{=} \mathcal{L}\{h(t)\} \overset{\text{def}}{=} \int_{-\infty}^{\infty} h(t)e^{-st}\,\mathrm{d}t$$

is exactly the way to get the eigenvalues from the impulse response. Of particular interest are pure sinusoids (i.e., exponential functions of the form $e^{j\omega t}$ where $\omega \in \mathbb{R}$ and $j \overset{\text{def}}{=} \sqrt{-1}$). These are generally called complex exponentials even though the argument is purely imaginary. The Fourier transform $H(j\omega) = \mathcal{F}\{h(t)\}$ gives the eigenvalues for pure complex sinusoids. Both of $H(s)$ and $H(j\omega)$ are called the *system function, system response*, or *transfer function.*

The Laplace transform is usually used in the context of one-sided signals, i.e. signals that are zero for all values of t less than some value. Usually, this "start time" is set to zero, for convenience and without loss of generality, with the transform integral being taken from zero to infinity (the transform shown above with lower limit of integration of negative infinity is formally known as the bilateral Laplace transform).

The Fourier transform is used for analyzing systems that process signals that are infinite in extent, such as modulated sinusoids, even though it cannot be directly applied to input and output signals that are not square integrable. The Laplace transform actually works directly for these signals if they are zero before a start time, even if they are not square integrable, for stable systems. The Fourier transform is often applied to spectra of infinite signals via the Wiener–Khinchin theorem even when Fourier transforms of the signals do not exist.

Due to the convolution property of both of these transforms, the convolution that gives the output of the system can be transformed to a multiplication in the transform domain, given signals for which the transforms exist

$$y(t) = (h * x)(t) \overset{\text{def}}{=} \int_{-\infty}^{\infty} h(t-\tau)x(\tau)\mathrm{d}\tau \overset{\text{def}}{=} \mathcal{L}^{-1}\{H(s)X(s)\}.$$

Not only is it often easier to do the transforms, multiplication, and inverse transform than the original convolution, but one can also gain insight into the behavior of the system from the system response. One can look at the modulus of the system function $|H(s)|$ to see whether the input $\exp(st)$ is *passed* (let through) the system or *rejected* or *attenuated* by the system (not let through).

Examples

- A simple example of an LTI operator is the derivative.

- $\frac{\mathrm{d}}{\mathrm{d}t}\left(c_1x_1(t)+c_2x_2(t)\right)=c_1x_1'(t)+c_2x_2'(t)$ (i.e., it is linear)

- $\frac{\mathrm{d}}{\mathrm{d}t}x(t-\tau)=x'(t-\tau)$ (i.e., it is time invariant)

When the Laplace transform of the derivative is taken, it transforms to a simple multiplication by the Laplace variable s.

$$\mathcal{L}\left\{\frac{\mathrm{d}}{\mathrm{d}t}x(t)\right\}=sX(s)$$

That the derivative has such a simple Laplace transform partly explains the utility of the transform.

- Another simple LTI operator is an averaging operator

$$\mathcal{A}\{x(t)\}\overset{\text{def}}{=}\int_{t-a}^{t+a}x(\lambda)\mathrm{d}\lambda.$$

By the linearity of integration,

$$\begin{aligned}\mathcal{A}\{c_1x_1(t)+c_2x_2(t)\}&=\int_{t-a}^{t+a}(c_1x_1(\lambda)+c_2x_2(\lambda))\mathrm{d}\lambda\\&=c_1\int_{t-a}^{t+a}x_1(\lambda)\mathrm{d}\lambda+c_2\int_{t-a}^{t+a}x_2(\lambda)\mathrm{d}\lambda\\&=c_1\mathcal{A}\{x_1(t)\}+c_2\mathcal{A}\{x_2(t)\}\end{aligned}$$

it is linear. Additionally, because

$$\begin{aligned}\mathcal{A}\{x(t-\tau)\}&=\int_{t-a}^{t+a}x(\lambda-\tau)\mathrm{d}\lambda\\&=\int_{(t-\tau)-a}^{(t-\tau)+a}x(\xi)\mathrm{d}\xi\\&=\mathcal{A}\{x\}(t-\tau)\end{aligned}$$

it is time invariant. In fact, $\mathcal{A}$ can be written as a convolution with the boxcar function $\Pi(t)$. That is,

$$\mathcal{A}\{x(t)\}=\int_{-\infty}^{\infty}\Pi\left(\frac{\lambda-t}{2a}\right)x(\lambda)\mathrm{d}\lambda,$$

where the boxcar function

$$\Pi(t)\overset{\text{def}}{=}\begin{cases}1 & \text{if } |t|<\frac{1}{2},\\ 0 & \text{if } |t|>\frac{1}{2}.\end{cases}$$

Important System Properties

Some of the most important properties of a system are causality and stability. Causality is a necessity if the independent variable is time, but not all systems have time as an independent variable. For example, a system that processes still images does not need to be causal. Non-causal systems can be built and can be useful in many circumstances. Even non-real systems can be built and are very useful in many contexts.

Causality

A system is causal if the output depends only on present and past, but not future inputs. A necessary and sufficient condition for causality is

$$h(t) = 0 \quad \forall t < 0,$$

where $h(t)$ is the impulse response. It is not possible in general to determine causality from the Laplace transform, because the inverse transform is not unique. When a region of convergence is specified, then causality can be determined.

Stability

A system is bounded-input, bounded-output stable (BIBO stable) if, for every bounded input, the output is finite. Mathematically, if every input satisfying

$$\| x(t) \|_\infty < \infty$$

leads to an output satisfying

$$\| y(t) \|_\infty < \infty$$

(that is, a finite maximum absolute value of $x(t)$ implies a finite maximum absolute value of $y(t)$), then the system is stable. A necessary and sufficient condition is that $h(t)$, the impulse response, is in L^1 (has a finite L^1 norm):

$$\| h(t) \|_1 = \int_{-\infty}^{\infty} |h(t)| \mathrm{d}t < \infty.$$

In the frequency domain, the region of convergence must contain the imaginary axis $s = j\omega$.

As an example, the ideal low-pass filter with impulse response equal to a sinc function is not BIBO stable, because the sinc function does not have a finite L^1 norm. Thus, for some bounded input, the output of the ideal low-pass filter is unbounded. In particular, if the input is zero for $t < 0$ and equal to a sinusoid at the cut-off frequency for $t < 0$, then the output will be unbounded for all times other than the zero crossings.

Discrete-time Systems

Almost everything in continuous-time systems has a counterpart in discrete-time systems.

Discrete-time Systems from Continuous-time Systems

In many contexts, a discrete time (DT) system is really part of a larger continuous time (CT) system. For example, a digital recording system takes an analog sound, digitizes it, possibly processes the digital signals, and plays back an analog sound for people to listen to.

Formally, the DT signals studied are almost always uniformly sampled versions of CT signals. If $x(t)$ is a CT signal, then an analog to digital converter will transform it to the DT signal:

$$x[n] \overset{\text{def}}{=} x(nT) \qquad \forall n \in \mathbb{Z},$$

where T is the sampling period. It is very important to limit the range of frequencies in the input signal for faithful representation in the DT signal, since then the sampling theorem guarantees that no information about the CT signal is lost. A DT signal can only contain a frequency range of $1/(2T)$; other frequencies are aliased to the same range.

Impulse Response and Convolution

Let $\{x[m-k];\ m\}$ represent the sequence $\{x[m-k];$ for all integer values of m$\}$..

And let the shorter notation $\{x\}$ represent $\{x[m];\ m\}$.

A discrete system transforms an input sequence, $\{x\}$ into an output sequence, $\{y\}$.In general, every element of the output can depend on every element of the input. Representing the transformation operator by O, we can write:

$$y[n] \overset{\text{def}}{=} O_n\{x\}.$$

Note that unless the transform itself changes with n, the output sequence is just constant, and the system is uninteresting. (Thus the subscript, n.) In a typical system, y[n] depends most heavily on the elements of x whose indices are near n.

For the special case of the Kronecker delta function, $x[m] = \delta[m]$, the output sequence is the impulse response:

$$h[n] \overset{\text{def}}{=} O_n\{\delta[m];\ m\}.$$

For a linear system, O must satisfy:

$$O_n\left\{\sum_{k=-\infty}^{\infty} c_k \cdot x_k[m];\ m\right\} = \sum_{k=-\infty}^{\infty} c_k \cdot O_n\{x_k\}. \qquad \textbf{(Eq.4)}$$

And the time-invariance requirement is:

$$\begin{aligned} O_n\{x[m-k];\ m\} &= y[n-k] \\ &\overset{\text{def}}{=} O_{n-k}\{x\}. \end{aligned} \qquad \textbf{(Eq.5)}$$

In such a system, the impulse response, $\{h\}$, characterizes the system completely. I.e., for any input sequence, the output sequence can be calculated in terms of the input and the impulse response. To see how that is done, consider the identity:

$$x[m] \equiv \sum_{k=-\infty}^{\infty} x[k] \cdot \delta[m-k],$$

which expresses $\{x\}$ in terms of a sum of weighted delta functions.

Therefore:

$$\begin{aligned} y[n] = O_n\{x\} &= O_n\left\{\sum_{k=-\infty}^{\infty} x[k] \cdot \delta[m-k];\ m\right\} \\ &= \sum_{k=-\infty}^{\infty} x[k] \cdot O_n\{\delta[m-k];\ m\}, \end{aligned}$$

where we have invoked Eq.4 for the case $c_k = x[k]$ and $x_k[m] = \delta[m-k]$.

And because of Eq.5, we may write:

$$\begin{aligned} O_n\{\delta[m-k];\ m\} &= O_{n-k}\{\delta[m];\ m\} \\ &\overset{\text{def}}{=} h[n-k]. \end{aligned}$$

Therefore:

$$y[n] = \sum_{k=-\infty}^{\infty} x[k] \cdot h[n-k]$$

$$= \sum_{k=-\infty}^{\infty} x[n-k] \cdot h[k], \qquad \text{(commutativity)}$$

which is the familiar discrete convolution formula. The operator O_n can therefore be interpreted as proportional to a weighted average of the function x[k]. The weighting function is h[-k], simply shifted by amount n. As n changes, the weighting function emphasizes different parts of the input function. Equivalently, the system's response to an impulse at n=0 is a "time" reversed copy of the unshifted weighting function. When h[k] is zero for all negative k, the system is said to be causal.

Exponentials as Eigenfunctions

An eigenfunction is a function for which the output of the operator is the same function, scaled by some constant. In symbols,

$$\mathcal{H}f = \lambda f,$$

where f is the eigenfunction and λ is the eigenvalue, a constant.

The exponential functions $z^n = e^{sTn}$, where $n \in \mathbb{Z}$,are eigenfunctions of a linear, time-invariant operator. $T \in \mathbb{R}$ is the sampling interval, and $z = e^{sT}$, $z, s \in \mathbb{C}$. A simple proof illustrates this concept.

Suppose the input is $x[n] = z^n$. The output of the system with impulse response $h[n]$ is then

$$\sum_{m=-\infty}^{\infty} h[n-m] z^m$$

which is equivalent to the following by the commutative property of convolution

$$\sum_{m=-\infty}^{\infty} h[m] z^{(n-m)} = z^n \sum_{m=-\infty}^{\infty} h[m] z^{-m} = z^n H(z)$$

where

$$H(z) \stackrel{\text{def}}{=} \sum_{m=-\infty}^{\infty} h[m] z^{-m}$$

is dependent only on the parameter z.

So z^n is an eigenfunction of an LTI system because the system response is the same as the input times the constant $H(z)$.

Z And Discrete-time Fourier Transforms

The eigenfunction property of exponentials is very useful for both analysis and insight into LTI systems. The Z transform

$$H(z) = \mathcal{Z}\{h[n]\} = \sum_{n=-\infty}^{\infty} h[n]z^{-n}$$

is exactly the way to get the eigenvalues from the impulse response. Of particular interest are pure sinusoids, i.e. exponentials of the form $e^{j\omega n}$, where $\omega \in \mathbb{R}$. These can also be written as z^n with $z = e^{j\omega}$. These are generally called complex exponentials even though the argument is purely imaginary. The Discrete-time Fourier transform (DTFT) $H(e^{j\omega}) = \mathcal{F}\{h[n]\}$ gives the eigenvalues of pure sinusoids. Both of $H(z)$ and $H(e^{j\omega})$ are called the *system function, system response*, or *transfer function'*.

The Z transform is usually used in the context of one-sided signals, i.e. signals that are zero for all values of t less than some value. Usually, this "start time" is set to zero, for convenience and without loss of generality. The Fourier transform is used for analyzing signals that are infinite in extent.

Due to the convolution property of both of these transforms, the convolution that gives the output of the system can be transformed to a multiplication in the transform domain. That is,

$$y[n] = (h * x)[n] = \sum_{m=-\infty}^{\infty} h[n-m]x[m] = \mathcal{Z}^{-1}\{H(z)X(z)\}.$$

Just as with the Laplace transform transfer function in continuous-time system analysis, the Z transform makes it easier to analyze systems and gain insight into their behavior. One can look at the modulus of the system function $|H(z)|$ to see whether the input z^n is *passed* (let through) by the system, or *rejected* or *attenuated* by the system (not let through).

Examples

- A simple example of an LTI operator is the delay operator $D\{x[n]\} \overset{\text{def}}{=} x[n-1]$.
- $D\left(c_1.x_1[n] + c_2.x_2[n]\right) = c_1.x_1[n-1] + c_2.x_2[n-1] = c_1.Dx_1[n] + c_2.Dx_2[n]$ (i.e., it is linear)
- $D\{x[n-m]\} = x[n-m-1] = x[(n-1)-m] = D\{x\}[n-m]$ (i.e., it is time invariant)

  The Z transform of the delay operator is a simple multiplication by z^{-1}. That is,

  $$\mathcal{Z}\{Dx[n]\} = z^{-1}X(z).$$

- Another simple LTI operator is the averaging operator

$$\mathcal{A}\{x[n]\} \stackrel{\text{def}}{=} \sum_{k=n-a}^{n+a} x[k].$$

Because of the linearity of sums,

$$\mathcal{A}\{c_1x_1[n]+c_2x_2[n]\} = \sum_{k=n-a}^{n+a} (c_1x_1[k]+c_2x_2[k])$$

$$c_1 \sum_{k=n-a}^{n+a} x_1[k]+c_2 \sum_{k=n-a}^{n+a} x_2[k] \quad , \quad = c_1\mathcal{A}\{x_1[n]\}+c_2\mathcal{A}\{x_2[n]\}$$

and so it is linear. Because,

$$\mathcal{A}\{x[n-m]\} = \sum_{k=n-a}^{n+a} x[k-m] \quad , \quad = \sum_{k'=(n-m)-a}^{(n-m)+a} x[k'] \quad , \quad = \mathcal{A}\{x\}[n-m]$$

it is also time invariant.

Important System Properties

The input-output characteristics of discrete-time LTI system are completely described by its impulse response $h[n]$. Some of the most important properties of a system are causality and stability. Unlike CT systems, non-causal DT systems can be realized. It is trivial to make an acausal FIR system causal by adding delays. It is even possible to make acausal IIR systems. Non-stable systems can be built and can be useful in many circumstances. Even non-real systems can be built and are very useful in many contexts.

Causality

A discrete-time LTI system is causal if the current value of the output depends on only the current value and past values of the input., A necessary and sufficient condition for causality is

$$h[n] = 0 \ \forall n < 0,$$

where $h[n]$ is the impulse response. It is not possible in general to determine causality from the Z transform, because the inverse transform is not unique. When a region of convergence is specified, then causality can be determined.

Stability

A system is bounded input, bounded output stable (BIBO stable) if, for every bounded input, the output is finite. Mathematically, if

$$\| x[n] \|_\infty < \infty$$

implies that

$$\| y[n] \|_\infty < \infty$$

(that is, if bounded input implies bounded output, in the sense that the maximum absolute values of $x[n]$ and $y[n]$ are finite), then the system is stable. A necessary and sufficient condition is that $h[n]$, , the impulse response, satisfies

$$\| h[n] \|_1 \overset{\text{def}}{=} \sum_{n=-\infty}^{\infty} |h[n]| < \infty.$$

In the frequency domain, the region of convergence must contain the unit circle (i.e., the locus satisfying $|z| = 1$ for complex z).

Delta Modulation

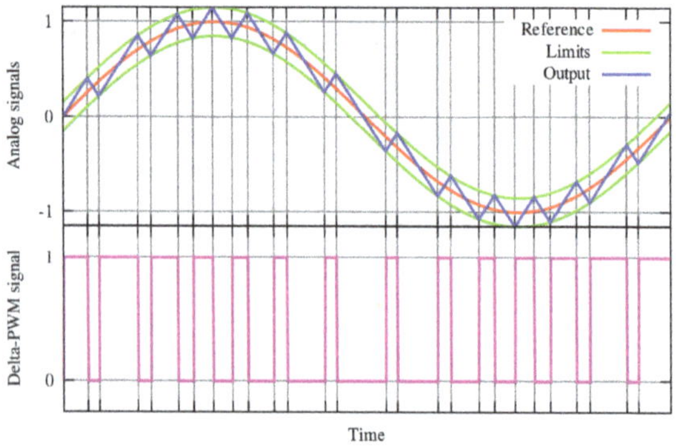

Principle of the delta PWM. The output signal (blue) is compared with the limits (green). The limits (green) correspond to the reference signal (red), offset by a given value. Every time the output signal reaches one of the limits, the PWM signal changes state.

A delta modulation (DM or Δ-modulation) is an analog-to-digital and digital-to-analog signal conversion technique used for transmission of voice information where quality is not of primary importance. DM is the simplest form of differential pulse-code modulation (DPCM) where the difference between successive samples are encoded into n-bit data streams. In delta modulation, the transmitted data are reduced to a 1-bit data stream. Its main features are:

- The analog signal is approximated with a series of segments.
- Each segment of the approximated signal is compared of successive bits is determined by this comparison.

- Only the change of information is sent, that is, only an increase or decrease of the signal amplitude from the previous sample is sent whereas a no-change condition causes the modulated signal to remain at the same 0 or 1 state of the previous sample.

To achieve high signal-to-noise ratio, delta modulation must use oversampling techniques, that is, the analog signal is sampled at a rate several times higher than the Nyquist rate.

Derived forms of delta modulation are continuously variable slope delta modulation, delta-sigma modulation, and differential modulation. Differential pulse-code modulation is the superset of DM.

Principle

Rather than quantizing the absolute value of the input analog waveform, delta modulation quantizes the difference between the current and the previous step, as shown in the block diagram in Fig.

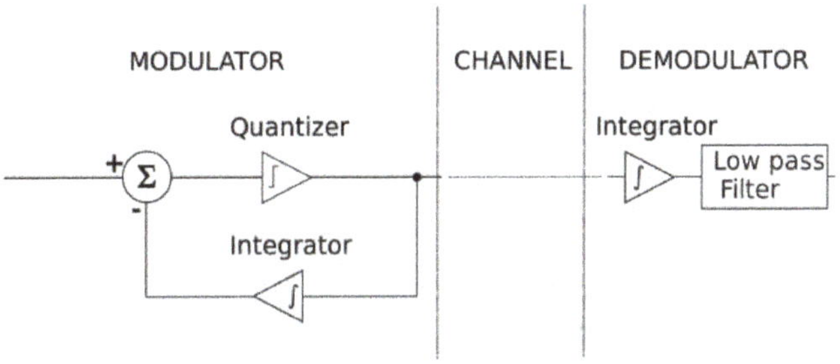

Block diagram of a Δ-modulator/demodulator

The modulator is made by a quantizer which converts the difference between the input signal and the average of the previous steps. In its simplest form, the quantizer can be realized with a comparator referenced to 0 (two levels quantizer), whose output is *1* or *0* if the input signal is positive or negative. It is also a bit-quantizer as it quantizes only a bit at a time. The demodulator is simply an integrator (like the one in the feedback loop) whose output rises or falls with each 1 or 0 received. The integrator itself constitutes a low-pass filter.

Transfer Characteristics

The transfer characteristics of a delta modulated system follows a signum function, as it quantizes only two levels and also one-bit at a time.

The two sources of noise in delta modulation are "slope overload", when step size is too small to track the original waveform, and "granularity", when step size is too large. But a 1971 study shows that slope overload is less objectionable compared to granularity than one might expect based solely on SNR measures.

Output Signal Power

In delta modulation there is a restriction on the amplitude of the input signal, because if the transmitted signal has a large derivative (abrupt changes) then the modulated signal can not follow the input signal and slope overload occurs. E.g. if the input signal is

$$m(t) = A\cos(\omega t),$$

the modulated signal (derivative of the input signal) which is transmitted by the modulator is

$$|\dot{m}(t)|_{max} = \omega A,$$

whereas the condition to avoid slope overload is

$$|\dot{m}(t)|_{max} = \omega A < \sigma f_s.$$

So the maximum amplitude of the input signal can be

$$A_{max} = \frac{\sigma f_s}{\omega},$$

where f_s is the sampling frequency and ω is the frequency of the input signal and σ is step size in quantization. So A_{max} is the maximum amplitude that DM can transmit without causing the slope overload and the power of transmitted signal depends on the maximum amplitude.

Bit-rate

If the communication channel is of limited bandwidth, there is the possibility of interference in either DM or PCM. Hence, 'DM' and 'PCM' operate at same bit-rate which is equal to N times the sampling frequency.

Adaptive Delta Modulation

Adaptive delta modulation (ADM) was first published by Dr. John E. Abate (AT&T Bell Laboratories Fellow) in his doctorial thesis at NJ Institute Of Technology in 1968. ADM was later selected as the standard for all NASA communications between mission control and space-craft.

Adaptive delta modulation or [continuously variable slope delta modulation] (CVSD) is a modification of DM in which the step size is not fixed. Rather, when several consecutive bits have the same direction value, the encoder and decoder assume that slope overload is occurring, and the step size becomes progressively larger.

Otherwise, the step size becomes gradually smaller over time. ADM reduces slope error, at the expense of increasing quantizing error.This error can be reduced by using a low-pass filter. ADM provides robust performance in the presence of bit errors meaning error detection and correction are not typically used in an ADM radio design, this allows for a reduction in host processor workload (allowing a low-cost processor to be used).

Applications

Contemporary applications of Delta Modulation includes, but is not limited to, recreating legacy synthesizer waveforms. With the increasing availability of FPGAs and game-related ASICs, sample rates are easily controlled so as to avoid slope overload and granularity issues. For example, the C64DTV used a 32 MHz sample rate, providing ample dynamic range to recreate the SID output to acceptable levels.

SBS Application 24 kbps Delta Modulation

Delta Modulation was used by Satellite Business Systems or SBS for its voice ports to provide long distance phone service to large domestic corporations with a significant inter-corporation communications need (such as IBM). This system was in service throughout the 1980s. The voice ports used digitally implemented 24 kbit/s delta modulation with Voice Activity Compression (VAC) and echo suppressors to control the half second echo path through the satellite. They performed formal listening tests to verify the 24 kbit/s delta modulator achieved full voice quality with no discernible degradation as compared to a high quality phone line or the standard 64 kbit/s μ-law companded PCM. This provided an eight to three improvement in satellite channel capacity. IBM developed the Satellite Communications Controller and the voice port functions.

The original proposal in 1974, used a state-of-the-art 24 kbit/s delta modulator with a single integrator and a Shindler Compander modified for gain error recovery. This proved to have less than full phone line speech quality. In 1977, one engineer with two assistants in the IBM Research Triangle Park, NC laboratory was assigned to improve the quality.

The final implementation replaced the integrator with a Predictor implemented with a two pole complex pair low-pass filter designed to approximate the long term average speech spectrum. The theory was that ideally the integrator should be a predictor designed to match the signal spectrum. A nearly perfect Shindler Compander replaced the modified version. It was found the modified compander resulted in a less than perfect step size at most signal levels and the fast gain error recovery increased the noise as determined by actual listening tests as compared to simple signal to noise measurements. The final compander achieved a very mild gain error recovery due to the natural truncation rounding error caused by twelve bit arithmetic.

The complete function of delta modulation, VAC and Echo Control for six ports was implemented in a single digital integrated circuit chip with twelve bit arithmetic. A single digital-to-analog converter (DAC) was shared by all six ports providing voltage compare functions for the modulators and feeding sample and hold circuits for the demodulator outputs. A single card held the chip, DAC and all the analog circuits for the phone line interface including transformers.

Sample Rate Conversion

Sample-rate conversion is the process of changing the sampling rate of a discrete signal to obtain a new discrete representation of the underlying continuous signal. Application areas include image scaling, and audio/visual systems, where different sampling-rates may be used for engineering, economic, or historical reasons.

For example, Compact Disc Digital Audio and Digital Audio Tape systems use different sampling rates, and American television, European television, and movies all use different frame rates. Sample rate conversion prevents changes in speed and pitch that would otherwise occur when transferring recorded material between such systems.

Within specific domains or for specific conversions, the following alternative terms for sample-rate conversion are also used: sampling-frequency conversion, resampling, upsampling, downsampling, interpolation, decimation, upscaling, downscaling. The term multi-rate digital signal processing is sometimes used to refer to systems that incorporate sample-rate conversion.

Techniques

Conceptual approaches to sample-rate conversion include: converting to an analog continuous signal, then re-sampling at the new rate, or calculating the values of the new samples directly from the old samples. The latter approach is more satisfactory since it introduces less noise and distortion. Two possible implementation methods are as follows:

1. If the ratio of the two sample-rates is (or can be approximated by) a fixed, rational number L/M: generate an intermediate signal by inserting L−1 0s between each of the original samples. Low-pass filter this signal at half of the lower of the two rates. Select every M^{th} sample from the filtered output, to obtain the result.
2. Treat the samples as geometric points and create any needed new points by interpolation. Choosing an interpolation method is a trade-off between implementation complexity and conversion quality (according to application requirements). Commonly used are: ZOH (for film/video frames), cubic (for image processing) and windowed sinc function (for audio).

The two methods are mathematically identical: picking an interpolation function in the second scheme is equivalent to picking the impulse response of the filter in the first scheme. Linear interpolation is equivalent to a triangular impulse response; windowed sinc approximates a brick-wall filter (it approaches the desirable "brick wall" filter as the number of points increase). The length of the impulse response of the filter in method 1 corresponds to the number of points used in interpolation in method 2.

In method 1, a slow pre-computation (such as the Remez algorithm) can be used to obtain an optimal (per application requirements) filter design. Method 2 will work in more general cases, e.g. where the ratio of sample-rates is not rational, or two real-time streams must be accommodated, or the sample-rates are time-varying.

Examples

Film and Television

The slow-scan TV signals from the Apollo moon missions were converted to the conventional TV rates for the viewers at home.

Movies (shot at 24 frames per second) are converted to television (roughly 50 or 60 fields per second). To convert a 24 frame/sec movie to 60 field/sec television, for example, alternate movie frames are shown 2 and 3 times, respectively. For 50 Hz systems such as PAL each frame is shown twice. Since 50 is not exactly 2×24, the movie will run 50/48 = 4% faster, and the audio pitch will be 4% higher, an effect known as PAL speed-up. This is often accepted for simplicity, but more complex methods are possible that preserve the running time and pitch. Every twelfth frame can be repeated 3 times rather than twice, or digital interpolation can be used in a video scaler.

Audio

Audio on Compact Disc has a sampling rate of 44.1 kHz; to transfer it to Digital Audio Tape, which uses 48 kHz, method 1 above can be used with L=160, M=147 (since 48000/44100 = 160/147). For the reverse conversion, the values of L and M are swapped. Per above, in both cases, the low-pass filter should be to 22.05 kHz.

Quantization (Signal Processing)

Quantization, in mathematics and digital signal processing, is the process of mapping a large set of input values to a (countable) smaller set. Rounding and truncation are typical examples of quantization processes. Quantization is involved to some degree in nearly all digital signal processing, as the process of representing a signal in digital form ordinarily involves rounding. Quantization also forms the core of essentially all lossy compression algorithms.

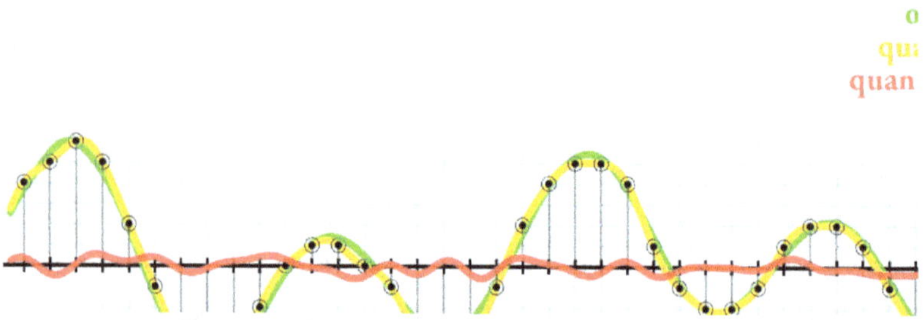

The simplest way to quantize a signal is to choose the digital amplitude value closest to the original analog amplitude. This example shows the original analog signal (green), the quantized signal (black dots), the signal reconstructed from the quantized signal (yellow) and the difference between the original signal and the reconstructed signal (red). The difference between the original signal and the reconstructed signal is the quantization error and, in this simple quantization scheme, is a deterministic function of the input signal.

The difference between an input value and its quantized value (such as round-off error) is referred to as quantization error. A device or algorithmic function that performs quantization is called a quantizer. An analog-to-digital converter is an example of a quantizer.

Basic Properties of Quantization

Because quantization is a many-to-few mapping, it is an inherently non-linear and irreversible process (i.e., because the same output value is shared by multiple input values, it is impossible in general to recover the exact input value when given only the output value).

The set of possible input values may be infinitely large, and may possibly be continuous and therefore uncountable (such as the set of all real numbers, or all real numbers within some limited range). The set of possible output values may be finite or countably infinite. The input and output sets involved in quantization can be defined in a rather general way. For example, *vector quantization* is the application of quantization to multi-dimensional (vector-valued) input data.

Basic Types of Quantization

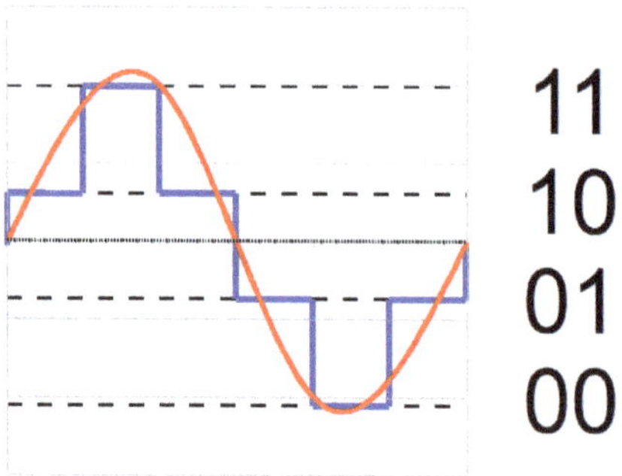

2-bit resolution with four levels of quantization compared to analog.

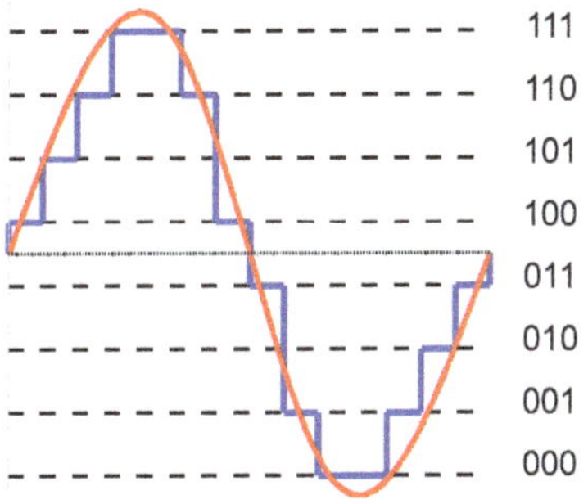

3-bit resolution with eight levels.

Analog-to-digital Converter (ADC)

Outside the realm of signal processing, this category may simply be called *rounding* or *scalar quantization*. An ADC can be modeled as two processes: sampling and quantization. Sampling converts a voltage signal (function of time) into a discrete-time signal (sequence of real numbers). Quantization replaces each real number with an approximation from a finite set of discrete values (levels), which is necessary for storage and processing by numerical methods. Most commonly, these discrete values are represented as fixed-point words (either proportional to the waveform values or companded) or floating-point words. Common word-lengths are 8-bit (256 levels), 16-bit (65,536 levels), 32-bit (4.3 billion levels), and so on, though any number of quantization levels is possible (not just powers of two). Quantizing a sequence of numbers produces a sequence of quantization errors which is sometimes modeled as an additive random signal called quantization noise because of its stochastic behavior. The more levels a quantizer uses, the lower is its quantization noise power.

In general, both ADC processes lose some information. So discrete-valued signals are only an approximation of the continuous-valued discrete-time signal, which is itself only an approximation of the original continuous-valued continuous-time signal. But both types of approximation errors can, in theory, be made arbitrarily small by good design.

Rate–distortion Optimization

Rate–distortion optimized quantization is encountered in source coding for "lossy" data compression algorithms, where the purpose is to manage distortion within the limits of the bit rate supported by a communication channel or storage medium. In this second setting, the amount of introduced distortion may be managed carefully by sophisticated techniques, and introducing some significant amount of distortion may be unavoidable. A quantizer designed for this purpose may be quite different and more elaborate in design than an ordinary rounding operation. It is in this domain that substantial rate–distortion theory analysis is likely to be applied. However, the same concepts actually apply in both use cases.

The analysis of quantization involves studying the amount of data (typically measured in digits or bits or bit *rate*) that is used to represent the output of the quantizer, and studying the loss of precision that is introduced by the quantization process (which is referred to as the *distortion*). The general field of such study of rate and distortion is known as *rate–distortion theory*.

Rounding Example

As an example, rounding a real number x to the nearest integer value forms a very basic type of quantizer – a *uniform* one. A typical (*mid-tread*) uniform quantizer with a quantization *step size* equal to some value Ä can be expressed as

$$Q(x) = \Delta \cdot \left\lfloor \frac{x}{\Delta} + \frac{1}{2} \right\rfloor = \Delta \cdot \text{floor}\left(\frac{x}{\Delta} + \frac{1}{2} \right),$$

where the notation $\lfloor\,\rfloor$ or $floor()$ depicts the floor function. For simple rounding to the nearest integer, the step size Δ is equal to 1. With $\Delta = 1$ or with Δ equal to any other integer value, this quantizer has real-valued inputs and integer-valued outputs, although this property is not a necessity – a quantizer may also have an integer input domain and may also have non-integer output values. The essential property of a quantizer is that it has a countable set of possible output values that has fewer members than the set of possible input values. The members of the set of output values may have integer, rational, or real values (or even other possible values as well, in general – such as vector values or complex numbers).

When the quantization step size is small (relative to the variation in the signal being measured), it is relatively simple to show that the mean squared error produced by such a rounding operation will be approximately $\Delta^2/12$. Mean squared error is also called the quantization noise power. Adding one bit to the quantizer halves the value of Δ, which reduces the noise power by the factor ¼. In terms of decibels, the noise power change is $10 \cdot \log_{10}\left(\frac{1}{4}\right) \approx -6$ dB.

Because the set of possible output values of a quantizer is countable, any quantizer can be decomposed into two distinct stages, which can be referred to as the *classification* stage (or *forward quantization* stage) and the *reconstruction* stage (or *inverse quantization* stage), where the classification stage maps the input value to an integer *quantization index* k and the reconstruction stage maps the index k to the *reconstruction value* y_k that is the output approximation of the input value. For the example uniform quantizer described above, the forward quantization stage can be expressed as

$$k = \left\lfloor \frac{x}{\Delta} + \frac{1}{2} \right\rfloor,$$

and the reconstruction stage for this example quantizer is simply

$$y_k = k \cdot \Delta.$$

This decomposition is useful for the design and analysis of quantization behavior, and it illustrates how the quantized data can be communicated over a communication channel – a *source encoder* can perform the forward quantization stage and send the index information through a communication channel (possibly applying entropy coding techniques to the quantization indices), and a *decoder* can perform the reconstruction stage to produce the output approximation of the original input data. In more elaborate quantization designs, both the forward and inverse quantization stages may be substantially more complex. In general, the forward quantization stage may use any function that maps the input data to the integer space of the quantization index data, and the inverse quantization stage can conceptually (or literally) be a table look-up operation to map each quantization index to a corresponding reconstruction value. This two-stage decomposition applies equally well to vector as well as scalar quantizers.

Mid-riser and mid-tread Uniform Quantizers

Most uniform quantizers for signed input data can be classified as being of one of two types: mid-riser and mid-tread. The terminology is based on what happens in the region around the value 0, and uses the analogy of viewing the input-output function of the quantizer as a stairway. Mid-tread quantizers have a zero-valued reconstruction level (corresponding to a *tread* of a stairway), while mid-riser quantizers have a zero-valued classification threshold (corresponding to a *riser* of a stairway).

The formulas for mid-tread uniform quantization are provided in the previous section.

The input-output formula for a mid-riser uniform quantizer is given by:

$$Q(x) = \Delta \cdot \left(\left\lfloor \frac{x}{\Delta} \right\rfloor + \frac{1}{2} \right),$$

where the classification rule is given by

$$k = \left\lfloor \frac{x}{\Delta} \right\rfloor$$

and the reconstruction rule is

$$y_k = \Delta \cdot \left(k + \tfrac{1}{2} \right).$$

Note that mid-riser uniform quantizers do not have a zero output value – their minimum output magnitude is half the step size. When the input data can be modeled as a random variable with a probability density function (pdf) that is smooth and symmetric around zero, mid-riser quantizers also always produce an output *entropy* of at least 1 bit per sample.

In contrast, mid-tread quantizers do have a zero output level, and can reach arbitrarily low bit rates per sample for input distributions that are symmetric and taper off at higher magnitudes. For some applications, having a zero output signal representation or supporting low output entropy may be a necessity. In such cases, using a mid-tread uniform quantizer may be appropriate while using a mid-riser one would not be.

In general, a mid-riser or mid-tread quantizer may not actually be a *uniform* quantizer – i.e., the size of the quantizer's classification intervals may not all be the same, or the spacing between its possible output values may not all be the same. The distinguishing characteristic of a mid-riser quantizer is that it has a classification threshold value that is exactly zero, and the distinguishing characteristic of a mid-tread quantizer is that is it has a reconstruction value that is exactly zero.

Dead-zone Quantizers

Another name for a mid-tread quantizer with symmetric behavior around 0 is dead-zone quantizer, and the classification region around the zero output value of such a quantizer is referred to as the *dead zone* or *deadband.* The dead zone can sometimes serve the same purpose as a noise gate or squelch function. Especially for compression applications, the dead-zone may be given a different width than that for the other steps. For an otherwise-uniform quantizer, the dead-zone width can be set to any value w by using the forward quantization rule

$$k = \operatorname{sgn}(x) \cdot \max\left(0, \left\lfloor \frac{|x| - w/2}{\Delta} + 1 \right\rfloor\right),$$

where the function sgn $()$ is the sign function (also known as the *signum* function). The general reconstruction rule for such a dead-zone quantizer is given by

$$y_k = \operatorname{sgn}(k) \cdot \left(\frac{w}{2} + \Delta \cdot (|k| - 1 + r_k) \right),$$

where r_k is a reconstruction offset value in the range of 0 to 1 as a fraction of the step size. Ordinarily, $0 \le r_k \le \frac{1}{2}$ when quantizing input data with a typical pdf that is symmetric around zero and reaches its peak value at zero (such as a Gaussian, Laplacian, or Generalized Gaussian pdf). Although r_k may depend on k in general, and can be chosen to fulfill the optimality condition described below, it is often simply set to a constant, such as $\frac{1}{2}$. (Note that in this definition, $y_0 = 0$ due to the definition of the sgn ( ) function, so r_0 has no effect.)

A very commonly used special case (e.g., the scheme typically used in financial accounting and elementary mathematics) is to set $w = \Delta$ and $r_k = \frac{1}{2}$ for all k.

Granular Distortion and Overload Distortion

Often the design of a quantizer involves supporting only a limited range of possible output values and performing clipping to limit the output to this range whenever the input exceeds the supported range. The error introduced by this clipping is referred to as *overload* distortion. Within the extreme limits of the supported range, the amount of spacing between the selectable output values of a quantizer is referred to as its *granularity*, and the error introduced by this spacing is referred to as *granular* distortion. It is common for the design of a quantizer to involve determining the proper balance between granular distortion and overload distortion. For a given supported number of possible output values, reducing the average granular distortion may involve increasing the average overload distortion, and vice versa. A technique for controlling the amplitude of the signal (or, equivalently, the quantization step size Δ) to achieve the appropriate balance is the use of *automatic gain control* (AGC). However, in some quantizer designs, the concepts of granular error and overload error may not apply (e.g., for a quantizer with a limited range of input data or with a countably infinite set of selectable output values).

The Additive Noise Model for Quantization Error

A common assumption for the analysis of quantization error is that it affects a signal processing system in a similar manner to that of additive white noise – having negligible correlation with the signal and an approximately flat power spectral density. The additive noise model is commonly used for the analysis of quantization error effects in digital filtering systems, and it can be very useful in such analysis. It has been shown to be a valid model in cases of high resolution quantization (small Δ relative to the signal strength) with smooth probability density functions. However, additive noise behaviour is not always a valid assumption, and care should be taken to avoid assuming that this model always applies. In actuality, the quantization error (for quantizers defined as described here) is deterministically related to the signal rather than being independent of it. Thus, periodic signals can create periodic quantization noise. And in some cases it can even cause limit cycles to appear in digital signal processing systems.

One way to ensure effective independence of the quantization error from the source signal is to perform *dithered quantization* (sometimes with *noise shaping*), which involves adding random (or pseudo-random) noise to the signal prior to quantization.

This can sometimes be beneficial for such purposes as improving the subjective quality of the result, however it can increase the total quantity of error introduced by the quantization process.

Quantization Error Models

In the typical case, the original signal is much larger than one least significant bit (LSB). When this is the case, the quantization error is not significantly correlated with the signal, and has an approximately uniform distribution. In the rounding case, the quantization error has a mean of zero and the RMS value is the standard deviation of this distribution, given by $\frac{1}{\sqrt{12}}\text{LSB} \approx 0.289\text{LSB}$. In the truncation case the error has a non-zero mean of $\frac{1}{2}\text{LSB}$ and the RMS value is $\frac{1}{\sqrt{3}}\text{LSB}$. In either case, the standard deviation, as a percentage of the full signal range, changes by a factor of 2 for each 1-bit change in the number of quantizer bits. The potential signal-to-quantization-noise power ratio therefore changes by 4, or $10 \cdot \log_{10}(4) = 6.02$ *decibels per bit.*

At lower amplitudes the quantization error becomes dependent on the input signal, resulting in distortion. This distortion is created after the anti-aliasing filter, and if these distortions are above 1/2 the sample rate they will alias back into the band of interest. In order to make the quantization error independent of the input signal, noise with an amplitude of 2 least significant bits is added to the signal. This slightly reduces signal to noise ratio, but, ideally, completely eliminates the distortion. It is known as dither.

Quantization Noise Model

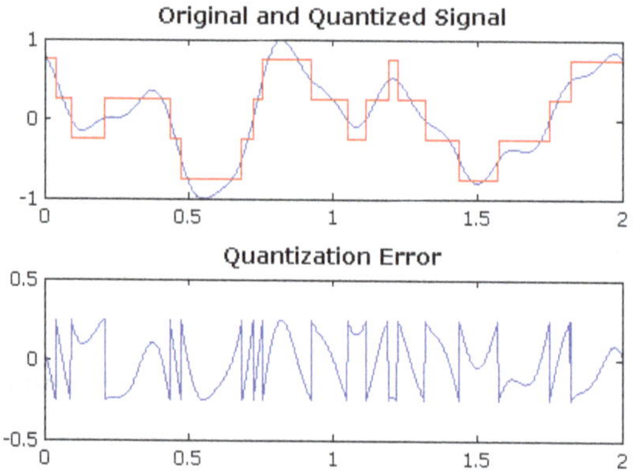

Quantization noise for a 2-bit ADC operating at infinite sample rate. The difference between the blue and red signals in the upper graph is the quantization error, which is "added" to the quantized signal and is the source of noise.

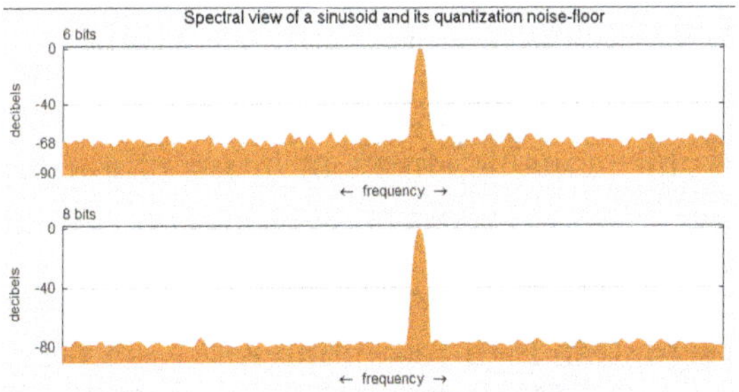

Comparison of quantizing a sinusoid to 64 levels (6 bits) and 256 levels (8 bits). The additive noise created by 6-bit quantization is 12 dB greater than the noise created by 8-bit quantization. When the spectral distribution is flat, as in this example, the 12 dB difference manifests as a measurable difference in the noise floors.

Quantization noise is a model of quantization error introduced by quantization in the analog-to-digital conversion (ADC) in telecommunication systems and signal processing. It is a rounding error between the analog input voltage to the ADC and the output digitized value. The noise is non-linear and signal-dependent. It can be modelled in several different ways.

In an ideal analog-to-digital converter, where the quantization error is uniformly distributed between −1/2 LSB and +1/2 LSB, and the signal has a uniform distribution covering all quantization levels, the Signal-to-quantization-noise ratio (SQNR) can be calculated from

$$\mathrm{SQNR} = 20\log_{10}(2^{Q}) \approx 6.02 \cdot Q \text{ dB}$$

Where Q is the number of quantization bits.

The most common test signals that fulfill this are full amplitude triangle waves and sawtooth waves.

For example, a 16-bit ADC has a maximum signal-to-noise ratio of 6.02 × 16 = 96.3 dB.

When the input signal is a full-amplitude sine wave the distribution of the signal is no longer uniform, and the corresponding equation is instead

$$\mathrm{SQNR} \approx 1.761 + 6.02 \cdot Q \text{ dB}$$

Here, the quantization noise is once again *assumed* to be uniformly distributed. When the input signal has a high amplitude and a wide frequency spectrum this is the case. In this case a 16-bit ADC has a maximum signal-to-noise ratio of 98.09 dB. The 1.761 difference in signal-to-noise only occurs due to the signal being a full-scale sine wave instead of a triangle/sawtooth.

Quantization noise power can be derived from

$$\mathrm{N} = \frac{(\delta \mathrm{v})^2}{12}\mathrm{W}$$

where δv is the voltage of the level.

(Typical real-life values are worse than this theoretical minimum, due to the addition of dither to reduce the objectionable effects of quantization, and to imperfections of the ADC circuitry.)

For complex signals in high-resolution ADCs this is an accurate model. For low-resolution ADCs, low-level signals in high-resolution ADCs, and for simple waveforms the quantization noise is not uniformly distributed, making this model inaccurate. In these cases the quantization noise distribution is strongly affected by the exact amplitude of the signal.

The calculations above, however, assume a completely filled input channel. If this is not the case - if the input signal is small - the relative quantization distortion can be very large. To circumvent this issue, analog compressors and expanders can be used, but these introduce large amounts of distortion as well, especially if the compressor does not match the expander. The application of such compressors and expanders is also known as companding.

Rate–distortion Quantizer Design

A scalar quantizer, which performs a quantization operation, can ordinarily be decomposed into two stages:

- Classification: A process that classifies the input signal range into M non-overlapping intervals $\{I_k\}_{k=1}^{M}$, by defining $M-1$ boundary (decision) values $\{b_k\}_{k=1}^{M-1}$, such that $I_k = [b_{k-1},\ b_k)$ for $k = 1, 2, \ldots, M$, with the extreme limits defined by $b_0 = -\infty$ and $b_M = \infty$. All the inputs x that fall in a given interval range I_k are associated with the same quantization index k.
- Reconstruction: Each interval I_k is represented by a reconstruction value y_k which implements the mapping $x \in I_k \Rightarrow y = y_k$.

These two stages together comprise the mathematical operation of $y = Q(x)$.

Entropy coding techniques can be applied to communicate the quantization indices from a source encoder that performs the classification stage to a decoder that performs the reconstruction stage. One way to do this is to associate each quantization index k with a binary codeword c_k. An important consideration is the number of bits used for each codeword, denoted here by $\text{length}(c_k)$.

As a result, the design of an M-level quantizer and an associated set of codewords for communicating its index values requires finding the values of $\{b_k\}_{k=1}^{M-1}$, $\{c_k\}_{k=1}^{M}$ and $\{y_k\}_{k=1}^{M}$ which optimally satisfy a selected set of design constraints such as the bit rate R and distortion D.

Assuming that an information source S produces random variables X with an associated probability density function $f(x)$, the probability S that the random variable falls within a particular quantization interval I_k is given by

$$p_k = P[x \in I_k] = \int_{b_{k-1}}^{b_k} f(x)dx.$$

The resulting bit rate R, in units of average bits per quantized value, for this quantizer can be derived as follows:

$$R = \sum_{k=1}^{M} p_k \cdot \text{length}(c_k) = \sum_{k=1}^{M} \text{length}(c_k) \int_{b_{k-1}}^{b_k} f(x)dx.$$

If it is assumed that distortion is measured by mean squared error, the distortion D, is given by:

$$D = E[(x - Q(x))^2] = \int_{-\infty}^{\infty} (x - Q(x))^2 f(x)dx = \sum_{k=1}^{M} \int_{b_{k-1}}^{b_k} (x - y_k)^2 f(x)dx.$$

Note that other distortion measures can also be considered, although mean squared error is a popular one.

A key observation is that rate R depends on the decision boundaries $\{b_k\}_{k=1}^{M-1}$ and the codeword lengths $(c_k)\}_{k=1}^{M}$, whereas the distortion D depends on the decision boundaries $\{b_k\}_{k=1}^{M-1}$ and the reconstruction levels $\{y_k\}_{k=1}^{M}$.

After defining these two performance metrics for the quantizer, a typical Rate–Distortion formulation for a quantizer design problem can be expressed in one of two ways:

1. Given a maximum distortion constraint $D \leq D_{\max}$, minimize the bit rate R
2. Given a maximum bit rate constraint $R \leq R_{\max}$, minimize the distortion D

Often the solution to these problems can be equivalently (or approximately) expressed and solved by converting the formulation to the unconstrained problem $\min\{D + \lambda \cdot R\}$ where the Lagrange multiplier λ is a non-negative constant that establishes the appropriate balance between rate and distortion. Solving the unconstrained problem is equivalent to finding a point on the convex hull of the family of solutions to an equivalent constrained formulation of the problem. However, finding a solution – especially a closed-form solution – to any of these three problem formulations can be difficult. Solutions that do not require multi-dimensional iterative optimization techniques have been published for only three probability distribution functions: the uniform, exponential, and Laplacian distributions. Iterative optimization approaches can be used to find solutions in other cases.

Note that the reconstruction values $\{y_k\}_{k=1}^{M}$ affect only the distortion – they do not affect the bit rate – and that each individual y_k makes a separate contribution d_k to the total distortion as shown below:

$$D = \sum_{k=1}^{M} d_k$$

where

$$d_k = \int_{b_{k-1}}^{b_k} (x - y_k)^2 f(x) dx$$

This observation can be used to ease the analysis – given the set of $\{b_k\}_{k=1}^{M-1}$ values, the value of each y_k can be optimized separately to minimize its contribution to the distortion D.

For the mean-square error distortion criterion, it can be easily shown that the optimal set of reconstruction values $\{y_k^*\}_{k=1}^{M}$ is given by setting the reconstruction value y_k within each interval I_k to the conditional expected value (also referred to as the *centroid*) within the interval, as given by:

$$y_k^* = \frac{1}{p_k} \int_{b_{k-1}}^{b_k} x f(x) dx .$$

The use of sufficiently well-designed entropy coding techniques can result in the use of a bit rate that is close to the true information content of the indices $\{k\}_{k=1}^{M}$, such that effectively

$$\text{length}(c_k) \approx -\log_2(p_k)$$

and therefore

$$R = \sum_{k=1}^{M} -p_k \cdot \log_2(p_k).$$

The use of this approximation can allow the entropy coding design problem to be separated from the design of the quantizer itself. Modern entropy coding techniques such as arithmetic coding can achieve bit rates that are very close to the true entropy of a source, given a set of known (or adaptively estimated) probabilities $\{p_k\}_{k=1}^{M}$.

In some designs, rather than optimizing for a particular number of classification regions M, the quantizer design problem may include optimization of the value of M as well. For some probabilistic source models, the best performance may be achieved when M approaches infinity.

Neglecting the Entropy Constraint: Lloyd–Max Quantization

In the above formulation, if the bit rate constraint is neglected by setting λ equal to 0, or equivalently if it is assumed that a fixed-length code (FLC) will be used to represent the quantized data instead of a variable-length code (or some other entropy coding technology such as arithmetic coding that is better than an FLC in the rate–distortion sense), the optimization problem reduces to minimization of distortion D alone.

The indices produced by an M -level quantizer can be coded using a fixed-length code using $R=\lceil \log_2 M \rceil$ bits/symbol. For example when $M=256$ levels, the FLC bit rate R is 8 bits/symbol. For this reason, such a quantizer has sometimes been called an 8-bit quantizer. However using an FLC eliminates the compression improvement that can be obtained by use of better entropy coding.

Assuming an FLC with M levels, the Rate–Distortion minimization problem can be reduced to distortion minimization alone. The reduced problem can be stated as follows: given a source X with pdf $f(x)$ and the constraint that the quantizer must use only M classification regions, find the decision boundaries $\{b_k\}_{k=1}^{M-1}$ and reconstruction levels $\{y_k\}_{k=1}^{M}$ to minimize the resulting distortion

$$D = E[(x-Q(x))^2] = \int_{-\infty}^{\infty} (x-Q(x))^2 f(x)dx = \sum_{k=1}^{M} \int_{b_{k-1}}^{b_k} (x-y_k)^2 f(x)dx = \sum_{k=1}^{M} d_k .$$

Finding an optimal solution to the above problem results in a quantizer sometimes called a MMSQE (minimum mean-square quantization error) solution, and the resulting pdf-optimized (non-uniform) quantizer is referred to as a *Lloyd–Max* quantizer, named after two people who independently developed iterative methods to solve the two sets of simultaneous equations resulting from $\partial D / \partial b_k = 0$ and $\partial D / \partial y_k = 0$, as follows:

$$\frac{\partial D}{\partial b_k} = 0 \Rightarrow b_k = \frac{y_k + y_{k+1}}{2},$$

which places each threshold at the midpoint between each pair of reconstruction values, and

$$\frac{\partial D}{\partial y_k} = 0 \Rightarrow y_k = \frac{\int_{b_{k-1}}^{b_k} xf(x)dx}{\int_{b_{k-1}}^{b_k} f(x)dx} = \frac{1}{p_k} \int_{b_{k-1}}^{b_k} xf(x)dx$$

which places each reconstruction value at the centroid (conditional expected value) of its associated classification interval.

Lloyd's Method I algorithm, originally described in 1957, can be generalized in a straightforward way for application to vector data. This generalization results in the

Linde–Buzo–Gray (LBG) or k-means classifier optimization methods. Moreover, the technique can be further generalized in a straightforward way to also include an entropy constraint for vector data.

Uniform Quantization and the 6 dB/Bit Approximation

The Lloyd–Max quantizer is actually a uniform quantizer when the input pdf is uniformly distributed over the range $[y_1 - \Delta/2, y_M + \Delta/2)$.. However, for a source that does not have a uniform distribution, the minimum-distortion quantizer may not be a uniform quantizer.

The analysis of a uniform quantizer applied to a uniformly distributed source can be summarized in what follows:

A symmetric source X can be modelled with $f(x) = \frac{1}{2X_{max}}$, , for $x \in [-X_{max}, X_{max}]$ and o elsewhere. The step size $\Delta = \frac{2X_{max}}{M}$ and the *signal to quantization noise ratio* (SQNR) of the quantizer is

$$\text{SQNR} = 10\log_{10}\frac{\sigma_x^2}{\sigma_q^2} = 10\log_{10}\frac{(M\Delta)^2/12}{\Delta^2/12} = 10\log_{10} M^2 = 20\log_{10} M .$$

For a fixed-length code using N bits, $M = 2^N$, resulting in $\text{SQNR} = 20\log_{10} 2^N =$ $N \cdot (20\log_{10} 2) = N \cdot 6.0206\text{dB}$,

or approximately 6 dB per bit. For example, for N =8 bits, M =256 levels and SQNR = 8*6 = 48 dB; and for N =16 bits, M =65536 and SQNR = 16*6 = 96 dB. The property of 6 dB improvement in SQNR for each extra bit used in quantization is a well-known figure of merit. However, it must be used with care: this derivation is only for a uniform quantizer applied to a uniform source.

For other source pdfs and other quantizer designs, the SQNR may be somewhat different from that predicted by 6 dB/bit, depending on the type of pdf, the type of source, the type of quantizer, and the bit rate range of operation.

However, it is common to assume that for many sources, the slope of a quantizer SQNR function can be approximated as 6 dB/bit when operating at a sufficiently high bit rate. At asymptotically high bit rates, cutting the step size in half increases the bit rate by approximately 1 bit per sample (because 1 bit is needed to indicate whether the value is in the left or right half of the prior double-sized interval) and reduces the mean squared error by a factor of 4 (i.e., 6 dB) based on the $\Delta^2/12$ approximation.

At asymptotically high bit rates, the 6 dB/bit approximation is supported for many source pdfs by rigorous theoretical analysis. Moreover, the structure of the optimal scalar quantizer (in the rate–distortion sense) approaches that of a uniform quantizer under these conditions.

Other Fields

Many physical quantities are actually quantized by physical entities. Examples of fields where this limitation applies include electronics (due to electrons), optics (due to photons), biology (due to DNA), physics (due to Planck limits) and chemistry (due to molecules). This is sometimes known as the "quantum noise limit" of systems in those fields. This is a different manifestation of "quantization error," in which theoretical models may be analog but physically occurs digitally. Around the quantum limit, the distinction between analog and digital quantities vanishes.

Discretization

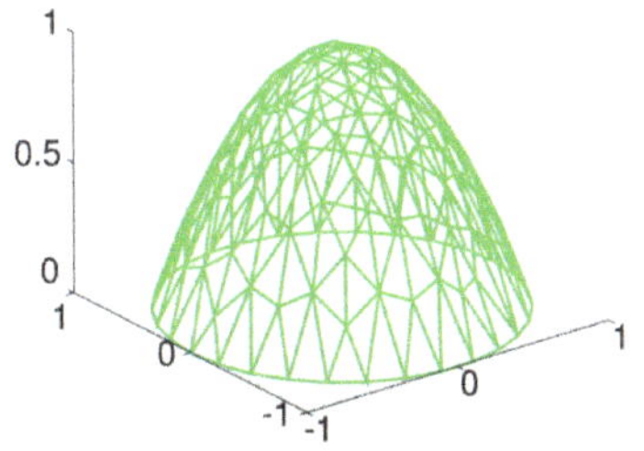

A solution to a discretized partial differential equation, obtained with the finite element method.

In mathematics, discretization concerns the process of transferring continuous functions, models, and equations into discrete counterparts. This process is usually carried out as a first step toward making them suitable for numerical evaluation and implementation on digital computers. Processing on a digital computer requires another process called quantization. Dichotomization is the special case of discretization in which the number of discrete classes is 2, which can approximate a continuous variable as a binary variable (creating a dichotomy for modeling purposes, as in binary classification).

- Euler–Maruyama method
- Zero-order hold

Discretization is also related to discrete mathematics, and is an important component of granular computing. In this context, *discretization* may also refer to modification of variable or category *granularity*, as when multiple discrete variables are aggregated or multiple discrete categories fused.

Whenever continuous data is discretized, there is always some amount of discretization error. The goal is to reduce the amount to a level considered negligible for the modeling purposes at hand.

Discretization of Linear State Space Models

Discretization is also concerned with the transformation of continuous differential equations into discrete difference equations, suitable for numerical computing.

The following continuous-time state space model

$$\dot{\mathbf{x}}(t) = \mathbf{A}\mathbf{x}(t) + \mathbf{B}\mathbf{u}(t) + \mathbf{w}(t)$$

$$\mathbf{y}(t) = \mathbf{C}\mathbf{x}(t) + \mathbf{D}\mathbf{u}(t) + \mathbf{v}(t)$$

where v and w are continuous zero-mean white noise sources with covariances

$$\mathbf{w}(t) \sim N(0, \mathbf{Q})$$

$$\mathbf{v}(t) \sim N(0, \mathbf{R})$$

can be discretized, assuming zero-order hold for the input u and continuous integration for the noise v, to

$$\mathbf{x}[k+1] = \mathbf{A}_d\mathbf{x}[k] + \mathbf{B}_d\mathbf{u}[k] + \mathbf{w}[k]$$

$$\mathbf{y}[k] = \mathbf{C}_d\mathbf{x}[k] + \mathbf{D}_d\mathbf{u}[k] + \mathbf{v}[k]$$

with covariances

$$\mathbf{w}[k] \sim N(0, \mathbf{Q}_d)$$

$$\mathbf{v}[k] \sim N(0, \mathbf{R}_d)$$

where

$$\mathbf{A}_d = e^{\mathbf{A}T} = \mathcal{L}^{-1}\{(s\mathbf{I} - \mathbf{A})^{-1}\}_{t=T}$$

$$\mathbf{B}_d = \left(\int_{\tau=0}^{T} e^{\mathbf{A}\tau} d\tau \right) \mathbf{B} = \mathbf{A}^{-1}(\mathbf{A}_d - I)\mathbf{B},$$

if $\mathbf{A}$ is nonsingular

$$\mathbf{C}_d = \mathbf{C}$$

$$\mathbf{D}_d = \mathbf{D}$$

$$\mathbf{Q}_d = \int_{\tau=0}^{T} e^{\mathbf{A}\tau}\mathbf{Q}e^{\mathbf{A}^T\tau}d\tau$$

$$\mathbf{R}_d = \frac{1}{T}\mathbf{R}$$

and T is the sample time, although $\mathbf{A}^T$ is the transposed matrix of $\mathbf{A}$.

A clever trick to compute A_d and B_d in one step is by utilizing the following property:

$$e^{\begin{bmatrix}\mathbf{A} & \mathbf{B}\\ \mathbf{0} & \mathbf{0}\end{bmatrix}T} = \begin{bmatrix}\mathbf{M}_{11} & \mathbf{M}_{12}\\ \mathbf{0} & \mathbf{I}\end{bmatrix}$$

and then having

$$\mathbf{A}_d = \mathbf{M}_{11}$$

$$\mathbf{B}_d = \mathbf{M}_{12}$$

Discretization of Process Noise

Numerical evaluation of $\mathbf{Q}_d$ is a bit trickier due to the matrix exponential integral. It can, however, be computed by first constructing a matrix, and computing the exponential of it (Van Loan, 1978):

$$\mathbf{F} = \begin{bmatrix}-\mathbf{A} & \mathbf{Q}\\ \mathbf{0} & \mathbf{A}^T\end{bmatrix}T$$

$$\mathbf{G} = e^{\mathbf{F}} = \begin{bmatrix}\cdots & \mathbf{A}_d^{-1}\mathbf{Q}_d\\ \mathbf{0} & \mathbf{A}_d^T\end{bmatrix}.$$

The discretized process noise is then evaluated by multiplying the transpose of the lower-right partition of G with the upper-right partition of G:

$$\mathbf{Q}_d = (\mathbf{A}_d^T)^T(\mathbf{A}_d^{-1}\mathbf{Q}_d).$$

Derivation

Starting with the continuous model

$$\dot{\mathbf{x}}(t) = \mathbf{A}\mathbf{x}(t) + \mathbf{B}\mathbf{u}(t)$$

we know that the matrix exponential is

$$\frac{d}{dt}e^{\mathbf{A}t} = \mathbf{A}e^{\mathbf{A}t} = e^{\mathbf{A}t}\mathbf{A}$$

and by premultiplying the model we get

$$e^{-\mathbf{A}t}\dot{\mathbf{x}}(t) = e^{-\mathbf{A}t}\mathbf{A}\mathbf{x}(t) + e^{-\mathbf{A}t}\mathbf{B}\mathbf{u}(t)$$

which we recognize as

$$\frac{d}{dt}(e^{-\mathbf{A}t}\mathbf{x}(t)) = e^{-\mathbf{A}t}\mathbf{B}\mathbf{u}(t)$$

and by integrating.

$$e^{-\mathbf{A}t}\mathbf{x}(t) - e^{0}\mathbf{x}(0) = \int_{0}^{t} e^{-\mathbf{A}\tau}\mathbf{B}\mathbf{u}(\tau)d\tau$$

$$\mathbf{x}(t) = e^{\mathbf{A}t}\mathbf{x}(0) + \int_{0}^{t} e^{\mathbf{A}(t-\tau)}\mathbf{B}\mathbf{u}(\tau)d\tau$$

which is an analytical solution to the continuous model.

Now we want to discretise the above expression. We assume that u is constant during each timestep.

$$\mathbf{x}[k] \overset{\text{def}}{=} \mathbf{x}(kT)$$

$$\mathbf{x}[k] = e^{\mathbf{A}kT}\mathbf{x}(0) + \int_{0}^{kT} e^{\mathbf{A}(kT-\tau)}\mathbf{B}\mathbf{u}(\tau)d\tau$$

$$\mathbf{x}[k+1] = e^{\mathbf{A}(k+1)T}\mathbf{x}(0) + \int_{0}^{(k+1)T} e^{\mathbf{A}((k+1)T-\tau)}\mathbf{B}\mathbf{u}(\tau)d\tau$$

$$\mathbf{x}[k+1] = e^{\mathbf{A}T}\left[e^{\mathbf{A}kT}\mathbf{x}(0) + \int_{0}^{kT} e^{\mathbf{A}(kT-\tau)}\mathbf{B}\mathbf{u}(\tau)d\tau\right] + \int_{kT}^{(k+1)T} e^{\mathbf{A}(kT+T-\tau)}\mathbf{B}\mathbf{u}(\tau)d\tau$$

We recognize the bracketed expression as $\mathbf{x}[k]$, and the second term can be simplified by substituting $v = kT + T - \tau$. We also assume that $\mathbf{u}$ is constant during the integral, which in turn yields

$$\begin{aligned}
\mathbf{x}[k+1] &= e^{\mathbf{A}T}\mathbf{x}[k] + \left(\int_{0}^{T} e^{\mathbf{A}v}dv\right)\mathbf{B}\mathbf{u}[k] \\
&= e^{\mathbf{A}T}\mathbf{x}[k] + \mathbf{A}^{-1}\left(e^{\mathbf{A}T} - \mathbf{I}\right)\mathbf{B}\mathbf{u}[k]
\end{aligned}$$

which is an exact solution to the discretization problem.

Approximations

Exact discretization may sometimes be intractable due to the heavy matrix exponential and integral operations involved. It is much easier to calculate an approximate discrete model, based on that for small timesteps $e^{\mathbf{A}T} \approx \mathbf{I} + \mathbf{A}T$.. The approximate solution then becomes:

$$\mathbf{x}[k+1] \approx (\mathbf{I} + \mathbf{A}T)\mathbf{x}[k] + T\mathbf{B}\mathbf{u}[k]$$

Other possible approximations are $e^{\mathbf{A}T} \approx \left(\mathbf{I} - \mathbf{A}T\right)^{-1}$ and $e^{\mathbf{A}T} \approx \left(\mathbf{I} + \frac{1}{2}\mathbf{A}T\right)\left(\mathbf{I} - \frac{1}{2}\mathbf{A}T\right)^{-1}$.

Each of them have different stability properties. The last one is known as the bilinear transform, or Tustin transform, and preserves the (in)stability of the continuous-time system.

Discretization of Continuous Features

In statistics and machine learning, discretization refers to the process of converting continuous features or variables to discretized or nominal features. This can be useful when creating probability mass functions.

References

- Phillips, C.l., Parr, J.M., & Riskin, E.A (2007). Signals, systems and Transforms. Prentice Hall. ISBN 0-13-041207-4.
- Sayood, Khalid (2005), Introduction to Data Compression, Third Edition, Morgan Kaufmann, ISBN 978-0-12-620862-7
- Jayant, Nikil S.; Noll, Peter (1984), Digital Coding of Waveforms: Principles and Applications to Speech and Video, Prentice–Hall, ISBN 978-0-13-211913-9
- Stein, Seymour; Jones, J. Jay (1967), Modern Communication Principles, McGraw–Hill, ISBN 978-0-07-061003-3
- Crutchfield, Steve (October 12, 2010), "The Joy of Convolution", Johns Hopkins University, retrieved November 21, 2010

4

Techniques and Methods of Digital Signal Processing

Digital signal processing has a number of techniques and methods; some of these are Z-transform, advanced Z-transform, matched Z- transform method, Zak transform, etc. The topics elaborated in this chapter will help in gaining a better perspective about the techniques and methods of digital signal processing.

Z-transform

In mathematics and signal processing, the Z-transform converts a discrete-time signal, which is a sequence of real or complex numbers, into a complex frequency domain representation.

It can be considered as a discrete-time equivalent of the Laplace transform. This similarity is explored in the theory of time scale calculus.

History

The basic idea now known as the Z-transform was known to Laplace, and it was re-introduced in 1947 by W. Hurewicz and others as a way to treat sampled-data control systems used with radar. It gives a tractable way to solve linear, constant-coefficient difference equations. It was later dubbed "the z-transform" by Ragazzini and Zadeh in the sampled-data control group at Columbia University in 1952.

The modified or advanced Z-transform was later developed and popularized by E. I. Jury.

The idea contained within the Z-transform is also known in mathematical literature as the method of generating functions which can be traced back as early as 1730 when it was introduced by de Moivre in conjunction with probability theory. From a mathematical view the Z-transform can also be viewed as a Laurent series where one views the sequence of numbers under consideration as the (Laurent) expansion of an analytic function.

Definition

The Z-transform, like many integral transforms, can be defined as either a *one-sided* or *two-sided* transform.

Bilateral Z-transform

The *bilateral* or *two-sided* Z-transform of a discrete-time signal *x[n]* is the formal power series *X(z)* defined as

$$X(z) = \mathcal{Z}\{x[n]\} = \sum_{n=-\infty}^{\infty} x[n]z^{-n}$$

where *n* is an integer and *z* is, in general, a complex number:

$$z = Ae^{j\phi} = A(\cos\phi + j\sin\phi)$$

where *A* is the magnitude of *z*, *j* is the imaginary unit, and ϕ is the *complex argument* (also referred to as *angle* or *phase*) in radians.

Unilateral Z-transform

Alternatively, in cases where *x[n]* is defined only for $n \geq 0$, the *single-sided* or *unilateral* Z-transform is defined as

$$\mathrm{X(z)} = \mathcal{Z}\{\mathrm{x[n]}\} = \sum_{\mathrm{n}=0}^{\infty} \mathrm{x[n]z^{-n}}.$$

In signal processing, this definition can be used to evaluate the Z-transform of the unit impulse response of a discrete-time causal system.

An important example of the unilateral Z-transform is the probability-generating function, where the component *x[n]* is the probability that a discrete random variable takes the value *n*, and the function *X(z)* is usually written as *X(s)*, in terms of $s = z^{-1}$. The properties of Z-transforms (below) have useful interpretations in the context of probability theory.

Geophysical Definition

In geophysics, the usual definition for the Z-transform is a power series in *z* as opposed to z^{-1}. This convention is used, for example, by Robinson and Treitel and by Kanasewich. The geophysical definition is:

$$X(z) = \mathcal{Z}\{x[n]\} = \sum_{n} x[n]z^{n}.$$

The two definitions are equivalent; however, the difference results in a number of changes. For example, the location of zeros and poles move from inside the unit circle using one definition, to outside the unit circle using the other definition. Thus, care is required to note which definition is being used by a particular author.

Inverse Z-transform

The *inverse* Z-transform is

$$x[n] = \mathcal{Z}^{-1}\{X(z)\} = \frac{1}{2\pi j}\oint_C X(z)z^{n-1}dz$$

where C is a counterclockwise closed path encircling the origin and entirely in the region of convergence (ROC). In the case where the ROC is causal, this means the path C must encircle all of the poles of $X(z)$.

A special case of this contour integral occurs when C is the unit circle (and can be used when the ROC includes the unit circle which is always guaranteed when $X(z)$ is stable, i.e. all the poles are within the unit circle). The inverse Z-transform simplifies to the inverse discrete-time Fourier transform:

$$x[n] = \frac{1}{2\pi}\int_{-\pi}^{+\pi} X(e^{j\omega})e^{j\omega n}d\omega.$$

The Z-transform with a finite range of n and a finite number of uniformly spaced z values can be computed efficiently via Bluestein's FFT algorithm. The discrete-time Fourier transform (DTFT)—not to be confused with the discrete Fourier transform (DFT)—is a special case of such a Z-transform obtained by restricting z to lie on the unit circle.

Region of Convergence

The region of convergence (ROC) is the set of points in the complex plane for which the Z-transform summation converges.

$$ROC = \{z : \left|\sum_{n=-\infty}^{\infty} x[n]z^{-n}\right| < \infty\}$$

Example 1 (no ROC)

Let $x[n] = (0.5)^n$. Expanding $x[n]$ on the interval $(-\infty, \infty)$ it becomes

$$x[n] = \left\{\cdots, 0.5^{-3}, 0.5^{-2}, 0.5^{-1}, 1, 0.5, 0.5^2, 0.5^3, \cdots\right\} = \left\{\cdots, 2^3, 2^2, 2, 1, 0.5, 0.5^2, 0.5^3, \cdots\right\}.$$

Looking at the sum

$$\sum_{n=-\infty}^{\infty} x[n]z^{-n} \to \infty.$$

Therefore, there are no values of z that satisfy this condition.

Example 2 (causal ROC)

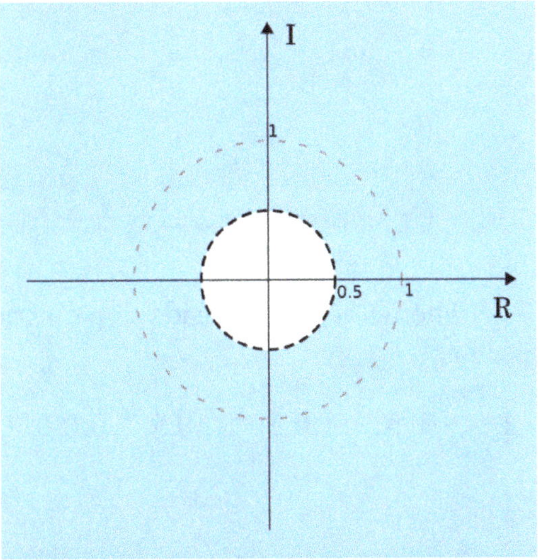

ROC shown in blue, the unit circle as a dotted grey circle (appears reddish to the eye) and the circle $|z| = 0.5$ is shown as a dashed black circle

Let $x[n] = 0.5^n u[n]$ (where u is the Heaviside step function). Expanding $x[n]$ on the interval $(-\infty, \infty)$ it becomes

$$x[n] = \{\cdots, 0, 0, 0, 1, 0.5, 0.5^2, 0.5^3, \cdots\}.$$

Looking at the sum

$$\sum_{n=-\infty}^{\infty} x[n]z^{-n} = \sum_{n=0}^{\infty} 0.5^n z^{-n} = \sum_{n=0}^{\infty} \left(\frac{0.5}{z}\right)^n = \frac{1}{1-0.5z^{-1}}.$$

The last equality arises from the infinite geometric series and the equality only holds if $|0.5z^{-1}| < 1$ which can be rewritten in terms of z as $|z| > 0.5$. Thus, the ROC is $|z| > 0.5$. In this case the ROC is the complex plane with a disc of radius 0.5 at the origin "punched out".

Example 3 (Anticausal ROC)

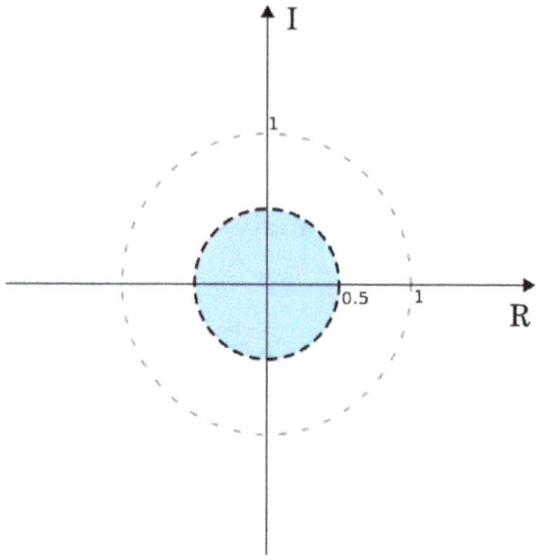

ROC shown in blue, the unit circle as a dotted grey circle and the circle $|z| = 0.5$ is shown as a dashed black circle

Let $x[n] = -(0.5)^n u[-n-1]$ (where u is the Heaviside step function). Expanding $x[n]$ on the interval $(-\infty, \infty)$ it becomes

$$x[n] = \left\{\cdots, -(0.5)^{-3}, -(0.5)^{-2}, -(0.5)^{-1}, 0, 0, 0, 0, \cdots\right\}.$$

Looking at the sum

$$\sum_{n=-\infty}^{\infty} x[n] z^{-n} = -\sum_{n=-\infty}^{-1} 0.5^n z^{-n} = -\sum_{m=1}^{\infty} \left(\frac{z}{0.5}\right)^m = -\frac{0.5^{-1} z}{1 - 0.5^{-1} z} = \frac{1}{1 - 0.5 z^{-1}}$$

Using the infinite geometric series, again, the equality only holds if $|0.5^{-1}z| < 1$ which can be rewritten in terms of z as $|z| < 0.5$. Thus, the ROC is $|z| < 0.5$. In this case the ROC is a disc centered at the origin and of radius 0.5.

What differentiates this example from the previous example is *only* the ROC. This is intentional to demonstrate that the transform result alone is insufficient.

Examples Conclusion

Examples 2 & 3 clearly show that the Z-transform $X(z)$ of $x[n]$ is unique when and only when specifying the ROC. Creating the pole–zero plot for the causal and anticausal case show that the ROC for either case does not include the pole that is at 0.5. This extends to cases with multiple poles: the ROC will *never* contain poles.

In example 2, the causal system yields an ROC that includes $|z| = \infty$ while the anticausal system in example 3 yields an ROC that includes $|z| = 0$.

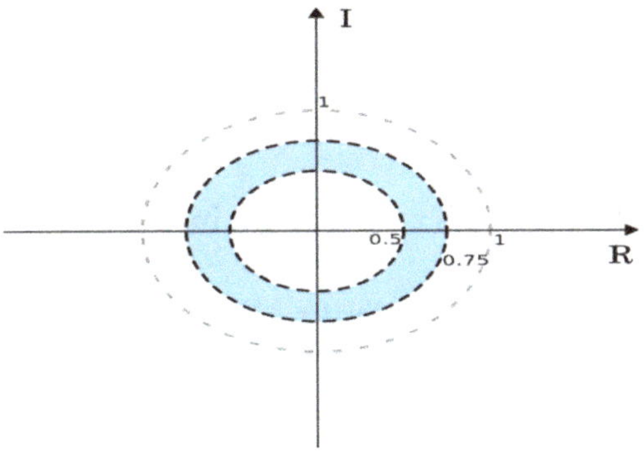

ROC shown as a blue ring 0.5 < |z| < 0.75

In systems with multiple poles it is possible to have an ROC that includes neither |z| = ∞ nor |z| = 0. The ROC creates a circular band. For example,

$$x[n] = 0.5^n u[n] - 0.75^n u[-n-1]$$

has poles at 0.5 and 0.75. The ROC will be 0.5 < |z| < 0.75, which includes neither the origin nor infinity. Such a system is called a mixed-causality system as it contains a causal term $(0.5)^n u[n]$ and an anticausal term $-(0.75)^n u[-n-1]$.

The stability of a system can also be determined by knowing the ROC alone. If the ROC contains the unit circle (i.e., |z| = 1) then the system is stable. In the above systems the causal system (Example 2) is stable because |z| > 0.5 contains the unit circle.

If you are provided a Z-transform of a system without an ROC (i.e., an ambiguous *x[n]*) you can determine a unique *x[n]* provided you desire the following:

- Stability
- Causality

If you need stability then the ROC must contain the unit circle. If you need a causal system then the ROC must contain infinity and the system function will be a right-sided sequence. If you need an anticausal system then the ROC must contain the origin and the system function will be a left-sided sequence. If you need both, stability and causality, all the poles of the system function must be inside the unit circle.

The unique *x[n]* can then be found.

Properties

Parseval's theorem

$$\sum_{n=-\infty}^{\infty} x_1[n]x_2^*[n] \quad = \quad \frac{1}{j2\pi}\oint_C X_1(v)X_2^*(\tfrac{1}{v^*})v^{-1}\mathrm{d}v$$

Initial value theorem: If $x[n]$ is causal, then

$$x[0] = \lim_{z \to \infty} X(z).$$

Final value theorem: If the poles of $(z-1)X(z)$ are inside the unit circle, then

$$x[\infty] = \lim_{z \to 1} (z-1)X(z).$$

Table of Common Z-transform Pairs

Here:

$$u : n \mapsto u[n] = \begin{cases} 1, & n \geq 0 \\ 0, & n < 0 \end{cases}$$

is the unit (or Heaviside) step function and

$$\delta : n \mapsto \delta[n] = \begin{cases} 1, & n = 0 \\ 0, & n \neq 0 \end{cases}$$

is the discrete-time unit impulse function (cf Dirac delta function which is a continuous-time version). The two functions are chosen together so that the unit step function is the accumulation (running total) of the unit impulse function.

Relationship to Fourier Series and Fourier Transform

For values of z in the region $|z|=1$, known as the unit circle, we can express the transform as a function of a single, real variable, ω, by defining $z=e^{j\omega}$. And the bi-lateral transform reduces to a Fourier series:

$$\sum_{n=-\infty}^{\infty} x[n]\, z^{-n} = \sum_{n=-\infty}^{\infty} x[n]\, e^{-j\omega n}, \qquad \textbf{\textit{(Eq.1)}}$$

which is also known as the discrete-time Fourier transform (DTFT) of the x[n] sequence. This 2π-periodic function is the periodic summation of a Fourier transform, which makes it a widely used analysis tool. To understand this, let X(f) be the Fourier transform of any function, x(t), whose samples at some interval, T, equal the x[n] sequence. Then the DTFT of the x[n] sequence can be written as:

$$\underbrace{\sum_{n=-\infty}^{\infty} \overbrace{x(nT)}^{x[n]}\, e^{-j2\pi fnT}}_{\mathbf{DTFT}} = \frac{1}{T} \sum_{k=-\infty}^{\infty} X(f - k/T).$$

When T has units of seconds, f has units of hertz. Comparison of the two series reveals that $\omega = 2\pi fT$ is a normalized frequency with units of *radians per sample*. The value ω=2π corresponds to $f = \frac{1}{T}$ Hz. And now, with the substitution $f = \frac{\omega}{2\pi T}$, Eq.1 can be expressed in terms of the Fourier transform, X(•):

$$\sum_{n=-\infty}^{\infty} x[n]\, e^{-j\omega n} = \frac{1}{T} \sum_{k=-\infty}^{\infty} \underbrace{X\left(\tfrac{\omega}{2\pi T} - \tfrac{k}{T}\right)}_{X\left(\frac{\omega - 2\pi k}{2\pi T}\right)}.$$

When sequence x(nT) represents the impulse response of an LTI system, these functions are also known as its frequency response. When the x(nT) sequence is periodic, its DTFT is divergent at one or more harmonic frequencies, and zero at all other frequencies. This is often represented by the use of amplitude-variant Dirac delta functions at the harmonic frequencies. Due to periodicity, there are only a finite number of unique amplitudes, which are readily computed by the much simpler discrete Fourier transform (DFT).

Relationship to Laplace Transform

Bilinear Transform

The bilinear transform can be used to convert continuous-time filters (represented in the Laplace domain) into discrete-time filters (represented in the Z-domain), and vice versa. The following substitution is used:

$$s = \frac{2}{T}\frac{(z-1)}{(z+1)}$$

to convert some function $H(s)$ in the Laplace domain to a function $H(z)$ in the Z-domain (Tustin transformation), or

$$z = \frac{2+sT}{2-sT}$$

from the Z-domain to the Laplace domain. Through the bilinear transformation, the complex s-plane (of the Laplace transform) is mapped to the complex z-plane (of the z-transform). While this mapping is (necessarily) nonlinear, it is useful in that it maps the entire $j\Omega$ axis of the s-plane onto the unit circle in the z-plane. As such, the Fourier transform (which is the Laplace transform evaluated on the $j\Omega$ axis) becomes the discrete-time Fourier transform. This assumes that the Fourier transform exists; i.e., that the $j\Omega$ axis is in the region of convergence of the Laplace transform.

Starred Transform

Given a one-sided Z-transform, X(z), of a time-sampled function, the corresponding starred transform produces a Laplace transform and restores the dependence on sampling parameter, T:

$$X^*(s) = X(z)\Big|_{z=e^{sT}}$$

The inverse Laplace transform is a mathematical abstraction known as an *impulse-sampled* function.

Linear Constant-coefficient Difference Equation

The linear constant-coefficient difference (LCCD) equation is a representation for a linear system based on the autoregressive moving-average equation.

$$\sum_{p=0}^{N} y[n-p]\alpha_p = \sum_{q=0}^{M} x[n-q]\beta_q$$

Both sides of the above equation can be divided by α_o, if it is not zero, normalizing α_o = 1 and the LCCD equation can be written

$$y[n] = \sum_{q=0}^{M} x[n-q]\beta_q - \sum_{p=1}^{N} y[n-p]\alpha_p .$$

This form of the LCCD equation is favorable to make it more explicit that the "current" output *y[n]* is a function of past outputs *y[n−p]*, current input *x[n]*, and previous inputs *x[n−q]*.

Transfer Function

Taking the Z-transform of the above equation (using linearity and time-shifting laws) yields

$$Y(z)\sum_{p=0}^{N} z^{-p}\alpha_p = X(z)\sum_{q=0}^{M} z^{-q}\beta_q$$

and rearranging results in

$$H(z) = \frac{Y(z)}{X(z)} = \frac{\sum_{q=0}^{M} z^{-q}\beta_q}{\sum_{p=0}^{N} z^{-p}\alpha_p} = \frac{\beta_0 + z^{-1}\beta_1 + z^{-2}\beta_2 + \cdots + z^{-M}\beta_M}{\alpha_0 + z^{-1}\alpha_1 + z^{-2}\alpha_2 + \cdots + z^{-N}\alpha_N}.$$

Zeros and Poles

From the fundamental theorem of algebra the numerator has M roots (corresponding to zeros of H) and the denominator has N roots (corresponding to poles). Rewriting the transfer function in terms of poles and zeros

$$H(z) = \frac{(1-q_1 z^{-1})(1-q_2 z^{-1})\cdots(1-q_M z^{-1})}{(1-p_1 z^{-1})(1-p_2 z^{-1})\cdots(1-p_N z^{-1})}$$

where q_k is the k-th zero and p_k is the k-th pole. The zeros and poles are commonly complex and when plotted on the complex plane (z-plane) it is called the pole–zero plot.

In addition, there may also exist zeros and poles at $z = 0$ and $z = \infty$. If we take these poles and zeros as well as multiple-order zeros and poles into consideration, the number of zeros and poles are always equal.

By factoring the denominator, partial fraction decomposition can be used, which can then be transformed back to the time domain. Doing so would result in the impulse response and the linear constant coefficient difference equation of the system.

Output Response

If such a system H(z) is driven by a signal X(z) then the output is Y(z) = H(z)X(z). By performing partial fraction decomposition on Y(z) and then taking the inverse Z-transform the output y[n] can be found. In practice, it is often useful to fractionally decompose $\frac{Y(z)}{z}$ before multiplying that quantity by z to generate a form of Y(z) which has terms with easily computable inverse Z-transforms.

Advanced Z-transform

In mathematics and signal processing, the advanced Z-transform is an extension of the Z-transform, to incorporate ideal delays that are not multiples of the sampling time. It takes the form

$$F(z,m) = \sum_{k=0}^{\infty} f(kT+m) z^{-k}$$

where

- T is the sampling period
- m (the "delay parameter") is a fraction of the sampling period $[0,T]$.

It is also known as the modified Z-transform.

The advanced Z-transform is widely applied, for example to accurately model processing delays in digital control.

Properties

If the delay parameter, m, is considered fixed then all the properties of the Z-transform hold for the advanced Z-transform.

Linearity

$$\mathcal{Z}\left\{\sum_{k=1}^{n} c_k f_k(t)\right\} = \sum_{k=1}^{n} c_k F(z,m).$$

Time Shift

$$\mathcal{Z}\{u(t-nT)f(t-nT)\} = z^{-n}F(z,m).$$

Damping

$$\mathcal{Z}\left\{f(t)e^{-at}\right\} = e^{-am}F(e^{aT}z,m).$$

Time Multiplication

$$\mathcal{Z}\left\{t^y f(t)\right\} = \left(-Tz\frac{d}{dz}+m\right)^y F(z,m).$$

Final Value Theorem

$$\lim_{k\to\infty} f(kT+m) = \lim_{z\to 1}(1-z^{-1})F(z,m).$$

Example

Consider the following example where $f(t)=\cos(\omega t)$:

$$\begin{aligned}
F(z,m) &= \mathcal{Z}\left\{\cos\left(\omega\left(kT+m\right)\right)\right\} \\
&= \mathcal{Z}\left\{\cos(\omega kT)\cos(\omega m) - \sin(\omega kT)\sin(\omega m)\right\} \\
&= \cos(\omega m)\mathcal{Z}\left\{\cos(\omega kT)\right\} - \sin(\omega m)\mathcal{Z}\left\{\sin(\omega kT)\right\} \\
&= \cos(\omega m)\frac{z\left(z-\cos(\omega T)\right)}{z^2-2z\cos(\omega T)+1} - \sin(\omega m)\frac{z\sin(\omega T)}{z^2-2z\cos(\omega T)+1} \\
&= \frac{z^2\cos(\omega m) - z\cos(\omega(T-m))}{z^2-2z\cos(\omega T)+1}.
\end{aligned}$$

If $m = 0$ then $F(z,m)$ reduces to the transform

$$F(z,0) = \frac{z^2 - z\cos(\omega T)}{z^2 - 2z\cos(\omega T) + 1},$$

which is clearly just the Z-transform of $f(t)$.

Matched Z-transform Method

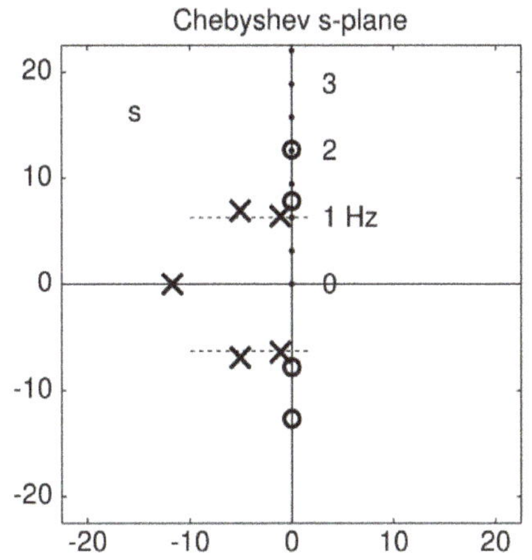

The s-plane poles and zeros of a 5th-order Chebyshev type II lowpass filter to be approximated as a discrete-time filter

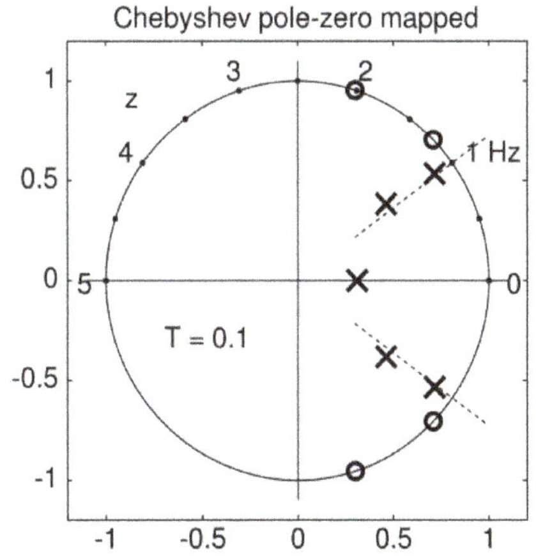

The z-plane poles and zeros of the discrete-time Chebyshev filter, as mapped into the z-plane using the matched Z-transform method with $T = 1/10$ second. The labeled frequency points and band-edge dotted lines have also been mapped through the function $z=e^{i\omega T}$, to show how frequencies along the $i\omega$ axis in the s-plane map onto the unit circle in the z-plane.

The matched Z-transform method, also called the pole–zero mapping or pole–zero matching method, is a technique for converting a continuous-time filter design to a discrete-time filter (digital filter) design.

The method works by mapping all poles and zeros of the s-plane design to z-plane locations $z=e^{sT}$, for a sample interval T.

Alternative methods include the bilinear transform and impulse invariance methods.

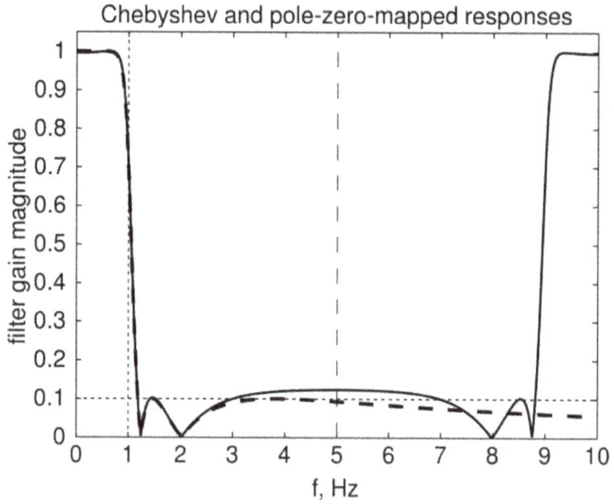

Responses of the filter (dashed), and its discrete-time approximation (solid), for nominal cutoff frequency of 1 Hz, sample rate 1/T = 10 Hz. The discrete-time filter does not reproduce the Chebyshev equiripple property in the stopband due to the interference from cyclic copies of the response.

Zak Transform

In mathematics, the Zak transform is a certain operation which takes as input a function of one variable and produces as output a function of two variables. The output function is called the Zak transform of the input function. The transform is defined as an infinite series in which each term is a product of a dilation of a translation by an integer of the function and an exponential function. In applications of Zak transform to signal processing the input function represents a signal and the transform will be a mixed time–frequency representation of the signal. The signal may be real valued or complex-valued, defined on a continuous set (for example, the real numbers) or a discrete set (for example, the integers or a finite subset of integers). The Zak transform is a generalization of the discrete Fourier transform.

The Zak transform had been discovered by several people in different fields and was called by different names. It was called the "Gel'fand mapping" because I.M. Gel'fand introduced it in his work on eigenfunction expansions. The transform was rediscovered independently by J. Zak in 1967 who called it the "k-q representation". There seems to be a general consent among experts in the field to call it the Zak transform, since Zak was indeed the first to systematically study that transform in a more general setting and recognize its usefulness.

Continuous-time Zak Transform: Definition

In defining the continuous-time Zak transform, the input function is a function of a real variable. So, let $f(t)$ be a function of a real variable t. The continuous-time Zak transform of $f(t)$ is a function of two real variables one of which is t. The other variable may be denoted by w. The continuous-time Zak transform has been defined variously.

Definition 1

Let a be a positive constant. The Zak transform of $f(t)$, denoted by $Z_a[f]$, is a function of t and w defined by

$$Z_a[f](t,w) = \sqrt{a}\sum_{k=-\infty}^{\infty} f(at+ak)e^{-2\pi kwi}..$$

Definition 2

The special case of Definition 1 obtained by taking $a = 1$ is sometimes taken as the definition of the Zak transform. In this special case, the Zak transform of $f(t)$ is denoted by $Z[f]$.

$$Z[f](t,w) = \sum_{k=-\infty}^{\infty} f(t+k)e^{-2\pi kwi}.$$

Definition 3

The notation $Z[f]$ is used to denote another form of the Zak transform. In this form, the Zak transform of $f(t)$ is defined as follows:

$$Z[f](t,v) = \sum_{k=-\infty}^{\infty} f(t+k)e^{-kvi}.$$

Definition 4

Let T be a positive constant. The Zak transform of $f(t)$, denoted by $Z_T[f]$, is a function of t and w defined by

$$Z_T[f](t,w) = \sqrt{T}\sum_{k=-\infty}^{\infty} f(t+kT)e^{-2\pi kwTi}.$$

Here t and w are assumed to satisfy the conditions $0 \le t \le T$ and $0 \le w \le 1/T$.

Example

The Zak transform of the function

$$\phi(t) = \begin{cases} 1, & 0 \le t < 1 \\ 0, & \text{otherwise} \end{cases}$$

is given by

$$Z[\phi](t,w) = e^{-2\pi\lceil -t \rceil wi}$$

where $-t$ denotes the smallest integer not less than $-t$ (the ceil function).

Properties of the Zak Transform

In the following it will be assumed that the Zak transform is as given in Definition 2.

1. Linearity

Let a and b be any real or complex numbers. Then

$$Z[af + bg](t,w) = aZ[f](t,w) + bZ[g](t,w)$$

2. Periodicity

$$Z[f](t,w+1) = Z[f](t,w)$$

3. Quasi-periodicity

$$Z[f](t+1,w) = e^{2\pi wi} Z[f](t,w)$$

4. Conjugation

$$Z[\overline{f}](t,w) = \overline{Z[f]}(t,-w)$$

5. Symmetry

If $f(t)$ is even then $Z[f](t,w) = Z[f](-t,-w)$

If $f(t)$ is odd then $Z[f](t,w) = -Z[f](-t,-w)$

6. Convolution

Let $\star$ denote convolution with respect to the variable t.

$$Z[f \star g](t,w) = Z[f](t,w) \star Z[g](t,w)$$

Inversion Formula

Given the Zak transform of a function, the function can be reconstructed using the following formula:

$$f(t) = \int_0^1 Z[f](t,w)\,dw.$$

Discrete Zak Transform: Definition

In defining the discrete Zak transform, the input function is a function of an integer variable. So, let *f*(*n*) be a function of an integer variable *n* (*n* taking all positive, zero and negative integers as values). The discrete Zak transform of *f*(*n*) is a function of two real variables one of which is the integer variable *n*. The other variable is a real variable which may be denoted by *w*. The discrete Zak transform has also been defined variously. However, only one of the definitions is given below.

Definition

The discrete Zak transform of the function *f*(*n*) where *n* is an integer variable, denoted by *Z*[*f*], is defined by

$$Z[f](n,w)=\sum_{k=-\infty}^{\infty} f(n+k)e^{-2\pi kwi}.$$

Inversion Formula

Given the discrete transform of a function *f*(*n*), the function can be reconstructed using the following formula:

$$f(n)=\int_0^1 Z[f](n,w)dw.$$

Applications

The Zak transform has been used successfully used in physics in quantum field theory, in electrical engineering in time-frequency representation of signals, and in digital data transmission. The Zak transform has also applications in mathematics. For example, it has been used in the Gabor representation problem.

Bilinear Transform

The bilinear transform (also known as Tustin's method) is used in digital signal processing and discrete-time control theory to transform continuous-time system representations to discrete-time and vice versa.

The bilinear transform is a special case of a conformal mapping (namely, the Möbius transformation), often used to convert a transfer function $H_a(s)$ of a linear, time-invariant (LTI) filter in the continuous-time domain (often called an analog filter) to a transfer function $H_d(z)$ of a linear, shift-invariant filter in the discrete-time domain (often called a digital filter although there are analog filters constructed with switched capacitors that are discrete-time filters). It maps positions on the $j\omega$ axis, $Re[s]=0$, in the s-plane to the unit circle, $|z|=1$, in the z-plane. Other bilinear transforms can be used to warp the frequency response of any discrete-time linear system (for example to approximate the non-linear frequency resolution of the human auditory system) and are implementable in the discrete domain by replacing a system's unit delays $\left(z^{-1}\right)$ with first order all-pass filters.

The transform preserves stability and maps every point of the frequency response of the continuous-time filter, $H_a(j\omega_a)$ to a corresponding point in the frequency response of the discrete-time filter, $H_d(e^{j\omega_d T})$ although to a somewhat different frequency, as shown in the Frequency warping section below. This means that for every feature that one sees in the frequency response of the analog filter, there is a corresponding feature, with identical gain and phase shift, in the frequency response of the digital filter but, perhaps, at a somewhat different frequency. This is barely noticeable at low frequencies but is quite evident at frequencies close to the Nyquist frequency.

Discrete-time Approximation

The bilinear transform is a first-order approximation of the natural logarithm function that is an exact mapping of the *z*-plane to the *s*-plane. When the Laplace transform is performed on a discrete-time signal (with each element of the discrete-time sequence attached to a correspondingly delayed unit impulse), the result is precisely the Z transform of the discrete-time sequence with the substitution of

$$\begin{aligned} z &= e^{sT} \\ &= \frac{e^{sT/2}}{e^{-sT/2}} \\ &\approx \frac{1+sT/2}{1-sT/2} \end{aligned}$$

where T is the numerical integration step size of the trapezoidal rule used in the bilinear transform derivation; or, in other words, the sampling period. The above bilinear approximation can be solved for s or a similar approximation for $s=(1/T)\ln(z)$ can be performed.

The inverse of this mapping (and its first-order bilinear approximation) is

$$\begin{aligned} s &= \frac{1}{T}\ln(z) \\ &= \frac{2}{T}\left[\frac{z-1}{z+1}+\frac{1}{3}\left(\frac{z-1}{z+1}\right)^3+\frac{1}{5}\left(\frac{z-1}{z+1}\right)^5+\frac{1}{7}\left(\frac{z-1}{z+1}\right)^7+\cdots\right] \end{aligned}$$

$$\approx \frac{2}{T}\frac{z-1}{z+1}$$
$$= \frac{2}{T}\frac{1-z^{-1}}{1+z^{-1}}$$

The bilinear transform essentially uses this first order approximation and substitutes into the continuous-time transfer function, $H_a(s)$

$$s \leftarrow \frac{2}{T}\frac{z-1}{z+1}.$$

That is

$$H_d(z) = H_a(s)\Big|_{s=\frac{2}{T}\frac{z-1}{z+1}} = H_a\left(\frac{2}{T}\frac{z-1}{z+1}\right).$$

Stability and Minimum-phase Property Preserved

A continuous-time causal filter is stable if the poles of its transfer function fall in the left half of the complex s-plane. A discrete-time causal filter is stable if the poles of its transfer function fall inside the unit circle in the complex z-plane. The bilinear transform maps the left half of the complex s-plane to the interior of the unit circle in the z-plane. Thus filters designed in the continuous-time domain that are stable are converted to filters in the discrete-time domain that preserve that stability.

Likewise, a continuous-time filter is minimum-phase if the zeros of its transfer function fall in the left half of the complex s-plane. A discrete-time filter is minimum-phase if the zeros of its transfer function fall inside the unit circle in the complex z-plane. Then the same mapping property assures that continuous-time filters that are minimum-phase are converted to discrete-time filters that preserve that property of being minimum-phase.

Example

As an example take a simple low-pass RC filter. This continuous-time filter has a transfer function

$$H_a(s) = \frac{1/sC}{R+1/sC}$$
$$= \frac{1}{1+RCs}.$$

If we wish to implement this filter as a digital filter, we can apply the bilinear transform by substituting for s the formula above; after some reworking, we get the following filter representation:

$$H_d(z) = H_a\left(\frac{2}{T}\frac{z-1}{z+1}\right)$$

$$= \frac{1}{1 + RC\left(\frac{2}{T}\frac{z-1}{z+1}\right)}$$

$$= \frac{1+z}{(1-2RC/T)+(1+2RC/T)z}$$

$$= \frac{1+z^{-1}}{(1+2RC/T)+(1-2RC/T)z^{-1}}.$$

The coefficients of the denominator are the 'feed-backward' coefficients and the coefficients of the numerator are the 'feed-forward' coefficients used to implement a real-time digital filter.

General Second-order Biquad Transformation

It is possible to relate the coefficients of a continuous-time, analog filter with those of a similar discrete-time digital filter created through the bilinear transform process. Transforming a general, second-order continuous-time filter with the given transfer function

$$H_a(s) = \frac{b_0 s^2 + b_1 s + b_2}{a_0 s^2 + a_1 s + a_2} = \frac{b_0 + b_1 s^{-1} + b_2 s^{-2}}{a_0 + a_1 s^{-1} + a_2 s^{-2}}$$

using the bilinear transform (without prewarping any frequency specification) requires the substitution of

$$s \leftarrow K\frac{1-z^{-1}}{1+z^{-1}}$$

where $K \triangleq \frac{2}{T}$.

This results in a discrete-time digital biquad filter with coefficients expressed in terms of the coefficients of the original continuous time filter:

$$H_d(z) = \frac{(b_0K^2 + b_1K + b_2) + (2b_2 - 2b_0K^2)z^{-1} + (b_0K^2 - b_1K + b_2)z^{-2}}{(a_0K^2 + a_1K + a_2) + (2a_2 - 2a_0K^2)z^{-1} + (a_0K^2 - a_1K + a_2)z^{-2}}$$

Normally the constant term in the denominator must be normalized to 1 before deriving the corresponding difference equation. This results in

$$H_d(z) = \frac{\frac{b_0K^2 + b_1K + b_2}{a_0K^2 + a_1K + a_2} + \frac{2b_2 - 2b_0K^2}{a_0K^2 + a_1K + a_2} z^{-1} + \frac{b_0K^2 - b_1K + b_2}{a_0K^2 + a_1K + a_2} z^{-2}}{1 + \frac{2a_2 - 2a_0K^2}{a_0K^2 + a_1K + a_2} z^{-1} + \frac{a_0K^2 - a_1K + a_2}{a_0K^2 + a_1K + a_2} z^{-2}}.$$

The difference equation (using the Direct Form I) is

$$\begin{aligned} y[n] = {} & \frac{b_0K^2 + b_1K + b_2}{a_0K^2 + a_1K + a_2} \cdot x[n] + \frac{2b_2 - 2b_0K^2}{a_0K^2 + a_1K + a_2} \cdot x[n-1] \\ & + \frac{b_0K^2 - b_1K + b_2}{a_0K^2 + a_1K + a_2} \cdot x[n-2] - \frac{2a_2 - 2a_0K^2}{a_0K^2 + a_1K + a_2} \cdot y[n-1] \\ & - \frac{a_0K^2 - a_1K + a_2}{a_0K^2 + a_1K + a_2} \cdot y[n-2]. \end{aligned}$$

Frequency Warping

To determine the frequency response of a continuous-time filter, the transfer function $H_a(s)$ is evaluated at $s = j\omega$ which is on the $j\omega$ axis. Likewise, to determine the frequency response of a discrete-time filter, the transfer function $H_d(z)$ is evaluated at $z = e^{j\omega T}$ which is on the unit circle, $|z| = 1$. When the actual frequency of ω_a is input to the discrete-time filter designed by use of the bilinear transform, it is desired to know at what frequency, ω_a, for the continuous-time filter that this ω is mapped to.

$$H_d(z) = H_a\left(\frac{2}{T}\frac{z-1}{z+1}\right)$$

$$H_d(e^{j\omega T}) = H_a\left(\frac{2}{T}\frac{e^{j\omega T}-1}{e^{j\omega T}+1}\right)$$

$$= H_a\left(\frac{2}{T} \cdot \frac{e^{j\omega T/2}\left(e^{j\omega T/2} - e^{-j\omega T/2}\right)}{e^{j\omega T/2}\left(e^{j\omega T/2} + e^{-j\omega T/2}\right)}\right)$$

$$= H_a\left(\frac{2}{T} \cdot \frac{\left(e^{j\omega T/2} - e^{-j\omega T/2}\right)}{\left(e^{j\omega T/2} + e^{-j\omega T/2}\right)}\right)$$

$$= H\left(j - \cdot \frac{\left(e^{j\omega T/2} \quad e^{\ j\omega T/2}\right)/(2j)}{\left(e^{j\omega T/2} \quad e^{\ j\omega T/2}\right)/2}\right)$$

$$= H_a\left(j\frac{2}{T}\cdot\frac{\sin(\omega T/2)}{\cos(\omega T/2)}\right)$$

$$= H_a\left(j\frac{2}{T}\cdot\tan\left(\omega T/2\right)\right)$$

This shows that every point on the unit circle in the discrete-time filter z-plane, $z = e^{j\omega T}$ is mapped to a point on the $j\omega$ axis on the continuous-time filter s-plane, $s = j\omega_a$. That is, the discrete-time to continuous-time frequency mapping of the bilinear transform is

$$\omega_a = \frac{2}{T}\tan\left(\omega\frac{T}{2}\right)$$

and the inverse mapping is

$$\omega = \frac{2}{T}\arctan\left(\omega_a\frac{T}{2}\right).$$

The discrete-time filter behaves at frequency ω the same way that the continuous-time filter behaves at frequency $(2/T)\tan(\omega T/2)$. Specifically, the gain and phase shift that the discrete-time filter has at frequency ω is the same gain and phase shift that the continuous-time filter has at frequency $(2/T)\tan(\omega T/2)$. This means that every feature, every "bump" that is visible in the frequency response of the continuous-time filter is also visible in the discrete-time filter, but at a different frequency. For low frequencies (that is, when $\omega \ll 2/T$ or $\omega_a 2/T$), $\omega \approx \omega_a$.

One can see that the entire continuous frequency range

$$-\infty < \omega_a < +\infty$$

is mapped onto the fundamental frequency interval

$$-\frac{\pi}{T} < \omega < +\frac{\pi}{T}.$$

The continuous-time filter frequency $\omega_a = 0$ corresponds to the discrete-time filter frequency $\omega = 0$ and the continuous-time filter frequency $\omega_a = \pm\infty$ correspond to the discrete-time filter frequency $\omega = \pm\pi/T$.

One can also see that there is a nonlinear relationship between ω_a and ω. This effect of the bilinear transform is called *frequency warping*. The continuous-time filter can be designed to compensate for this frequency warping by setting $\omega_a = \frac{2}{T}\tan\left(\omega\frac{T}{2}\right)$ for every frequency specification that the designer has control over (such as corner frequency or center frequency). This is called *pre-warping* the filter design.

When designing a digital filter as an approximation of a continuous time filter, the frequency response (both amplitude and phase) of the digital filter can be made to match the frequency response of the continuous filter at frequency ω_0 if the following transform is substituted into the continuous filter transfer function. This is a modified version of Tustin's transform shown above. However, note that this transform becomes the above transform as $\omega_0 \to 0.\omega_0 \to 0.$. That is to say, the above transform causes the digital filter response to match the analog filter response at DC.

$$s \leftarrow \frac{\omega_0}{\tan \frac{\omega_0 T}{2}} \frac{z-1}{z+1}.$$

The main advantage of the warping phenomenon is the absence of aliasing distortion of the frequency response characteristic, such as observed with Impulse invariance. It is necessary, however, to compensate for the frequency warping by pre-warping the given frequency specifications of the continuous-time system. These pre-warped specifications may then be used in the bilinear transform to obtain the desired discrete-time system.

Discrete Fourier Transform

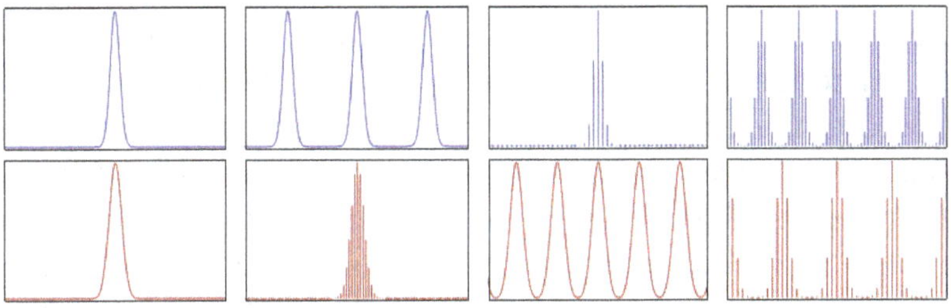

Relationship between the (continuous) Fourier transform and the discrete Fourier transform. Left column: A continuous function (top) and its Fourier transform (bottom). Center-left column: Periodic summation of the original function (top). Fourier transform (bottom) is zero except at discrete points. The inverse transform is a sum of sinusoids called Fourier series. Center-right column: Original function is discretized (multiplied by a Dirac comb) (top). Its Fourier transform (bottom) is a periodic summation (DTFT) of the original transform. Right column: The DFT (bottom) computes discrete samples of the continuous DTFT. The inverse DFT (top) is a periodic summation of the original samples. The FFT algorithm computes one cycle of the DFT and its inverse is one cycle of the DFT inverse.

In mathematics, the discrete Fourier transform (DFT) converts a finite sequence of equally-spaced samples of a function into an equivalent-length sequence of equally-spaced samples of the discrete-time Fourier transform (DTFT), which is a complex-valued function of frequency. The interval at which the DTFT is sampled is the reciprocal of the duration of the input sequence. An inverse DFT is a Fourier series, using the DTFT samples as coefficients of complex sinusoids at the corresponding DTFT frequencies. It has the same sample-values as the original input sequence. The DFT is therefore said to be a frequency domain representation of the original input sequence. If the original sequence spans all the non-zero values of a function, its DTFT is continuous (and periodic), and the DFT provides discrete samples of one cycle. If the original sequence is one cycle of a periodic function, the DFT provides all the non-zero values of one DTFT cycle.

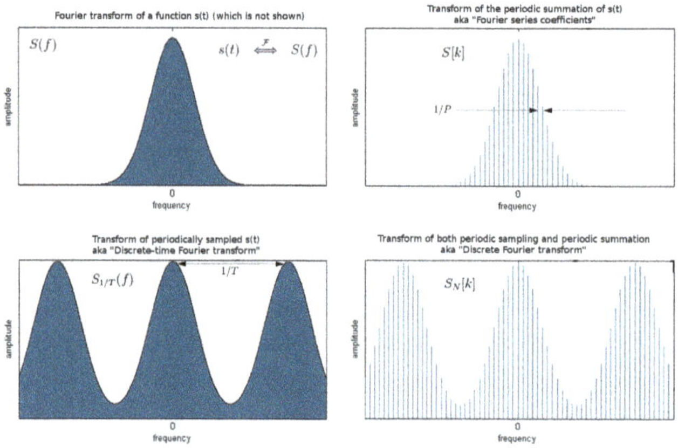

Depiction of a Fourier transform (upper left) and its periodic summation (DTFT) in the lower left corner. The spectral sequences at (a) upper right and (b) lower right are respectively computed from (a) one cycle of the periodic summation of s(t) and (b) one cycle of the periodic summation of the s(nT) sequence. The respective formulas are (a) the Fourier series integral and (b) the DFT summation. Its similarities to the original transform, S(f), and its relative computational ease are often the motivation for computing a DFT sequence.

The DFT is the most important discrete transform, used to perform Fourier analysis in many practical applications. In digital signal processing, the function is any quantity or signal that varies over time, such as the pressure of a sound wave, a radio signal, or daily temperature readings, sampled over a finite time interval (often defined by a window function). In image processing, the samples can be the values of pixels along a row or column of a raster image. The DFT is also used to efficiently solve partial differential equations, and to perform other operations such as convolutions or multiplying large integers.

Since it deals with a finite amount of data, it can be implemented in computers by numerical algorithms or even dedicated hardware. These implementations usually employ efficient fast Fourier transform (FFT) algorithms; so much so that the terms "FFT" and "DFT" are often used interchangeably. Prior to its current usage, the "FFT" initialism may have also been used for the ambiguous term "finite Fourier transform".

Definition

The sequence of *N* complex numbers $x_0, x_1, \ldots, x_{N-1}$ is transformed into an *N*-periodic sequence of complex numbers:

$$X_k \stackrel{\text{def}}{=} \sum_{n=0}^{N-1} x_n \cdot e^{-2\pi ikn/N}, \quad k \in \mathbb{Z} \text{ (integers)} \qquad \textit{(Eq.1)}$$

Because of periodicity, the customary domain of k actually computed is $[0, N-1]$. That is always the case when the DFT is implemented via the Fast Fourier transform algorithm. But other common domains are $[-N/2, N/2-1]$ (N even) and $[-(N-1)/2, (N-1)/2]$ (N odd), as when the left and right halves of an FFT output sequence are swapped.

The transform is sometimes denoted by the symbol $\mathcal{F}$, as in $\mathbf{X} = \mathcal{F}\{\mathbf{x}\}$ or $\mathcal{F}(\mathbf{x})$ or $\mathcal{F}\mathbf{x}$

Eq.1 can be interpreted or derived in various ways, for example:

- It completely describes the discrete-time Fourier transform (DTFT) of an N-periodic sequence, which comprises only discrete frequency components. (Using the DTFT with periodic data)
- It can also provide uniformly spaced samples of the continuous DTFT of a finite length sequence. (Sampling the DTFT)
- It is the cross correlation of the *input* sequence, x_n, and a complex sinusoid at frequency k/N. Thus it acts like a matched filter for that frequency.
- It is the discrete analogy of the formula for the coefficients of a Fourier series:

$$x_n = \frac{1}{N}\sum_{k=0}^{N-1} X_k \cdot e^{2\pi ikn/N}, \quad n \in \mathbb{Z} \qquad \textit{(Eq.2)}$$

which is also N-periodic. In the domain $n \in [0, N-1]$, this is the inverse transform of Eq.1. In this interpretation, each X_k is a complex number that encodes both amplitude and phase of a complex sinusoidal component ($e^{2\pi ikn/N}$) of function x_n. The sinusoid's frequency is k cycles per N samples. Its amplitude and phase are:

$$|X_k|/N = \sqrt{\text{Re}(X_k)^2 + Im(X_k)^2}\ /\ N$$

$$\arg(X_k) = \text{atan}\,2\big(\text{Im}(X_k), \text{Re}(X_k)\big) = -i\ln\left(\frac{X_k}{|X_k|}\right),$$

where atan2 is the two-argument form of the arctan function.

The normalization factor multiplying the DFT and IDFT (here 1 and $1/N$) and the signs of the exponents are merely conventions, and differ in some treatments. The only requirements of these conventions are that the DFT and IDFT have opposite-sign exponents and that the product of their normalization factors be $1/N$. A normalization of $\sqrt{1/N}$ for both the DFT and IDFT, for instance, makes the transforms unitary.

In the following discussion the terms "sequence" and "vector" will be considered interchangeable.

Using Euler's formula, the DFT formulae can be converted to the trigonometric forms sometimes used in engineering and computer science:

Fourier Transform:

$$X_k = \sum_{n=0}^{N-1} x_n \cdot (\cos(-2\pi k \frac{n}{N}) + i \sin(-2\pi k \frac{n}{N})), \quad k \in \qquad \textbf{(Eq.3)}$$

Inverse Fourier Transform:

$$x_n = \frac{1}{N} \sum_{k=0}^{N-1} X_k \cdot (\cos(2\pi k \frac{n}{N}) + i \sin(2\pi k \frac{n}{N})), \quad n \in \mathbb{Z} \qquad \textbf{(Eq.4)}$$

N = number of time samples we have

n = current sample we're considering (0, ..., $N-1$)

x_n = value of the signal at time n

k = current frequency we're considering (0 Hertz up to N-1 Hertz)

X_k = amount of frequency k in the signal (Amplitude and Phase, a complex number)

Properties

Completeness

The discrete Fourier transform is an invertible, linear transformation

$$\mathcal{F} : \mathbb{C}^N \to \mathbb{C}^N$$

with $\mathbb{C}$ denoting the set of complex numbers. In other words, for any $N > 0$, an N-dimensional complex vector has a DFT and an IDFT which are in turn N-dimensional complex vectors.

Orthogonality

The vectors $u_k = \left[e^{\frac{2\pi i}{N} kn} \mid n = 0, 1, \ldots, N-1 \right]^T$ form an orthogonal basis over the set of N-dimensional complex vectors:

$$u_k^T u_{k'}^* = \sum_{n=0}^{N-1} \left(e^{\frac{2\pi i}{N} kn} \right) \left(e^{\frac{2\pi i}{N} (-k')n} \right) = \sum_{n=0}^{N-1} e^{\frac{2\pi i}{N} (k-k')n} = N \delta_{kk'}$$

where $\delta_{kk'}$ is the Kronecker delta. (In the last step, the summation is trivial if $k = k'$, where it is 1+1+⋯=N, and otherwise is a geometric series that can be explicitly summed to obtain zero.) This orthogonality condition can be used to derive the formula for the IDFT from the definition of the DFT, and is equivalent to the unitarity property below.

The Plancherel Theorem and Parseval's Theorem

If X_k and Y_k are the DFTs of x_n and y_n respectively then the Parseval's theorem states:

$$\sum_{n=0}^{N-1} x_n y_n^* = \frac{1}{N}\sum_{k=0}^{N-1} X_k Y_k^*$$

where the star denotes complex conjugation. Plancherel theorem is a special case of the Parseval's theorem and states:

$$\sum_{n=0}^{N-1} |x_n|^2 = \frac{1}{N}\sum_{k=0}^{N-1} |X_k|^2.$$

These theorems are also equivalent to the unitary condition below.

Periodicity

The periodicity can be shown directly from the definition:

$$X_{k+N} \stackrel{\text{def}}{=} \sum_{n=0}^{N-1} x_n e^{-\frac{2\pi i}{N}(k+N)n} = \sum_{n=0}^{N-1} x_n e^{-\frac{2\pi i}{N}kn} \underbrace{e^{-2\pi i n}}_{1} = \sum_{n=0}^{N-1} x_n e^{-\frac{2\pi i}{N}kn} = X_k.$$

Similarly, it can be shown that the IDFT formula leads to a periodic extension.

Shift Theorem

Multiplying x_n by a *linear phase* $e^{\frac{2\pi i}{N}nm}$ for some integer m corresponds to a *circular shift* of the output X_k: X_k is replaced by X_{k-m}, where the subscript is interpreted modulo N (i.e., periodically). Similarly, a circular shift of the input x_n corresponds to multiplying the output X_k by a linear phase. Mathematically, if $\{x_n\}$ represents the vector x then

if $\mathcal{F}(\{x_n\})_k = X_k$

then $\mathcal{F}(\{x_n \cdot e^{\frac{2\pi i}{N}nm}\})_k = X_{k-m}$

and $\mathcal{F}(\{x_{n-m}\})_k = X_k \cdot e^{-\frac{2\pi i}{N}km}$

Circular Convolution Theorem and Cross-correlation Theorem

The convolution theorem for the discrete-time Fourier transform indicates that a convolution of two infinite sequences can be obtained as the inverse transform of the product of the individual transforms. An important simplification occurs when the sequences are of finite length, N. In terms of the DFT and inverse DFT, it can be written as follows:

$$\mathcal{F}^{-1}\{\mathbf{X}\cdot\mathbf{Y}\}_n = \sum_{l=0}^{N-1} x_l \cdot (y_N)_{n-l} \stackrel{\text{def}}{=} (\mathbf{x} * \mathbf{y}_\mathbf{N})_n,$$

which is the convolution of the $\mathbf{x}$ sequence with a $\mathbf{y}$ sequence extended by periodic summation:

$$(\mathbf{y}_\mathbf{N})_n \stackrel{\text{def}}{=} \sum_{p=-\infty}^{\infty} y_{(n-pN)} = y_{n(modN)}.$$

Similarly, the cross-correlation of $\mathbf{x}$ and $\mathbf{y}_\mathbf{N}$ is given by:

$$\mathcal{F}^{-1}\{\mathbf{X}^*\cdot\mathbf{Y}\}_n = \sum_{l=0}^{N-1} x_l^* \cdot (y_N)_{n+l} \stackrel{\text{def}}{=} (\mathbf{x} \star \mathbf{y}_\mathbf{N})_n.$$

When either sequence contains a string of zeros, of length L, L+1 of the circular convolution outputs are equivalent to values of $\mathbf{x} * \mathbf{y}$. Methods have also been developed to use this property as part of an efficient process that constructs $\mathbf{x} * \mathbf{y}$ with an $\mathbf{x}$ or $\mathbf{y}$ sequence potentially much longer than the practical transform size (N). Two such methods are called overlap-save and overlap-add. The efficiency results from the fact that a direct evaluation of either summation (above) requires $O(N^2)$ operations for an output sequence of length N. An indirect method, using transforms, can take advantage of the $O(N \log N)$ efficiency of the fast Fourier transform (FFT) to achieve much better performance. Furthermore, convolutions can be used to efficiently compute DFTs via Rader's FFT algorithm and Bluestein's FFT algorithm.

Convolution Theorem Duality

It can also be shown that:

$$\mathcal{F}\{\mathbf{x}\cdot\mathbf{y}\}_k \stackrel{\text{def}}{=} \sum_{n=0}^{N-1} x_n \cdot y_n \cdot e^{-\frac{2\pi i}{N}kn}$$

$$= \frac{1}{N}(\mathbf{X} * \mathbf{Y}_\mathbf{N})_k,$$ which is the circular convolution of $\mathbf{X}$ and $\mathbf{Y}$.

Trigonometric Interpolation Polynomial

The trigonometric interpolation polynomial

$$p(t)=\frac{1}{N}\left[X_0+X_1e^{2\pi it}+\cdots+X_{N/2-1}e^{(N/2-1)2\pi it}+X_{N/2}\cos(N\pi t)+X_{N/2+1}e^{(-N/2+1)2\pi it}+\cdots+X_{N-1}e^{-2\pi it}\right]$$

for N even ,

$$p(t)=\frac{1}{N}\left[X_0+X_1e^{2\pi it}+\cdots+X_{\lfloor N/2\rfloor}e^{\lfloor N/2\rfloor 2\pi it}+X_{\lfloor N/2\rfloor+1}e^{-\lfloor N/2\rfloor 2\pi it}+\cdots+X_{N-1}e^{-2\pi it}\right]$$

for N odd,

where the coefficients X_k are given by the DFT of x_n above, satisfies the interpolation property $p(n>$ for $n=0,\ldots,N-1$..

For even N, notice that the Nyquist component $\frac{X_{N/2}}{N}\cos(N\pi t)$ is handled specially.

This interpolation is *not unique*: aliasing implies that one could add N to any of the complex-sinusoid frequencies (e.g. changing e^{-it} to $e^{i(N-1)t}$) without changing the interpolation property, but giving *different* values in between the x_n points. The choice above, however, is typical because it has two useful properties. First, it consists of sinusoids whose frequencies have the smallest possible magnitudes: the interpolation is bandlimited. Second, if the x_n are real numbers, then $p(t)$ is real as well.

In contrast, the most obvious trigonometric interpolation polynomial is the one in which the frequencies range from 0 to $N-1$ (instead of roughly $-N/2$ to $+N>$ as above), similar to the inverse DFT formula. This interpolation does *not* minimize the slope, and is *not* generally real-valued for real x_n; its use is a common mistake.

The Unitary DFT

Another way of looking at the DFT is to note that in the above discussion, the DFT can be expressed as the DFT matrix, a Vandermonde matrix, introduced by Sylvester in 1867,

$$\mathbf{F}=\begin{bmatrix}\omega_N^{0\cdot 0} & \omega_N^{0\cdot 1} & \cdots & \omega_N^{0\cdot(N-1)}\\ \omega_N^{1\cdot 0} & \omega_N^{1\cdot 1} & \cdots & \omega_N^{1\cdot(N-1)}\\ \vdots & \vdots & \ddots & \vdots\\ \omega_N^{(N-1)\cdot 0} & \omega_N^{(N-1)\cdot 1} & \cdots & \omega_N^{(N-1)\cdot(N-1)}\end{bmatrix}$$

where

$$\omega_N=e^{-2\pi i/N}$$

is a primitive Nth root of unity.

The inverse transform is then given by the inverse of the above matrix,

$$\mathbf{F}^{-1} = \frac{1}{N}\mathbf{F}^*$$

With unitary normalization constants $1/\sqrt{N}$, the DFT becomes a unitary transformation, defined by a unitary matrix:

$$\mathbf{U} = \mathbf{F}/\sqrt{N}$$

$$\mathbf{U}^{-1} = \mathbf{U}^*$$

$$|\det(\mathbf{U})| = 1$$

where *det()* is the determinant function. The determinant is the product of the eigenvalues, which are always ± 1 or $\pm i$ as described below. In a real vector space, a unitary transformation can be thought of as simply a rigid rotation of the coordinate system, and all of the properties of a rigid rotation can be found in the unitary DFT.

The orthogonality of the DFT is now expressed as an orthonormality condition (which arises in many areas of mathematics as described in root of unity):

$$\sum_{m=0}^{N-1} U_{km} U^*_{mn} = \delta_{kn}$$

If X is defined as the unitary DFT of the vector x, then

$$X_k = \sum_{n=0}^{N-1} U_{kn} x_n$$

and the Plancherel theorem is expressed as

$$\sum_{n=0}^{N-1} x_n y_n^* = \sum_{k=0}^{N-1} X_k Y_k^*$$

If we view the DFT as just a coordinate transformation which simply specifies the components of a vector in a new coordinate system, then the above is just the statement that the dot product of two vectors is preserved under a unitary DFT transformation. For the special case $\mathbf{x} = \mathbf{y}$, this implies that the length of a vector is preserved as well—this is just Parseval's theorem,

$$\sum_{n=0}^{N-1} |x_n|^2 = \sum_{k=0}^{N-1} |X_k|^2$$

A consequence of the circular convolution theorem is that the DFT matrix *F* diagonalizes any circulant matrix.

Expressing the Inverse DFT in Terms of the DFT

A useful property of the DFT is that the inverse DFT can be easily expressed in terms of the (forward) DFT, via several well-known "tricks". (For example, in computations, it is often convenient to only implement a fast Fourier transform corresponding to one transform direction and then to get the other transform direction from the first.)

First, we can compute the inverse DFT by reversing all but one of the inputs (Duhamel *et al.*, 1988):

$$\mathcal{F}^{-1}(\{x_n\}) = \mathcal{F}(\{x_{N-n}\}) / N$$

(As usual, the subscripts are interpreted modulo *N*; thus, for $n = 0$, we have $x_{N-0} = x_0$.)

Second, one can also conjugate the inputs and outputs:

$$\mathcal{F}^{-1}(\mathbf{x}) = \mathcal{F}(\mathbf{x}^*)^* / N$$

Third, a variant of this conjugation trick, which is sometimes preferable because it requires no modification of the data values, involves swapping real and imaginary parts (which can be done on a computer simply by modifying pointers). Define swap(x_n) as x_n with its real and imaginary parts swapped—that is, if $x_n = a + bi$ then swap(x_n) is $b + ai$. Equivalently, swap(x_n) equals ix_n^*. Then

$$\mathcal{F}^{-1}(\mathbf{x}) = \text{swap}(\mathcal{F}(\text{swap}(\mathbf{x}))) / N$$

That is, the inverse transform is the same as the forward transform with the real and imaginary parts swapped for both input and output, up to a normalization (Duhamel *et al.*, 1988).

The conjugation trick can also be used to define a new transform, closely related to the DFT, that is involutory—that is, which is its own inverse. In particular, $T(\mathbf{x}) = \mathcal{F}(\mathbf{x}^*) / \sqrt{N}$ is clearly its own inverse: $T(T(\mathbf{x})) = \mathbf{x}$. A closely related involutory transformation (by a factor of $(1+i)/\sqrt{2}$) is $H(\mathbf{x}) = \mathcal{F}((1+i)\mathbf{x}^*) / \sqrt{2N}$, since the $(1+i)$ factors in $H(H(\mathbf{x}))$ cancel the 2. For real inputs $\mathbf{x}$, the real part of $H(\mathbf{x})$ is none other than the discrete Hartley transform, which is also involutory.

Eigenvalues and Eigenvectors

The eigenvalues of the DFT matrix are simple and well-known, whereas the eigenvectors are complicated, not unique, and are the subject of ongoing research.

Consider the unitary form **U** defined above for the DFT of length N, where

$$\mathbf{U}_{m,n} = \frac{1}{\sqrt{N}}\omega_N^{(m-1)(n-1)} = \frac{1}{\sqrt{N}}e^{-\frac{2\pi i}{N}(m-1)(n-1)}.$$

This matrix satisfies the matrix polynomial equation:

$$\mathbf{U}^4 = \mathbf{I}.$$

This can be seen from the inverse properties above: operating **U** twice gives the original data in reverse order, so operating **U** four times gives back the original data and is thus the identity matrix. This means that the eigenvalues λ satisfy the equation:

$$\lambda^4 = 1.$$

Therefore, the eigenvalues of **U** are the fourth roots of unity: λ is +1, −1, $+i$, or $-i$.

Since there are only four distinct eigenvalues for this $N \times N$ matrix, they have some multiplicity. The multiplicity gives the number of linearly independent eigenvectors corresponding to each eigenvalue. (Note that there are N independent eigenvectors; a unitary matrix is never defective.)

The problem of their multiplicity was solved by McClellan and Parks (1972), although it was later shown to have been equivalent to a problem solved by Gauss (Dickinson and Steiglitz, 1982). The multiplicity depends on the value of N modulo 4, and is given by the following table:

size N	$\lambda = +1$	$\lambda = -1$	$\lambda = -i$	$\lambda = +i$
$4m$	$m + 1$	m	m	$m - 1$
$4m + 1$	$m + 1$	m	m	m
$4m + 2$	$m + 1$	$m + 1$	m	m
$4m + 3$	$m + 1$	$m + 1$	$m + 1$	m

Multiplicities of the eigenvalues λ of the unitary DFT matrix **U** as a function of the transform size N (in terms of an integer m).

Otherwise stated, the characteristic polynomial of **U** is:

$$\det(\lambda I - \mathbf{U}) = (\lambda - 1)^{\left\lfloor \frac{N+4}{4} \right\rfloor}(\lambda + 1)^{\left\lfloor \frac{N+2}{4} \right\rfloor}(\lambda + i)^{\left\lfloor \frac{N+1}{4} \right\rfloor}(\lambda - i)^{\left\lfloor \frac{N-1}{4} \right\rfloor}.$$

No simple analytical formula for general eigenvectors is known. Moreover, the eigenvectors are not unique because any linear combination of eigenvectors for the same eigenvalue is also an eigenvector for that eigenvalue. Various researchers have proposed different choices of eigenvectors, selected to satisfy useful properties like orthogonality and to have "simple" forms (e.g., McClellan and Parks, 1972; Dickinson and Steiglitz, 1982; Grünbaum, 1982; Atakishiyev and Wolf, 1997; Candan *et al.*, 2000; Hanna *et al.*, 2004; Gurevich and Hadani, 2008).

A straightforward approach is to discretize an eigenfunction of the continuous Fourier transform, of which the most famous is the Gaussian function. Since periodic summation of the function means discretizing its frequency spectrum and discretization means periodic summation of the spectrum, the discretized and periodically summed Gaussian function yields an eigenvector of the discrete transform:

- $$F(m)=\sum_{k\in\mathbb{Z}}\exp\left(-\frac{\pi\cdot(m+N\cdot k)^2}{N}\right).$$

  A closed form expression for the series is not known, but it converges rapidly.

Two other simple closed-form analytical eigenvectors for special DFT period N were found (Kong, 2008):

For DFT period $N = 2L + 1 = 4K + 1$, where K is an integer, the following is an eigenvector of DFT:

- $$F(m)=\prod_{s=K+1}^{L}\left[\cos\left(\frac{2\pi}{N}m\right)-\cos\left(\frac{2\pi}{N}s\right)\right]$$

  For DFT period $N = 2L = 4K$, where K is an integer, the following is an eigenvector of DFT:

- $$F(m)=\sin\left(\frac{2\pi}{N}m\right)\prod_{s=K+1}^{L-1}\left[\cos\left(\frac{2\pi}{N}m\right)-\cos\left(\frac{2\pi}{N}s\right)\right]$$

The choice of eigenvectors of the DFT matrix has become important in recent years in order to define a discrete analogue of the fractional Fourier transform—the DFT matrix can be taken to fractional powers by exponentiating the eigenvalues (e.g., Rubio and Santhanam, 2005). For the continuous Fourier transform, the natural orthogonal eigenfunctions are the Hermite functions, so various discrete analogues of these have been employed as the eigenvectors of the DFT, such as the Kravchuk polynomials (Atakishiyev and Wolf, 1997). The "best" choice of eigenvectors to define a fractional discrete Fourier transform remains an open question, however.

Uncertainty Principle

If the random variable X_k is constrained by

$$\sum_{n=0}^{N-1} |X_n|^2 = 1 ,$$

then

$$P_n = |X_n|^2$$

may be considered to represent a discrete probability mass function of n, with an associated probability mass function constructed from the transformed variable,

$$Q_m = N\,|x_m|^2 .$$

For the case of continuous functions $P(x)$ and $Q(k)$, the Heisenberg uncertainty principle states that

$$D_0(X)D_0(x) \geq \frac{1}{16\pi^2}$$

where $D_0(X)$ and $D_0(x)$ are the variances of $|X|^2$ and $|x|^2$ respectively, with the equality attained in the case of a suitably normalized Gaussian distribution. Although the variances may be analogously defined for the DFT, an analogous uncertainty principle is not useful, because the uncertainty will not be shift-invariant. Still, a meaningful uncertainty principle has been introduced by Massar and Spindel.

However, the Hirschman entropic uncertainty will have a useful analog for the case of the DFT. The Hirschman uncertainty principle is expressed in terms of the Shannon entropy of the two probability functions.

In the discrete case, the Shannon entropies are defined as

$$H(X) = -\sum_{n=0}^{N-1} P_n \ln P_n$$

and

$$H(x) = -\sum_{m=0}^{N-1} Q_m \ln Q_m ,$$

and the entropic uncertainty principle becomes

$$H(X) + H(x) \geq \ln(N) .$$

The equality is obtained for P_n equal to translations and modulations of a suitably normalized Kronecker comb of period A where A is any exact integer divisor of N. The probability mass function Q_m will then be proportional to a suitably translated Kronecker comb of period $B = N / A$.

There is also a well-known deterministic uncertainty principle that uses signal sparsity (or the number of non-zero coefficients). Let x_0 and $\| X \|_0$ be the number of non-zero elements of the time and frequency sequences $x_0, x_1, \ldots, x_{N-1}$ and $X_0, X_1, \ldots, X_{N-1}$, respectively. Then,

$$N \leq x_0 \cdot X_0.$$

As an immediate consequence of the inequality of arithmetic and geometric means, one also has $2\sqrt{N} \leq \| x \|_0 + \| X \|_0$. Both uncertainty principles were shown to be tight for specifically-chosen "picket-fence" sequences (discrete impulse trains), and find practical use for signal recovery applications.

The Real-input DFT

If $x_0, \ldots, x_{N-1}$ are real numbers, as they often are in practical applications, then the DFT obeys the symmetry:

$$X_{N-k} \equiv X_{-k} = X_k^*, \text{ where } X^* \text{ denotes complex conjugation.}$$

It follows that X_o and $X_{N/2}$ are real-valued, and the remainder of the DFT is completely specified by just *N/2-1* complex numbers.

Generalized DFT (Shifted and Non-linear Phase)

It is possible to shift the transform sampling in time and/or frequency domain by some real shifts *a* and *b*, respectively. This is sometimes known as a generalized DFT (or GDFT), also called the shifted DFT or offset DFT, and has analogous properties to the ordinary DFT:

$$X_k = \sum_{n=0}^{N-1} x_n e^{-\frac{2\pi i}{N}(k+b)(n+a)} \qquad k = 0, \ldots, N-1.$$

Most often, shifts of $1/2$ (half a sample) are used. While the ordinary DFT corresponds to a periodic signal in both time and frequency domains, $a = 1/2$ produces a signal that is anti-periodic in frequency domain ($X_{k+N} = -X_k$) and vice versa for $b = 1/2$. Thus, the specific case of $a = b = 1/2$ is known as an *odd-time odd-frequency* discrete Fourier transform (or O^2 DFT). Such shifted transforms are most often used for symmetric data, to represent different boundary symmetries, and for real-symmetric data they correspond to different forms of the discrete cosine and sine transforms.

Another interesting choice is $a = b = -(N-1) >,$, which is called the centered DFT (or CDFT). The centered DFT has the useful property that, when N is a multiple of four, all four of its eigenvalues (see above) have equal multiplicities (Rubio and Santhanam, 2005)

The term GDFT is also used for the non-linear phase extensions of DFT. Hence, GDFT method provides a generalization for constant amplitude orthogonal block transforms including linear and non-linear phase types. GDFT is a framework to improve time and frequency domain properties of the traditional DFT, e.g. auto/cross-correlations, by the addition of the properly designed phase shaping function (non-linear, in general) to the original linear phase functions (Akansu and Agirman-Tosun, 2010).

The discrete Fourier transform can be viewed as a special case of the z-transform, evaluated on the unit circle in the complex plane; more general z-transforms correspond to *complex* shifts *a* and *b* above.

Multidimensional DFT

The ordinary DFT transforms a one-dimensional sequence or array x_n that is a function of exactly one discrete variable *n*. The multidimensional DFT of a multidimensional array $x_{n_1,n_2,\ldots,n_d}$ that is a function of d discrete variables $n_\ell = 0,1,\ldots,N_\ell - 1$ for ℓ in $1,2,\ldots,d$ is defined by:

$$X_{k_1,k_2,\ldots,k_d} = \sum_{n_1=0}^{N_1-1}\left(\omega_{N_1}^{k_1 n_1}\sum_{n_2=0}^{N_2-1}\left(\omega_{N_2}^{k_2 n_2}\cdots\sum_{n_d=0}^{N_d-1}\omega_{N_d}^{k_d n_d}\cdot x_{n_1,n_2,\ldots,n_d}\right)\right),$$

where $\omega_{N_\ell} = \exp(-2\pi i / N_\ell)$ as above and the d output indices run from $k_\ell = 0,1,\ldots,N_\ell - 1$. . This is more compactly expressed in vector notation, where we define $\mathbf{n} = (n_1, n_2, \ldots, n_d)$ and $\mathbf{k} = (k_1, k_2, \ldots, k_d)$ as *d*-dimensional vectors of indices from 0 to $\mathbf{N} - 1$, which we define as $\mathbf{N} - 1 = (N_1 - 1, N_2 - 1, \ldots, N_d - 1)$

$$X_{\mathbf{k}} = \sum_{\mathbf{n}=0}^{\mathbf{N}-1} e^{-2\pi i \mathbf{k}\cdot(\mathbf{n}/\mathbf{N})} x_{\mathbf{n}},$$

where the division $\mathbf{n}/\mathbf{N}$ is defined as $\mathbf{n}/\mathbf{N} = (n_1/N_1, \ldots, n_d/N_d)$ to be performed element-wise, and the sum denotes the set of nested summations above.

The inverse of the multi-dimensional DFT is, analogous to the one-dimensional case, given by:

$$x_{\mathbf{n}} = \frac{1}{\prod_{\ell=1}^{d} N_{\ell}} \sum_{\mathbf{k}=\mathbf{0}}^{\mathbf{N}-1} e^{2\pi i \mathbf{n}\cdot(\mathbf{k}/\mathbf{N})} X_{\mathbf{k}}.$$

As the one-dimensional DFT expresses the input x_n as a superposition of sinusoids, the multidimensional DFT expresses the input as a superposition of plane waves, or multidimensional sinusoids. The direction of oscillation in space is $\mathbf{k}/\mathbf{N}$. The amplitudes are $X_{\mathbf{k}}$. This decomposition is of great importance for everything from digital image processing (two-dimensional) to solving partial differential equations. The solution is broken up into plane waves.

The multidimensional DFT can be computed by the composition of a sequence of one-dimensional DFTs along each dimension. In the two-dimensional case x_{n_1,n_2} the N_1 independent DFTs of the rows (i.e., along n_2) are computed first to form a new array y_{n_1,k_2}. Then the independent DFTs of y along the columns (along n_1) are computed to form the final result X_{k_1,k_2}. Alternatively the columns can be computed first and then the rows. The order is immaterial because the nested summations above commute.

An algorithm to compute a one-dimensional DFT is thus sufficient to efficiently compute a multidimensional DFT. This approach is known as the *row-column* algorithm. There are also intrinsically multidimensional FFT algorithms.

The Real-input multidimensional DFT

For input data $x_{n_1,n_2,\ldots,n_d}$ consisting of real numbers, the DFT outputs have a conjugate symmetry similar to the one-dimensional case above:

$$X_{k_1,k_2,\ldots,k_d} = X^*_{N_1-k_1,N_2-k_2,\ldots,N_d-k_d},$$

where the star again denotes complex conjugation and the ℓ-th subscript is again interpreted modulo N_ℓ (for $\ell = 1,2,\ldots,d$).

Applications

The DFT has seen wide usage across a large number of fields; we only sketch a few examples below. All applications of the DFT depend crucially on the availability of a fast algorithm to compute discrete Fourier transforms and their inverses, a fast Fourier transform.

Spectral Analysis

When the DFT is used for signal spectral analysis, the $\{x_n\}$ sequence usually represents a finite set of uniformly spaced time-samples of some signal $x(t)$, where t represents time. The conversion from continuous time to samples (discrete-time) changes the underlying Fourier transform of x(t) into a discrete-time Fourier transform (DTFT), which generally entails a type of distortion called aliasing. Choice of an appropriate sample-rate is the key to minimizing that distortion. Similarly, the conversion from a very long (or infinite) sequence to a manageable size entails a type of distortion called *leakage*, which is manifested as a loss of detail (a.k.a. resolution) in the DTFT. Choice of an appropriate sub-sequence length is the primary key to minimizing that effect. When the available data (and time to process it) is more than the amount needed to attain the desired frequency resolution, a standard technique is to perform multiple DFTs, for example to create a spectrogram. If the desired result is a power spectrum and noise or randomness is present in the data, averaging the magnitude components of the multiple DFTs is a useful procedure to reduce the variance of the spectrum (also called a periodogram in this context); two examples of such techniques are the Welch method and the Bartlett method; the general subject of estimating the power spectrum of a noisy signal is called spectral estimation.

A final source of distortion (or perhaps *illusion*) is the DFT itself, because it is just a discrete sampling of the DTFT, which is a function of a continuous frequency domain. That can be mitigated by increasing the resolution of the DFT. That procedure is illustrated at Sampling the DTFT.

The procedure is sometimes referred to as *zero-padding*, which is a particular implementation used in conjunction with the fast Fourier transform (FFT) algorithm. The inefficiency of performing multiplications and additions with zero-valued "samples" is more than offset by the inherent efficiency of the FFT.

As already noted, leakage imposes a limit on the inherent resolution of the DTFT. So there is a practical limit to the benefit that can be obtained from a fine-grained DFT.

Data Compression

The field of digital signal processing relies heavily on operations in the frequency domain (i.e. on the Fourier transform). For example, several lossy image and sound compression methods employ the discrete Fourier transform: the signal is cut into short segments, each is transformed, and then the Fourier coefficients of high frequencies, which are assumed to be unnoticeable, are discarded. The decompressor computes the inverse transform based on this reduced number of Fourier coefficients. (Compression applications often use a specialized form of the DFT, the discrete cosine transform or sometimes the modified discrete cosine transform.) Some relatively recent compression algorithms, however, use wavelet transforms, which give a more uniform compromise between time and frequency domain than obtained by chopping data into segments and transforming each segment. In the case of JPEG2000, this avoids the spurious image features that appear when images are highly compressed with the original JPEG.

Partial Differential Equations

Discrete Fourier transforms are often used to solve partial differential equations, where again the DFT is used as an approximation for the Fourier series (which is recovered in the limit of infinite *N*). The advantage of this approach is that it expands the signal in complex exponentials e^{inx}, which are eigenfunctions of differentiation: $\mathrm{d}\left(e^{inx}\right)/\,\mathrm{d}x = ine^{inx}$.. Thus, in the Fourier representation, differentiation is simple—we just multiply by in. (Note, however, that the choice of n is not unique due to aliasing; for the method to be convergent, a choice similar to that in the trigonometric interpolation section above should be used.) A linear differential equation with constant coefficients is transformed into an easily solvable algebraic equation. One then uses the inverse DFT to transform the result back into the ordinary spatial representation. Such an approach is called a spectral method.

Polynomial Multiplication

Suppose we wish to compute the polynomial product $c(x) = a(x) \cdot b(x)$. The ordinary product expression for the coefficients of c involves a linear (acyclic) convolution, where indices do not "wrap around." This can be rewritten as a cyclic convolution by taking the coefficient vectors for $a(x)$ and $b(x)$ with constant term first, then appending zeros so that the resultant coefficient vectors a and b have dimension $d > \deg(a(x)) + \deg(b(x))$. Then,

$$\mathbf{c} = \mathbf{a} * \mathbf{b}$$

Where c is the vector of coefficients for $c(x)$, and the convolution operator $*$ is defined so

$$c_n = \sum_{m=0}^{d-1} a_m b_{n-m \bmod d} \qquad n = 0,1\ldots,d-1$$

But convolution becomes multiplication under the DFT:

$$\mathcal{F}(\mathbf{c}) = \mathcal{F}(\mathbf{a})\mathcal{F}(\mathbf{b})$$

Here the vector product is taken elementwise. Thus the coefficients of the product polynomial $c(x)$ are just the terms 0, ..., $\deg(a(x)) + \deg(b(x))$ of the coefficient vector

With a fast Fourier transform, the resulting algorithm takes O ($N \log N$) arithmetic operations. Due to its simplicity and speed, the Cooley–Tukey FFT algorithm, which is limited to composite sizes, is often chosen for the transform operation. In this case, d should be chosen as the smallest integer greater than the sum of the input polynomial degrees that is factorizable into small prime factors (e.g. 2, 3, and 5, depending upon the FFT implementation).

Multiplication of Large Integers

The fastest known algorithms for the multiplication of very large integers use the polynomial multiplication method outlined above. Integers can be treated as the value of a polynomial evaluated specifically at the number base, with the coefficients of the polynomial corresponding to the digits in that base. After polynomial multiplication, a relatively low-complexity carry-propagation step completes the multiplication.

Convolution

When data is convolved with a function with wide support, such as for downsampling by a large sampling ratio, because of the Convolution theorem and the FFT algorithm, it may be faster to transform it, multiply pointwise by the transform of the filter and then reverse transform it. Alternatively, a good filter is obtained by simply truncating the transformed data and re-transforming the shortened data set.

Generalizations

Representation Theory

The DFT can be interpreted as the complex-valued representation theory of the finite cyclic group. In other words, a sequence of n complex numbers can be thought of as an element of n-dimensional complex space C^n or equivalently a function f from the finite cyclic group of order n to the complex numbers, $Z_n \to C$. So f is a class function on the finite cyclic group, and thus can be expressed as a linear combination of the irreducible characters of this group, which are the roots of unity.

From this point of view, one may generalize the DFT to representation theory generally, or more narrowly to the representation theory of finite groups.

More narrowly still, one may generalize the DFT by either changing the target (taking values in a field other than the complex numbers), or the domain (a group other than a finite cyclic group), as detailed in the sequel.

Other Fields

Many of the properties of the DFT only depend on the fact that $e^{-\frac{2\pi i}{N}}$ is a primitive root of unity, sometimes denoted ω_N or W_N (so that $\omega_N^N = 1$). Such properties include the completeness, orthogonality, Plancherel/Parseval, periodicity, shift, convolution, and unitarity properties above, as well as many FFT algorithms. For this reason, the discrete Fourier transform can be defined by using roots of unity in fields other than the complex numbers, and such generalizations are commonly called *number-theoretic transforms* (NTTs) in the case of finite fields.

Other Finite Groups

The standard DFT acts on a sequence $x_0, x_1, \ldots, x_{N-1}$ of complex numbers, which can be viewed as a function {0, 1, ..., $N - 1$} → C. The multidimensional DFT acts on multidimensional sequences, which can be viewed as functions

$$\{0,1,\ldots,N_1-1\}\times\cdots\times\{0,1,\ldots,N_d-1\}\to\mathbb{C}.$$

This suggests the generalization to Fourier transforms on arbitrary finite groups, which act on functions $G \to$ C where G is a finite group. In this framework, the standard DFT is seen as the Fourier transform on a cyclic group, while the multidimensional DFT is a Fourier transform on a direct sum of cyclic groups.

Alternatives

There are various alternatives to the DFT for various applications, prominent among which are wavelets. The analog of the DFT is the discrete wavelet transform (DWT). From the point of view of time–frequency analysis, a key limitation of the Fourier transform is that it does not include *location* information, only *frequency* information, and thus has difficulty in representing transients. As wavelets have location as well as frequency, they are better able to represent location, at the expense of greater difficulty representing frequency.

Discrete-time Fourier Transform

In mathematics, the discrete-time Fourier transform (DTFT) is a form of Fourier analysis that is applicable to the uniformly-spaced samples of a continuous function. The term *discrete-time* refers to the fact that the transform operates on discrete data (samples) whose interval often has units of time. From only the samples, it produces a function of frequency that is a periodic summation of the continuous Fourier transform of the original continuous function. Under certain theoretical conditions, described by the sampling theorem, the original continuous function can be recovered perfectly from the DTFT and thus from the original discrete samples. The DTFT itself is a continuous function of frequency, but discrete samples of it can be readily calculated via the discrete Fourier transform (DFT) (see Sampling the DTFT), which is by far the most common method of modern Fourier analysis.

Both transforms are invertible. The inverse DTFT is the original sampled data sequence. The inverse DFT is a periodic summation of the original sequence. The fast Fourier transform (FFT) is an algorithm for computing one cycle of the DFT, and its inverse produces one cycle of the inverse DFT.

Definition

The discrete-time Fourier transform of a discrete set of real or complex numbers $x[n]$, for all integers n, is a Fourier series, which produces a periodic function of a frequency variable. When the frequency variable, ω, has normalized units of *radians/sample*, the periodicity is 2π, and the Fourier series is:

$$X_{2\pi}(\omega) = \sum_{n=-\infty}^{\infty} x[n]e^{-i\omega n}. \qquad \textbf{(Eq.1)}$$

The utility of this frequency domain function is rooted in the Poisson summation formula. Let X(f) be the Fourier transform of any function, x(t), whose samples at some interval T (*seconds*) are equal (or proportional to) the x[n] sequence, i.e. $T \cdot x(nT) = x[n]$. Then the periodic function represented by the Fourier series is a periodic summation of X(f). In terms of frequency f in hertz (*cycles/sec*):

$$X_{1/T}(f) = X_{2\pi}(2\pi fT) \overset{\text{def}}{=} \sum_{n=-\infty}^{\infty} \underbrace{T \cdot x(nT)}_{x[n]} e^{-i2\pi fTn} \overset{\text{Poisson f.}}{=} \sum_{k=-\infty}^{\infty} X(f - k/T). \qquad \textbf{(Eq.2)}$$

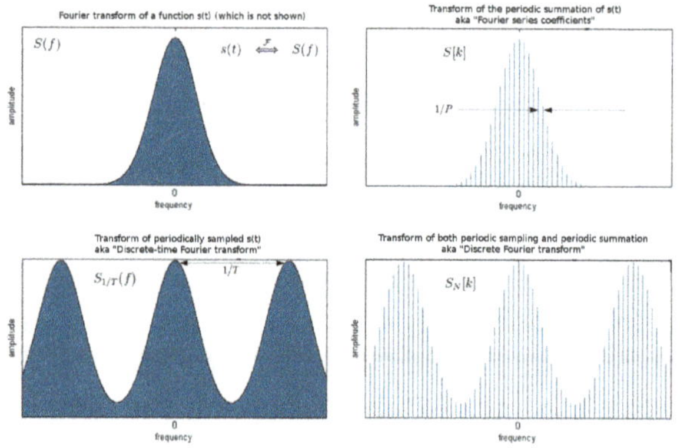

Fig 1. Depiction of a Fourier transform (upper left) and its periodic summation (DTFT) in the lower left corner. The lower right corner depicts samples of the DTFT that are computed by a discrete Fourier transform (DFT).

The integer k has units of *cycles/sample*, and $1/T$ is the sample-rate, f_s (*samples/sec*). So $X_{1/T}(f)$ comprises exact copies of $X(f)$ that are shifted by multiples of f_s hertz and combined by addition. For sufficiently large f_s the k=0 term can be observed in the region $[-f_s/2, f_s/2]$ with little or no distortion (aliasing) from the other terms. In Fig.1, the extremities of the distribution in the upper left corner are masked by aliasing in the periodic summation (lower left).

We also note that $e^{-i2\pi fTn}$ is the Fourier transform of $\delta(t-nT)$. Therefore, an alternative definition of DTFT is:

$$X_{1/T}(f) = \mathcal{F}\left\{\sum_{n=-\infty}^{\infty} x[n]\cdot\delta(t-nT)\right\} \qquad \textbf{(\textit{Eq.3})}$$

The modulated Dirac comb function is a mathematical abstraction sometimes referred to as *impulse sampling*.

Inverse Transform

An operation that recovers the discrete data sequence from the DTFT function is called an *inverse DTFT*. For instance, the inverse continuous Fourier transform of both sides of Eq.3 produces the sequence in the form of a modulated Dirac comb function:

$$\sum_{n=-\infty}^{\infty} x[n]\cdot\delta(t-nT) = \mathcal{F}^{-1}\left\{X_{1/T}(f)\right\} \stackrel{\text{def}}{=} \int_{-\infty}^{\infty} X_{1/T}(f)\cdot e^{i2\pi ft}\,df.$$

However, noting that $X_{1/T}(f)$ is periodic, all the necessary information is contained within any interval of length $1/T$. In both Eq.1 and Eq.2, the summations over n are a Fourier series, with coefficients x[n]. The standard formulas for the Fourier coefficients are also the inverse transforms:

$$x[n] = T\int_{\frac{1}{T}} X_{1/T}(f)\cdot e^{i2\pi fnT}\,df \qquad \text{(integral over any interval of length } 1/T\text{)}$$

$$= \frac{1}{2\pi}\int_{2\pi} X_{2\pi}(\omega)\cdot e^{i\omega n}\,d\omega \qquad \text{(integral over any interval of length } 2\pi\text{)}$$

Periodic Data

When the input data sequence $x[n]$ is N-periodic, Eq.2 can be computationally reduced to a discrete Fourier transform (DFT), because:

- All the available information is contained within N samples.
- $X_{1/T}(f)$ converges to zero everywhere except integer multiples of $\frac{1}{NT}$, known as harmonic frequencies.

- The DTFT is periodic, so the maximum number of unique harmonic amplitudes is $\frac{1}{T}/\frac{1}{NT} = N.$

The kernel $x[n]\cdot e^{-i2\pi fTn}$ is N-periodic at the harmonic frequencies, $f = \frac{k}{NT}$. Introducing the notation $\sum_N$ to represent a sum over any n-sequence of length N, we can write:

$$X_{1/T}\left(\frac{k}{NT}\right) = \sum_{m=-\infty}^{\infty}\left(\sum_N x[n-mN]\cdot e^{-i2\pi\frac{k}{N}(n-mN)}\right)$$

$$= \sum_{m=-\infty}^{\infty}\left(\sum_N x[n]\cdot e^{-i2\pi\frac{k}{N}n}\right) = T\underbrace{\left(\sum_N x(nT)\cdot e^{-i2\pi\frac{k}{N}n}\right)}_{X[k]\ \textbf{(DFT)}}\cdot\left(\sum_{m=-\infty}^{\infty}1\right).$$

Therefore, the DTFT diverges at the harmonic frequencies, but at different frequency-dependent rates. And those rates are given by the DFT of one cycle of the $x[n]$ sequence. In terms of a Dirac comb function, this is represented by:

$$X_{1/T}(f) = \sum_{k=-\infty}^{\infty}(T\cdot X[k])\cdot\frac{1}{NT}\delta\left(f-\frac{k}{NT}\right) = \underbrace{\frac{1}{N}\sum_{k=-\infty}^{\infty}X[k]\cdot\delta\left(f-\frac{k}{NT}\right)}_{\text{DTFT of a periodic sequence}}.$$

Sampling the DTFT

When the DTFT is continuous, a common practice is to compute an arbitrary number of samples (N) of one cycle of the periodic function $X_{1/T}$:

$$\underbrace{X_{1/T}\left(\frac{k}{NT}\right)}_{X_k} = \sum_{n=-\infty}^{\infty}x[n]\cdot e^{-i2\pi\frac{kn}{N}} \qquad k = 0,\ldots,N-1$$

$$= \underbrace{\sum_N x_N[n]\cdot e^{-i2\pi\frac{kn}{N}},}_{DFT} \quad \text{(sum over any } n\text{-sequence of length } N)$$

where x_N is a periodic summation:

$$x_N[n] \stackrel{\text{def}}{=} \sum_{m=-\infty}^{\infty}x[n-mN].$$

The x_N sequence is the inverse DFT. Thus, our sampling of the DTFT causes the inverse transform to become periodic. The array of $|X_k|^2$ is known as a periodogram, where parameter N is known to Matlab users as *NFFT*.

In order to evaluate one cycle of x_N numerically, we require a finite-length x[n] sequence. For instance, a long sequence might be truncated by a window function of length L resulting in two cases worthy of special mention: $L \leq N$ and $L = I \cdot N$, for some integer I (typically 6 or 8). For notational simplicity, consider the $x[n]$ values below to represent the modified values.

When $L = I \cdot N$ a cycle of x_N reduces to a summation of I *blocks* of length N. This goes by various names, such as:

- *window-presum FFT*
- *Weight, OverLap, Add (WOLA)*
- *polyphase FFT*
- *multiple block windowing*
- *time-aliasing.*

An interesting way to understand/motivate the technique is to recall that decimation of sampled data in one domain (time or frequency) produces aliasing in the other, and vice versa. The x_N summation is mathematically equivalent to aliasing, leading to decimation in frequency, leaving only DTFT samples least affected by spectral leakage. That is usually a priority when implementing an FFT filter-bank (channelizer). With a conventional window function of length L, scalloping loss would be unacceptable. So multi-block windows are created using FIR filter design tools. Their frequency profile is flat at the highest point and falls off quickly at the midpoint between the remaining DTFT samples. The larger the value of parameter I the better the potential performance. We note that the same results can be obtained by computing and decimating an L-length DFT, but that is not computationally efficient.

When $L \leq N$ the DFT is usually written in this more familiar form:

$$X_k = \sum_{n=0}^{N-1} x[n] \cdot e^{-i2\pi\frac{kn}{N}}.$$

In order to take advantage of a fast Fourier transform algorithm for computing the DFT, the summation is usually performed over all N terms, even though N-L of them are zeros. Therefore, the case $L < N$ is often referred to as "zero-padding".

Spectral leakage, which increases as L decreases, is detrimental to certain important performance metrics, such as resolution of multiple frequency components and the amount of noise measured by each DTFT sample. But those things don't always matter, for instance when the x[n] sequence is a noiseless sinusoid (or a constant), shaped by a window function. Then it is a common practice to use *zero-padding* to graphically display and compare the detailed leakage patterns of window functions. To illustrate that for a rectangular window, consider the sequence:

$$x[n] = e^{i2\pi\frac{1}{8}n}, \quad \text{and } L = 64.$$

The two figures below are plots of the magnitude of two different sized DFTs, as indicated in their labels. In both cases, the dominant component is at the signal frequency: $f = 1/8 = 0.125$. Also visible on the right is the spectral leakage pattern of the $L = 64$ rectangular window. The illusion on the left is a result of sampling the DTFT at all of its zero-crossings. Rather than the DTFT of a finite-length sequence, it gives the impression of an infinitely long sinusoidal sequence. Contributing factors to the illusion are the use of a rectangular window, and the choice of a frequency ($1/8 = 8/64$) with exactly 8 (an integer) cycles per 64 samples.

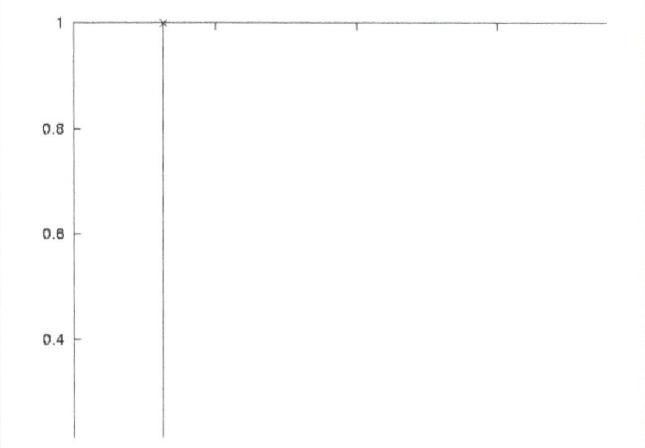

DFT for L = 64 and N = 64

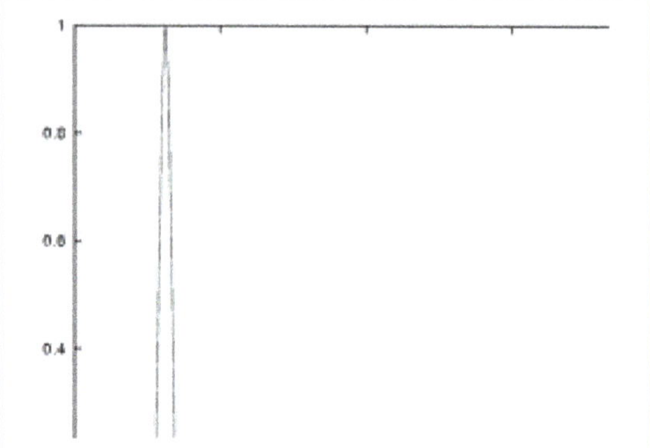

DFT for L = 64 and N = 256

Convolution

The convolution theorem for sequences is:

$$x * y = \text{DTFT}^{-1}\left[\text{DTFT}\{x\} \cdot \text{DTFT}\{y\}\right].$$

An important special case is the circular convolution of sequences x and y defined by $x_N * y$ where x_N is a periodic summation. The discrete-frequency nature of DTFT$\{x_N\}$ "selects" only discrete values from the continuous function DTFT$\{y\}$, which results in considerable simplification of the inverse transform. As shown at Convolution theorem#Functions of discrete variable sequences:

$$x_N * y = \text{DTFT}^{-1}\left[\text{DTFT}\{x_N\} \cdot \text{DTFT}\{y\}\right] = \text{DFT}^{-1}\left[\text{DFT}\{x_N\} \cdot \text{DFT}\{y_N\}\right].$$

For x and y sequences whose non-zero duration is less than or equal to N, a final simplification is:

$$x_N * y = \text{DFT}^{-1}\left[\text{DFT}\{x\} \cdot \text{DFT}\{y\}\right].$$

The significance of this result is expounded at Circular convolution and Fast convolution algorithms.

Relationship to the Z-Transform

The bilateral Z-transform is defined by:

$$\underbrace{\mathcal{Z}\{x[n]\}}_{\hat{X}(z)} = \sum_{n=-\infty}^{\infty} x[n]z^{-n},$$ where z is a complex variable.

We denote this function as $\hat{X}(z)$ to avoid confusion with the Fourier transform. For the special case that z is constrained to values of the form $e^{i\omega}$:

$$\hat{X}(e^{i\omega}) = \sum_{n=-\infty}^{\infty} x[n]e^{-i\omega n} = X_{2\pi}(\omega) = X_{1/T}\left(\tfrac{\omega}{2\pi T}\right) = \sum_{k=-\infty}^{\infty} X\left(\tfrac{\omega}{2\pi T} - k/T\right)$$

$$= \sum_{k=-\infty}^{\infty} X\left(\tfrac{\omega-2\pi k}{2\pi T}\right)$$

Note that when parameter T changes, the terms of $X_{2\pi}(\omega)$ remain a constant separation (2π) apart, and their width scales up or down. The terms of $X_{1/T}(f)$ remain a constant width and their separation (1/T) scales up or down.

Table of Discrete-time Fourier Transforms

Some common transform pairs are shown in the table below. The following notation applies:

- $\omega = 2\pi fT$ is a real number representing continuous angular frequency (in radians per sample). (f is in cycles/sec, and T is in sec/sample.) In all cases in the table, the DTFT is 2π-periodic (in ω).
- $X_{2\pi}(\omega)$ designates a function defined on $-\infty < \omega < \infty$.
- $X(\omega)$ designates a function defined on $-\pi < \omega \le \pi$, and zero elsewhere. Then:

$$X_{2\pi}(\omega) \stackrel{\text{def}}{=} \sum_{k=-\infty}^{\infty} X(\omega - 2\pi k).$$

- $\delta(\omega)$ is the Dirac delta function
- sinc(t) is the normalized sinc function
- rect(t) is the rectangle function
- tri(t) is the triangle function
- n is an integer representing the discrete-time domain (in samples)

- $u[n]$ is the discrete-time unit step function
- $\delta[n]$ is the Kronecker delta $\delta_{n,0}$

Discrete Wavelet Transform

An example of the 2D discrete wavelet transform that is used in JPEG2000. The original image is high-pass filtered, yielding the three large images, each describing local changes in brightness (details) in the original image. It is then low-pass filtered and downscaled, yielding an approximation image; this image is high-pass filtered to produce the three smaller detail images, and low-pass filtered to produce the final approximation image in the upper-left.

In numerical analysis and functional analysis, a discrete wavelet transform (DWT) is any wavelet transform for which the wavelets are discretely sampled. As with other wavelet transforms, a key advantage it has over Fourier transforms is temporal resolution: it captures both frequency *and* location information (location in time).

Examples

Haar Wavelets

The first DWT was invented by Hungarian mathematician Alfréd Haar. For an input represented by a list of 2^n numbers, the Haar wavelet transform may be considered to pair up input values, storing the difference and passing the sum. This process is repeated recursively, pairing up the sums to provide the next scale, which leads to $2^n - 1$ differences and a final sum.

Daubechies Wavelets

The most commonly used set of discrete wavelet transforms was formulated by the Belgian mathematician Ingrid Daubechies in 1988. This formulation is based on the use of recurrence relations to generate progressively finer discrete samplings of an implicit mother wavelet function; each resolution is twice that of the previous scale. In her seminal paper, Daubechies derives a family of wavelets, the first of which is the Haar wavelet. Interest in this field has exploded since then, and many variations of Daubechies' original wavelets were developed.

The Dual-tree Complex Wavelet Transform (DCWT)

The dual-tree complex wavelet transform (CWT) is a relatively recent enhancement to the discrete wavelet transform (DWT), with important additional properties: It is nearly shift invariant and directionally selective in two and higher dimensions. It achieves this with a redundancy factor of only 2^d substantially lower than the undecimated DWT. The multidimensional (M-D) dual-tree CWT is nonseparable but is based on a computationally efficient, separable filter bank (FB).

Others

Other forms of discrete wavelet transform include the non- or undecimated wavelet transform (where downsampling is omitted), the Newland transform (where an orthonormal basis of wavelets is formed from appropriately constructed top-hat filters in frequency space). Wavelet packet transforms are also related to the discrete wavelet transform. Complex wavelet transform is another form.

Properties

The Haar DWT illustrates the desirable properties of wavelets in general. First, it can be performed in $O(n)$ operations; second, it captures not only a notion of the frequency content of the input, by examining it at different scales, but also temporal content, i.e. the times at which these frequencies occur. Combined, these two properties make the Fast wavelet transform (FWT) an alternative to the conventional fast Fourier transform (FFT).

Time Issues

Due to the rate-change operators in the filter bank, the discrete WT is not time-invariant but actually very sensitive to the alignment of the signal in time. To address the time-varying problem of wavelet transforms, Mallat and Zhong proposed a new algorithm for wavelet representation of a signal, which is invariant to time shifts. According to this algorithm, which is called a TI-DWT, only the scale parameter is sampled along the dyadic sequence 2^j (j∈Z) and the wavelet transform is calculated for each point in time.

Applications

The discrete wavelet transform has a huge number of applications in science, engineering, mathematics and computer science. Most notably, it is used for signal coding, to represent a discrete signal in a more redundant form, often as a preconditioning for data compression. Practical applications can also be found in signal processing of accelerations for gait analysis, in digital communications and many others.

It is shown that discrete wavelet transform (discrete in scale and shift, and continuous in time) is successfully implemented as analog filter bank in biomedical signal processing for design of low-power pacemakers and also in ultra-wideband (UWB) wireless communications.

Comparison with Fourier Transform

To illustrate the differences and similarities between the discrete wavelet transform with the discrete Fourier transform, consider the DWT and DFT of the following sequence: (1,0,0,0), a unit impulse.

The DFT has Orthogonal Basis (DFT matrix):

$$\begin{bmatrix} 1 & 1 & 1 & 1 \\ 1 & -i & -1 & i \\ 1 & -1 & 1 & -1 \\ 1 & i & -1 & -i \end{bmatrix}$$

while the DWT with Haar wavelets for length 4 data has orthogonal basis in the rows of:

$$\begin{bmatrix} 1 & 1 & 1 & 1 \\ 1 & 1 & -1 & -1 \\ 1 & -1 & 0 & 0 \\ 0 & 0 & 1 & -1 \end{bmatrix}$$

(To simplify notation, whole numbers are used, so the bases are orthogonal but not orthonormal.)

Preliminary observations include:

- Sinusoidal waves differ only in their frequency. The first does not complete any cycles, the second completes one full cycle, the third completes two cycles, and the fourth completes three cycles (which is equivalent to completing one cycle in the opposite direction). Differences in phase can be represented by multiplying a given basis vector by a complex constant.
- Wavelets, by contrast, have both frequency and location. As before, the first completes zero cycles, and the second completes one cycle. However, the second and third both have the same frequency, twice that of the first. Rather than differing in frequency, they differ in *location* — the third is nonzero over the first two elements, and the second is nonzero over the second two elements.

Decomposing the sequence with respect to these bases yields:

$$(1,0,0,0) = \frac{1}{4}(1,1,1,1) + \frac{1}{4}(1,1,-1,-1) + \frac{1}{2}(1,-1,0,0) \qquad \text{Haar DWT}$$

$$(1,0,0,0) = \frac{1}{4}(1,1,1,1) + \frac{1}{4}(1,i,-1,-i) + \frac{1}{4}(1,-1,1,-1) + \frac{1}{4}(1,-i,-1,i) \qquad \text{DFT}$$

The DWT demonstrates the localization: the (1,1,1,1) term gives the average signal value, the (1,1,−1,−1) places the signal in the left side of the domain, and the (1,−1,0,0) places it at the left side of the left side, and truncating at any stage yields a downsampled version of the signal:

$$\left(\frac{1}{4},\frac{1}{4},\frac{1}{4},\frac{1}{4}\right)$$

$$\left(\frac{1}{2},\frac{1}{2},0,0\right) \qquad \text{2-term truncation}$$

$$(1,0,0,0)$$

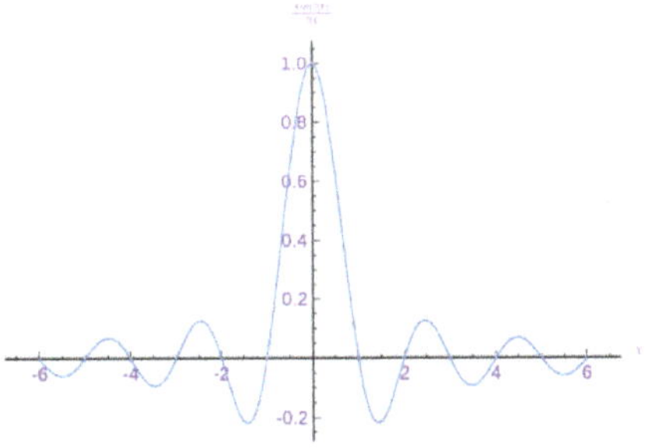

The sinc function, showing the time domain artifacts (undershoot and ringing) of truncating a Fourier series.

The DFT, by contrast, expresses the sequence by the interference of waves of various frequencies – thus truncating the series yields a low-pass filtered version of the series:

$$\left(\frac{1}{4},\frac{1}{4},\frac{1}{4},\frac{1}{4}\right)$$

$$\left(\frac{3}{4},\frac{1}{4},-\frac{1}{4},\frac{1}{4}\right) \qquad \text{2-term truncation}$$

$$(1,0,0,0)$$

Notably, the middle approximation (2-term) differs. From the frequency domain perspective, this is a better approximation, but from the time domain perspective it has drawbacks – it exhibits undershoot – one of the values is negative, though the original series is non-negative everywhere – and ringing, where the right side is non-zero, unlike in the wavelet transform. On the other hand, the Fourier approximation correctly shows a peak, and all points are within $1/4$ of their correct value, though all points have error. The wavelet approximation, by contrast, places a peak on the left half, but has no peak at the first point, and while it is exactly correct for half the values (reflecting location), it has an error of $1/2$ for the other values.

This illustrates the kinds of trade-offs between these transforms, and how in some respects the DWT provides preferable behavior, particularly for the modeling of transients.

Definition

One Level of the Transform

The DWT of a signal x is calculated by passing it through a series of filters. First the samples are passed through a low pass filter with impulse response g resulting in a convolution of the two:

$$y[n] = (x * g)[n] = \sum_{k=-\infty}^{\infty} x[k]g[n-k]$$

The signal is also decomposed simultaneously using a high-pass filter h. The outputs giving the detail coefficients (from the high-pass filter) and approximation coefficients (from the low-pass). It is important that the two filters are related to each other and they are known as a quadrature mirror filter.

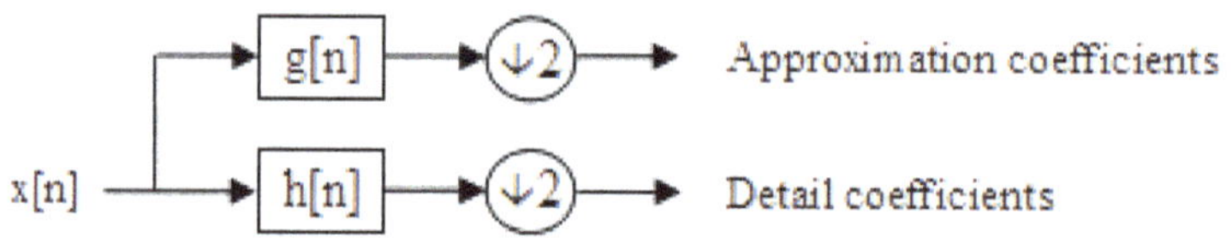

Block diagram of filter analysis

However, since half the frequencies of the signal have now been removed, half the samples can be discarded according to Nyquist's rule. The filter outputs are then subsampled by 2. In the next two formulas, the notation is the opposite: g- denotes high pass and h- low pass as is Mallat's and the common notation:

$$y_{\text{low}}[n] = \sum_{k=-\infty}^{\infty} x[k]g[2n-k]$$

$$y_{\text{high}}[n] = \sum_{k=-\infty}^{\infty} x[k]h[2n-k]$$

This decomposition has halved the time resolution since only half of each filter output characterises the signal. However, each output has half the frequency band of the input so the frequency resolution has been doubled.

With the subsampling operator $\downarrow$

$$(y \downarrow k)[n] = y[kn]$$

the above summation can be written more concisely.

$$y_{\text{low}} = (x * g) \downarrow 2$$

$$y_{\text{high}} = (x * h) \downarrow 2$$

However computing a complete convolution $x * g$ with subsequent downsampling would waste computation time.

The Lifting scheme is an optimization where these two computations are interleaved.

Cascading and filter banks

This decomposition is repeated to further increase the frequency resolution and the approximation coefficients decomposed with high and low pass filters and then down-sampled. This is represented as a binary tree with nodes representing a sub-space with a different time-frequency localisation. The tree is known as a filter bank.

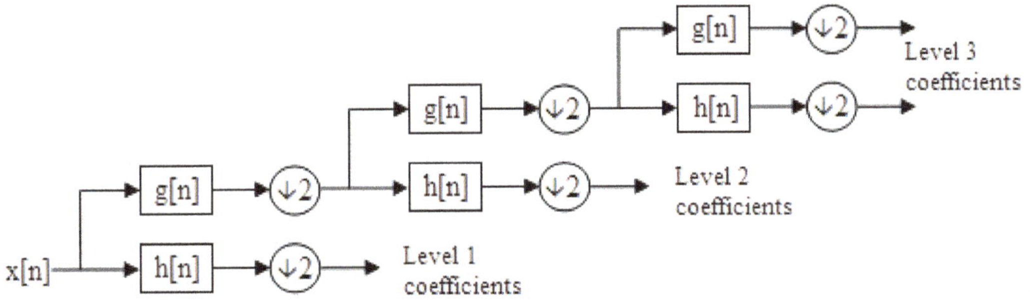

A 3 level filter bank

At each level in the above diagram the signal is decomposed into low and high frequencies. Due to the decomposition process the input signal must be a multiple of 2^n where n is the number of levels.

For example a signal with 32 samples, frequency range 0 to f_n and 3 levels of decomposition, 4 output scales are produced:

Level	Frequencies	Samples
3	to	4
	to	4
2	to	8
1	to	16

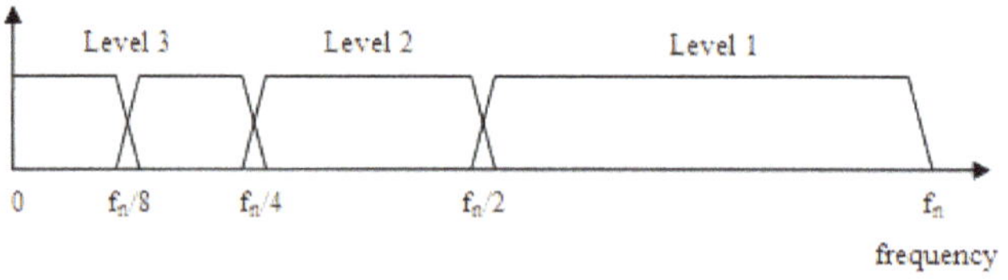

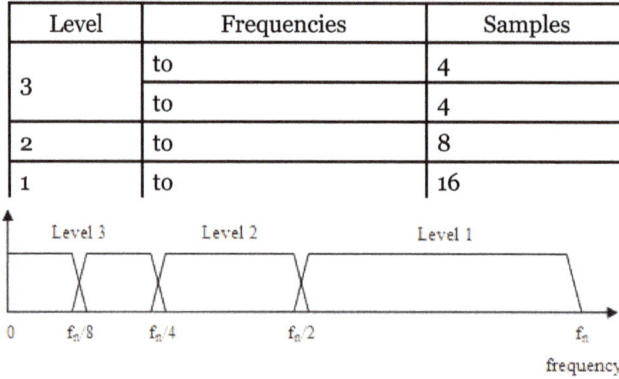

Frequency domain representation of the DWT

Relationship to the Mother Wavelet

The filterbank implementation of wavelets can be interpreted as computing the wavelet coefficients of a discrete set of child wavelets for a given mother wavelet $\psi(t)$. In the case of the discrete wavelet transform, the mother wavelet is shifted and scaled by powers of two

$$\psi_{j,k}(t) = \frac{1}{\sqrt{2^j}}\psi\left(\frac{t-k2^j}{2^j}\right)$$

where j is the scale parameter and k is the shift parameter, both which are integers.

Recall that the wavelet coefficient γ of a signal $x(t)$ is the projection of $x(t)$ onto a wavelet, and let $x(t)$ be a signal of length 2^N. In the case of a child wavelet in the discrete family above,

$$\gamma_{jk} = \int_{-\infty}^{\infty} x(t)\frac{1}{\sqrt{2^j}}\psi\left(\frac{t-k2^j}{2^j}\right)dt$$

Now fix j at a particular scale, so that γ_{jk} is a function of k only. In light of the above equation, γ_{jk} can be viewed as a convolution of $x(t)$ with a dilated, reflected, and normalized version of the mother wavelet, $h(t) = \frac{1}{\sqrt{2^j}}\psi\left(\frac{-t}{2^j}\right)$, sampled at the points $1, 2^j, 2^{2j}, \ldots, 2^N$. But this is precisely what the detail coefficients give at level j of the discrete wavelet transform. Therefore, for an appropriate choice of $h[n]$ and $g[n]$, the detail coefficients of the filter bank correspond exactly to a wavelet coefficient of a discrete set of child wavelets for a given mother wavelet $\psi(t)$.

As an example, consider the discrete Haar wavelet, whose mother wavelet is $\psi = [1,-1]$. Then the dilated, reflected, and normalized version of this wavelet is $h[n] = \frac{1}{\sqrt{2}}[-1,1]$, which is, indeed, the highpass decomposition filter for the discrete Haar wavelet transform.

Time Complexity

The filterbank implementation of the Discrete Wavelet Transform takes only O(N) in certain cases, as compared to O($N \log N$) for the fast Fourier transform.

Note that if $g[n]$ and $h[n]$ are both a constant length (i.e. their length is independent of N), then $x*h$ and $x*g$ each take O(N) time. The wavelet filterbank does each of these two O(N) convolutions, then splits the signal into two branches of size N/2. But it only recursively splits the upper branch convolved with $g[n]$ (as contrasted with the FFT, which recursively splits both the upper branch and the lower branch). This leads to the following recurrence relation

$$T(N) = 2N + T\left(\frac{N}{2}\right)$$

which leads to an O(N) time for the entire operation, as can be shown by a geometric series expansion of the above relation.

As an example, the discrete Haar wavelet transform is linear, since in that case $h[n]$ and $g[n]$ are constant length 2.

$$h[n] = \left[\frac{-\sqrt{2}}{2}, \frac{\sqrt{2}}{2}\right] g[n] = \left[\frac{\sqrt{2}}{2}, \frac{\sqrt{2}}{2}\right]$$

Other Transforms

The Adam7 algorithm, used for interlacing in the Portable Network Graphics (PNG) format, is a multiscale model of the data which is similar to a DWT with Haar wavelets.

Unlike the DWT, it has a specific scale – it starts from an 8×8 block, and it downsamples the image, rather than decimating (low-pass filtering, then downsampling). It thus offers worse frequency behavior, showing artifacts (pixelation) at the early stages, in return for simpler implementation.

Code Example

In its simplest form, the DWT is remarkably easy to compute.

The Haar wavelet in Java:

```
public static int[] discreteHaarWaveletTransform(int[] input) {
    // This function assumes that input.length=2^n, n>1
```

```
    int[] output = new int[input.length];
    for (int length = input.length >> 1; ; length >>= 1) {
      // length = input.length / 2^n, WITH n INCREASING to log_2(input.length)
      for (int i = 0; i < length; ++i) {
        int sum = input[i * 2] + input[i * 2 + 1];
        int difference = input[i * 2] - input[i * 2 + 1];
        output[i] = sum;
        output[length + i] = difference;
      }
      if (length == 1) {
        return output;
      }
      //Swap arrays to do next iteration
      System.arraycopy(output, 0, input, 0, length << 1);
    }
}
```

Complete Java code for a 1-D and 2-D DWT using Haar, Daubechies, Coiflet, and Legendre wavelets is available from the open source project: JWave. Furthermore, a fast lifting implementation of the discrete biorthogonal CDF 9/7 wavelet transform in C, used in the JPEG 2000 image compression standard can be found here (archived 5 March 2012).

Example of Above Code

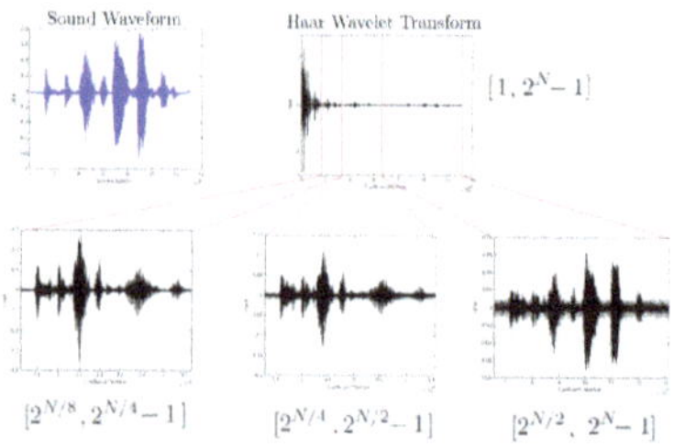

An example of computing the discrete Haar wavelet coefficients for a sound signal of someone saying "I Love Wavelets." The original waveform is shown in blue in the upper left, and the wavelet coefficients are shown in black in the upper right. Along the bottom is shown three zoomed-in regions of the wavelet coefficients for different ranges.

This figure shows an example of applying the above code to compute the Haar wavelet coefficients on a sound waveform. This example highlights two key properties of the wavelet transform:

- Natural signals often have some degree of smoothness, which makes them sparse in the wavelet domain. There are far fewer significant components in the wavelet domain in this example than there are in the time domain, and most of the significant components are towards the coarser coefficients on the left. Hence, natural signals are compressible in the wavelet domain.
- The wavelet transform is a multiresolution, bandpass representation of a signal. This can be seen directly from the filterbank definition of the discrete wavelet transform given in this article. For a signal of length 2^N, the coefficients in the range $[2^{\frac{N}{j+1}}, 2^{\frac{N}{j}} - 1]$ represent a version of the original signal which is in the pass-band $\left[\frac{\pi}{2^j}, \frac{\pi}{2^{j-1}}\right]$. This is why zooming in on these ranges of the wavelet coefficients looks so similar in structure to the original signal. Ranges which are closer to the left (larger j in the above notation), are coarser representations of the signal, while ranges to the right represent finer details.

Discrete Cosine Transform

A discrete cosine transform (DCT) expresses a finite sequence of data points in terms of a sum of cosine functions oscillating at different frequencies. DCTs are important to numerous applications in science and engineering, from lossy compression of audio (e.g. MP3) and images (e.g. JPEG) (where small high-frequency components can be discarded), to spectral methods for the numerical solution of partial differential equations. The use of cosine rather than sine functions is critical for compression, since it turns out (as described below) that fewer cosine functions are needed to approximate a typical signal, whereas for differential equations the cosines express a particular choice of boundary conditions.

In particular, a DCT is a Fourier-related transform similar to the discrete Fourier transform (DFT), but using only real numbers. The DCTs are generally related to Fourier Series coefficients of a periodically and symmetrically extended sequence whereas DFTs are related to Fourier Series coefficients of a periodically extended sequence. DCTs are equivalent to DFTs of roughly twice the length, operating on real data with even symmetry (since the Fourier transform of a real and even function is real and even), whereas in some variants the input and/or output data are shifted by half a sample. There are eight standard DCT variants, of which four are common.

The most common variant of discrete cosine transform is the type-II DCT, which is often called simply “the DCT”. Its inverse, the type-III DCT, is correspondingly often called simply "the inverse DCT" or "the IDCT". Two related transforms are the discrete sine transform (DST), which is equivalent to a DFT of real and *odd* functions, and the modified discrete cosine transform (MDCT), which is based on a DCT of *overlapping* data. Multidimensional DCTs (MD DCTs) are developed to extend the concept of DCT on MD Signals. There are several algorithms to compute MD DCT. A new variety of fast alogrithms are also developed to reduce the computational complexity of implementing DCT.

Applications

The DCT, and in particular the DCT-II, is often used in signal and image processing, especially for lossy compression, because it has a strong "energy compaction" property: in typical applications, most of the signal information tends to be concentrated in a few low-frequency components of the DCT. For strongly correlated Markov processes, the DCT can approach the compaction efficiency of the Karhunen-Loève transform (which is optimal in the decorrelation sense). As explained below, this stems from the boundary conditions implicit in the cosine functions.

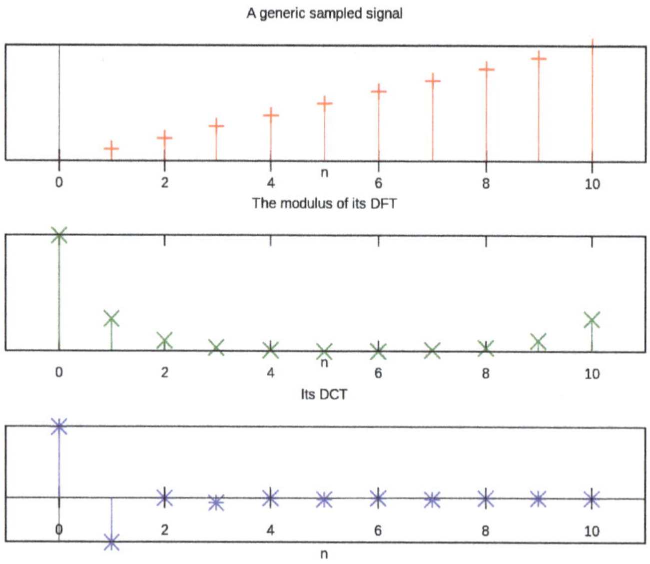

DCT-II (bottom) compared to the DFT (middle) of an input signal (top).

A related transform, the *modified* discrete cosine transform, or MDCT (based on the DCT-IV), is used in AAC, Vorbis, WMA, and MP3 audio compression.

DCTs are also widely employed in solving partial differential equations by spectral methods, where the different variants of the DCT correspond to slightly different even/odd boundary conditions at the two ends of the array.

DCTs are also closely related to Chebyshev polynomials, and fast DCT algorithms (below) are used in Chebyshev approximation of arbitrary functions by series of Chebyshev polynomials, for example in Clenshaw–Curtis quadrature.

JPEG

The DCT is used in JPEG image compression, MJPEG, MPEG, DV, Daala, and Theora video compression. There, the two-dimensional DCT-II of $N \times N$ blocks are computed and the results are quantized and entropy coded. In this case, N is typically 8 and the DCT-II formula is applied to each row and column of the block. The result is an 8×8 transform coefficient array in which the $(0,0)$ element (top-left) is the DC (zero-frequency) component and entries with increasing vertical and horizontal index values represent higher vertical and horizontal spatial frequencies.

Informal Overview

Like any Fourier-related transform, discrete cosine transforms (DCTs) express a function or a signal in terms of a sum of sinusoids with different frequencies and amplitudes. Like the discrete Fourier transform (DFT), a DCT operates on a function at a finite number of discrete data points. The obvious distinction between a DCT and a DFT is that the former uses only cosine functions, while the latter uses both cosines and sines (in the form of complex exponentials). However, this visible difference is merely a consequence of a deeper distinction: a DCT implies different boundary conditions from the DFT or other related transforms.

The Fourier-related transforms that operate on a function over a finite domain, such as the DFT or DCT or a Fourier series, can be thought of as implicitly defining an *extension* of that function outside the domain. That is, once you write a function $f(x)$ as a sum of sinusoids, you can evaluate that sum at any x, even for x where the original $f(x)$ was not specified. The DFT, like the Fourier series, implies a periodic extension of the original function. A DCT, like a cosine transform, implies an even extension of the original function.

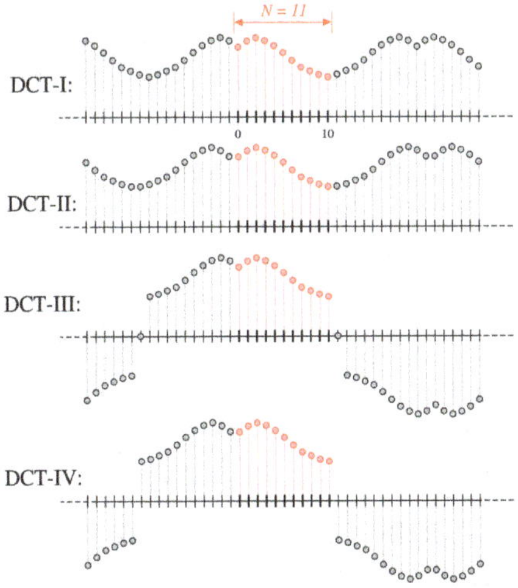

Illustration of the implicit even/odd extensions of DCT input data, for N=11 data points (red dots), for the four most common types of DCT (types I-IV).

However, because DCTs operate on *finite*, *discrete* sequences, two issues arise that do not apply for the continuous cosine transform. First, one has to specify whether the function is even or odd at *both* the left and right boundaries of the domain (i.e. the min-n and max-n boundaries in the definitions below, respectively). Second, one has to specify around *what point* the function is even or odd. In particular, consider a sequence *abcd* of four equally spaced data points, and say that we specify an even *left* boundary. There are two sensible possibilities: either the data are even about the sample a, in which case the even extension is *dcbabcd*, or the data are even about the point *halfway* between a and the previous point, in which case the even extension is *dcbaabcd* (a is repeated).

These choices lead to all the standard variations of DCTs and also discrete sine transforms (DSTs). Each boundary can be either even or odd (2 choices per boundary) and can be symmetric about a data point or the point halfway between two data points (2 choices per boundary), for a total of $2 \times 2 \times 2 \times 2 = 16$ possibilities. Half of these possibilities, those where the *left* boundary is even, correspond to the 8 types of DCT; the other half are the 8 types of DST.

These different boundary conditions strongly affect the applications of the transform and lead to uniquely useful properties for the various DCT types. Most directly, when using Fourier-related transforms to solve partial differential equations by spectral methods, the boundary conditions are directly specified as a part of the problem being solved. Or, for the MDCT (based on the type-IV DCT), the boundary conditions are intimately involved in the MDCT's critical property of time-domain aliasing cancellation. In a more subtle fashion, the boundary conditions are responsible for the "energy compactification" properties that make DCTs useful for image and audio compression, because the boundaries affect the rate of convergence of any Fourier-like series.

In particular, it is well known that any discontinuities in a function reduce the rate of convergence of the Fourier series, so that more sinusoids are needed to represent the function with a given accuracy. The same principle governs the usefulness of the DFT and other transforms for signal compression; the smoother a function is, the fewer terms in its DFT or DCT are required to represent it accurately, and the more it can be compressed. (Here, we think of the DFT or DCT as approximations for the Fourier series or cosine series of a function, respectively, in order to talk about its "smoothness".) However, the implicit periodicity of the DFT means that discontinuities usually occur at the boundaries: any random segment of a signal is unlikely to have the same value at both the left and right boundaries. (A similar problem arises for the DST, in which the odd left boundary condition implies a discontinuity for any function that does not happen to be zero at that boundary.) In contrast, a DCT where *both* boundaries are even *always* yields a continuous extension at the boundaries (although the slope is generally discontinuous). This is why DCTs, and in particular DCTs of types I, II, V, and VI (the types that have two even boundaries) generally perform better for signal compression than DFTs and DSTs. In practice, a type-II DCT is usually preferred for such applications, in part for reasons of computational convenience.

Formal Definition

Formally, the discrete cosine transform is a linear, invertible function $f:\mathbb{R}^N \to \mathbb{R}^N$(where $\mathbb{R}$ denotes the set of real numbers), or equivalently an invertible $N \times N$ square matrix. There are several variants of the DCT with slightly modified definitions. The N real numbers x_0, ..., x_{N-1} are transformed into the N real numbers X_0, ..., X_{N-1} according to one of the formulas:

DCT-I

$$X_k = \frac{1}{2}(x_0 + (-1)^k x_{N-1}) + \sum_{n=1}^{N-2} x_n \cos\left[\frac{\pi}{N-1} nk\right] \qquad k = 0,\ldots,N-1.$$

Some authors further multiply the x_0 and x_{N-1} terms by √2, and correspondingly multiply the X_0 and X_{N-1} terms by 1/√2. This makes the DCT-I matrix orthogonal, if one further multiplies by an overall scale factor of $\sqrt{2/(N-1)}$, but breaks the direct correspondence with a real-even DFT.

The DCT-I is exactly equivalent (up to an overall scale factor of 2), to a DFT of $2N-2$ real numbers with even symmetry. For example, a DCT-I of N=5 real numbers *abcde* is exactly equivalent to a DFT of eight real numbers *abcdedcb* (even symmetry), divided by two. (In contrast, DCT types II-IV involve a half-sample shift in the equivalent DFT.)

Note, however, that the DCT-I is not defined for N less than 2. (All other DCT types are defined for any positive N.)

Thus, the DCT-I corresponds to the boundary conditions: x_n is even around n=0 and even around n=N-1; similarly for X_k.

DCT-II

$$X_k = \sum_{n=0}^{N-1} x_n \cos\left[\frac{\pi}{N}\left(n+\frac{1}{2}\right)k\right] \qquad k = 0,\ldots,N-1.$$

The DCT-II is probably the most commonly used form, and is often simply referred to as "the DCT".

This transform is exactly equivalent (up to an overall scale factor of 2) to a DFT of $4N$ real inputs of even symmetry where the even-indexed elements are zero. That is, it is half of the DFT of the $4N$ inputs y_n, where $y_{2n}=0$, $y_{2n+1}=x_n$ for $0 \le n < N$, $y_{2N}=0$, and $y_{4N-n}=y_n$ for $0 < n < 2N$.

Some authors further multiply the X_0 term by $1/\sqrt{2}$ and multiply the resulting matrix by an overall scale factor of $\sqrt{2/N}$ (see below for the corresponding change in DCT-III). This makes the DCT-II matrix orthogonal, but breaks the direct correspondence with a real-even DFT of half-shifted input. This is the normalization used by Matlab, for example. In many applications, such as JPEG, the scaling is arbitrary because scale factors can be combined with a subsequent computational step (e.g. the quantization step in JPEG), and a scaling that can be chosen that allows the DCT to be computed with fewer multiplications.

The DCT-II implies the boundary conditions: x_n is even around n=-1/2 and even around n=N-1/2; X_k is even around k=0 and odd around k=N.

DCT-III

$$X_k = \frac{1}{2}x_0 + \sum_{n=1}^{N-1} x_n \cos\left[\frac{\pi}{N} n \left(k+\frac{1}{2}\right)\right] \qquad k = 0,\dots,N-1.$$

Because it is the inverse of DCT-II (up to a scale factor, see below), this form is sometimes simply referred to as "the inverse DCT" ("IDCT").

Some authors divide the x_0 term by $\sqrt{2}$ instead of by 2 (resulting in an overall $x_0/\sqrt{2}$ term) and multiply the resulting matrix by an overall scale factor of $\sqrt{2/N}$ (see above for the corresponding change in DCT-II), so that the DCT-II and DCT-III are transposes of one another. This makes the DCT-III matrix orthogonal, but breaks the direct correspondence with a real-even DFT of half-shifted output.

The DCT-III implies the boundary conditions: x_n is even around n=0 and odd around n=N; X_k is even around k=-1/2 and odd around k=N-1/2.

DCT-IV

$$X_k = \sum_{n=0}^{N-1} x_n \cos\left[\frac{\pi}{N}\left(n+\frac{1}{2}\right)\left(k+\frac{1}{2}\right)\right] \qquad k = 0,\dots,N-1.$$

The DCT-IV matrix becomes orthogonal (and thus, being clearly symmetric, its own inverse) if one further multiplies by an overall scale factor of $\sqrt{2/N}$.

A variant of the DCT-IV, where data from different transforms are *overlapped*, is called the modified discrete cosine transform (MDCT) (Malvar, 1992).

The DCT-IV implies the boundary conditions: x_n is even around n=-1/2 and odd around n=N-1/2; similarly for X_k.

DCT V-VIII

DCTs of types I-IV treat both boundaries consistently regarding the point of symmetry: they are even/odd around either a data point for both boundaries or halfway between two data points for both boundaries. By contrast, DCTs of types V-VIII imply boundaries that are even/odd around a data point for one boundary and halfway between two data points for the other boundary.

In other words, DCT types I-IV are equivalent to real-even DFTs of even order (regardless of whether N is even or odd), since the corresponding DFT is of length $2(N-1)$ (for DCT-I) or $4N$ (for DCT-II/III) or $8N$ (for DCT-IV). The four additional types of discrete cosine transform (Martucci, 1994) correspond essentially to real-even DFTs of logically odd order, which have factors of $N\pm\frac{1}{2}$ in the denominators of the cosine arguments.

However, these variants seem to be rarely used in practice. One reason, perhaps, is that FFT algorithms for odd-length DFTs are generally more complicated than FFT algorithms for even-length DFTs (e.g. the simplest radix-2 algorithms are only for even lengths), and this increased intricacy carries over to the DCTs as described below.

(The trivial real-even array, a length-one DFT (odd length) of a single number a, corresponds to a DCT-V of length N=1.)

Inverse Transforms

Using the normalization conventions above, the inverse of DCT-I is DCT-I multiplied by $2/(N\text{-}1)$. The inverse of DCT-IV is DCT-IV multiplied by $2/N$. The inverse of DCT-II is DCT-III multiplied by $2/N$ and vice versa.

Like for the DFT, the normalization factor in front of these transform definitions is merely a convention and differs between treatments. For example, some authors multiply the transforms by $\sqrt{2/N}$ so that the inverse does not require any additional multiplicative factor. Combined with appropriate factors of $\sqrt{2}$ (see above), this can be used to make the transform matrix orthogonal.

Multidimensional Dcts

Multidimensional variants of the various DCT types follow straightforwardly from the one-dimensional definitions: they are simply a separable product (equivalently, a composition) of DCTs along each dimension.

M-D DCT-II

For example, a two-dimensional DCT-II of an image or a matrix is simply the one-dimensional DCT-II, from above, performed along the rows and then along the columns (or vice versa). That is, the 2D DCT-II is given by the formula (omitting normalization and other scale factors, as above):

$$\begin{aligned} X_{k_1,k_2} &= \sum_{n_1=0}^{N_1-1}\left(\sum_{n_2=0}^{N_2-1} x_{n_1,n_2} \cos\left[\frac{\pi}{N_2}\left(n_2+\frac{1}{2}\right)k_2\right]\right)\cos\left[\frac{\pi}{N_1}\left(n_1+\frac{1}{2}\right)k_1\right] \\ &= \sum_{n_1=0}^{N_1-1}\sum_{n_2=0}^{N_2-1} x_{n_1,n_2} \cos\left[\frac{\pi}{N_1}\left(n_1+\frac{1}{2}\right)k_1\right]\cos\left[\frac{\pi}{N_2}\left(n_2+\frac{1}{2}\right)k_2\right]. \end{aligned}$$

The inverse of a multi-dimensional DCT is just a separable product of the inverse(s) of the corresponding one-dimensional DCT(s) (see above), e.g. the one-dimensional inverses applied along one dimension at a time in a row-column algorithm.

The 3-D DCT-II is only the extension of 2-D DCT-II in three dimensional space and mathematically can be calculated by the formula

$$X_{k_1,k_2,k_3} = \sum_{n_1=0}^{N_1-1}\sum_{n_2=0}^{N_2-1}\sum_{n_3=0}^{N-3-1} x_{n_1,n_2,n_3} \cos\left[\frac{\pi}{N_1}\left(n_1+\frac{1}{2}\right)k_1\right]\cos\left[\frac{\pi}{N_2}\left(n_2+\frac{1}{2}\right)k_2\right]\cos\left[\frac{\pi}{N_3}\left(n_3+\frac{1}{2}\right)k_3\right], k_1,k_2,k_3 = 0,1,2,3....N-1.$$

The inverse of 3-D DCT-II is 3-D DCT-III and can be computed from the formula given by

$$x_{n_1,n_2,n_3} = \sum_{k_1=0}^{N_1-1}\sum_{k_2=0}^{N_2-1}\sum_{k_3=0}^{N-3-1} X_{k_1,k_2,k_3} \cos\left[\frac{\pi}{N_1}\left(n_1+\frac{1}{2}\right)k_1\right]\cos\left[\frac{\pi}{N_2}\left(n_2+\frac{1}{2}\right)k_2\right]\cos\left[\frac{\pi}{N_3}\left(n_3+\frac{1}{2}\right)k_3\right], n_1,n_2,n_3 = 0,1,2,3....N-1.$$

Technically, computing a two-, three- (or -multi) dimensional DCT by sequences of one-dimensional DCTs along each dimension is known as a *row-column* algorithm. As with multidimensional FFT algorithms, however, there exist other methods to compute the same thing while performing the computations in a different order (i.e. interleaving/combining the algorithms for the different dimensions). Owing to the rapid growth in the applications based on the 3-D DCT, several fast algorithms are developed for the computation of 3-D DCT-II. Vector-Radix algorithms are applied for computing M-D DCT to reduce the computational complexity and to increase the computational speed. To compute 3-D DCT-II efficiently, a fast algorithm, Vector-Radix Decimation in Frequency (VR DIF) algorithm was developed.

3-D DCT-II VR DIF

In order to apply the VR DIF algorithm the input data is to be formlated and rearranged as follows. The transform sixe $N \, x \, N \, x \, N$ is assumed to be 2.

$$\tilde{x}(n_1,n_2,n_3)=x(2n_1,2n_2,2n_3)$$
$$\tilde{x}(n_1,n_2,N-n_3-1)=x(2n_1,2n_2,2n_3+1)$$
$$\tilde{x}(n_1,N-n_2-1,n_3)=x(2n_1,2n_2+1,2n_3)$$
$$\tilde{x}(n_1,N-n_2-1,N-n_3-1)=x(2n_1,2n_2+1,2n_3+1)$$
$$\tilde{x}(N-n_1-1,n_2,n_3)=x(2n_1+1,2n_2,2n_3)$$
$$\tilde{x}(N-n_1-1,n_2,N-n_3-1)=x(2n_1+1,2n_2,2n_3+1)$$
$$\tilde{x}(N-n_1-1,N-n_2-1,n_3)=x(2n_1+1,2n_2+1,2n_3)$$
$$\tilde{x}(N-n_1-1,N-n_2-1,N-n_3-1=x(2n_1+1,2n_2+1,2n_3+1)$$

$$where\,0\le n_1,n_2,n_3\le\frac{N}{2}-1$$

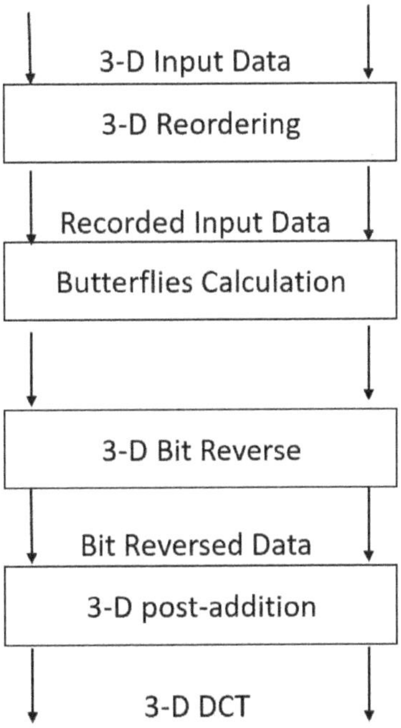

The four basic stages of computing 3-D DCT-II using VR DIF Algorithm.

The figure to the adjacent shows the four stages that are involved in calculating 3-D DCT-II using VR DIF algorithm. The first stage is the 3-D reordering using the index mapping illustrated by the above equations. The second stage is the butterfly calculation. Each butterfly calculates eight points together as shown in the figure just below., where $c(\phi_i)=cos(\phi_i)$

The original 3-D DCT-II now can be written as

$$X(k_1,k_2,k_3)=\sum_{n_1=1}^{N-1}\sum_{n_2=1}^{N-1}\sum_{n_3=1}^{N-1}\tilde{x}(n_1,n_2,n_3)cos(\phi k_1)cos(\phi k_2)cos(\phi k_3)$$

where $\phi_i = \frac{\pi}{2N}(4N_i + 1)$, and i= 1,2,3.

If the even and the odd parts of k_1, k_2 and k_3 and are considered, the general formula for the calculation of the 3-D DCT-II can be expressed as

$$X(k_1,k_2,k_3) = \sum_{n_1=1}^{N/2-1}\sum_{n_2=1}^{N/2-1}\sum_{n_1=1}^{N/2-1} \tilde{x}_{ijl}(n_1,n_2,n_3)cos(\phi(2k_1+i)cos(\phi(2k_2+j)cos(\phi(2k_3+l))$$

where $\tilde{x}_{ijl}(n_1,n_2,n_3) = \tilde{x}(n_1,n_2,n_3) + (-1)^l \tilde{x}(n_1,n_2,n_3+\frac{n}{2})$

$$+(-1)^j \tilde{x}(n_1,n_2+\frac{n}{2},n_3) + (-1)^{j+l}\tilde{x}(n_1,n_2+\frac{n}{2},n_3+\frac{n}{2})$$

$$+(-1)^i \tilde{x}(n_1+\frac{n}{2},n_2,n_3) + (-1)^{i+j}\tilde{x}(n_1+\frac{n}{2}+\frac{n}{2},n_2,n_3)$$

$$+(-1)^{i+l}\tilde{x}(n_1+\frac{n}{2},n_2,n_3+\frac{n}{3})$$

$$+(-1)^{i+j+l}\tilde{x}(n_1+\frac{n}{2},n_2+\frac{n}{2},n_3+\frac{n}{2}),\ \text{i,j,l} = 0 \text{ or } 1.$$

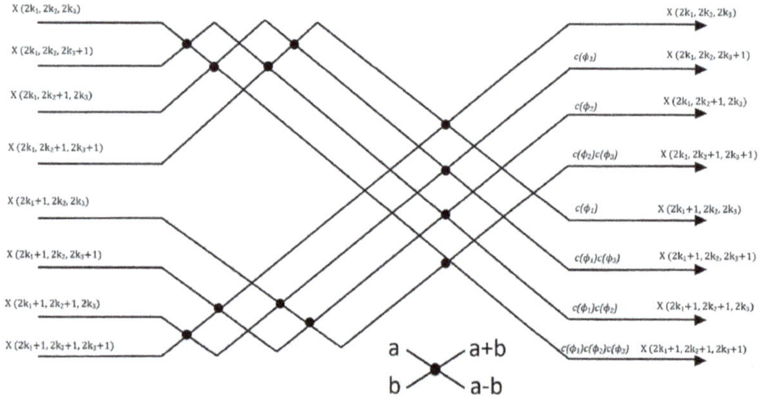

The single butterfly stage of VR DIF algorithm.

Arithmetic complexity

The whole 3-D DCT calculation needs $[\log_2 N]$ stages, and each stage involves $N^3/8$ butterflies. The whole 3-D DCT requires $[(N^3/8)\log_2 N]$ butterflies to be computed. Each butterfly requires seven real multiplications (including trivial multiplications) and 24 real additions (including trivial additions). Therefore, the total number of real multiplications needed for this stage is $[(7/8)N^3 log_2 N]$, and the total number of real additions i.e. including the post-additions (recursive additions) which can be calculated directly after the butterfly stage or after the bit-reverse stage are given by

$$\underbrace{[\frac{3}{2}N^3 log_2 N]}_{Real}+\underbrace{[\frac{3}{2}N^3 log_2 N-3N^3+3N^2]}_{Recursive}=[\frac{9}{2}N^3 log_2 N-3N^3+3N^2]$$

The conventional method to calculate MD-DCT-II is using a Row-Column-Frame (RCF) approach which is computationally complex and less productive on most advanced recent hardware platforms. The number of multiplications required to compute VR DIF Algorithm when compared to RCF algorithm are quite a few in number. The number of Multiplications and additions involved in RCF approach are given by $[\frac{3}{2}N^3 log_2 N]$ and $[\frac{9}{2}N^3 log_2 N-3N^3+3N^2]$ respectively. From Table 1, it can be seen that the total number.

TABLE 1 Comparison of VR DIF & RCF Algorithms for computing 3D-DCT-II

Transform Size	3D VR Mults	RCF Mults	3D VR Adds	RCF Adds
8 x 8 x 8	2.625	4.5	10.875	10.875
16 x 16 x 16	3.5	6	15.188	15.188
32 x 32 x 32	4.375	7.5	19.594	19.594
64 x 64 x 64	5.25	9	24.047	24.047

Of multiplications associated with the 3-D DCT VR algorithm is less than that associated with the RCF approach by more than 40%. In addition, the RCF approach involves matrix transpose and more indexing and data swapping than the new VR algorithm. This makes the 3-D DCT VR algorithm more efficient and better suited for 3-D applications that involve the 3-D DCT-II such as video compression and other 3-D image processing applications. The main consideration in choosing a fast algorithm is to avoid computational and structural complexities. As the technology of computers and DSPs advances, the execution time of arithmetic operations (multiplications and additions) is becoming very fast, and regular computational structure becomes the most important factor. Therefore, although the above proposed 3-D VR algorithm does not achieve the theoretical lower bound on the number of multiplications, it has a simpler computational structure as compared to other 3-D DCT algorithms. It can be implemented in place using a single butterfly and possesses the properties of the Cooley–Tukey FFT algorithm in 3-D. Hence, the 3-D VR presents a good choice for reducing arithmetic operations in the calculation of the 3-D DCT-II while keeping the simple structure that characterize butterfly style Cooley–Tukey FFT algorithms.

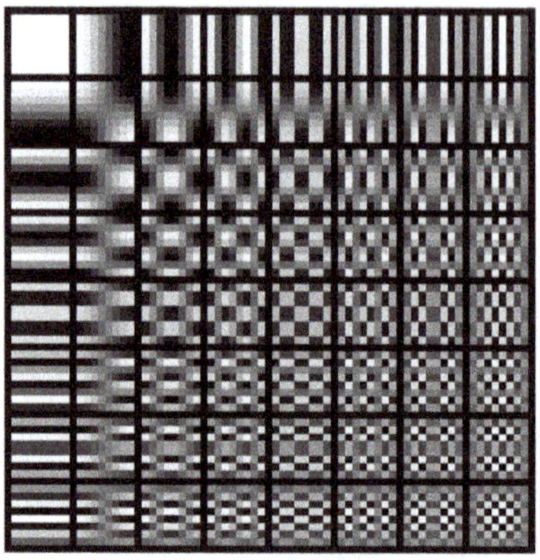

Two-dimensional DCT frequencies from the JPEG DCT

The image to the right shows a combination of horizontal and vertical frequencies for an 8 x 8 $N_1 = N_2 = 8$ two-dimensional DCT. Each step from left to right and top to bottom is an increase in frequency by 1/2 cycle.

For example, moving right one from the top-left square yields a half-cycle increase in the horizontal frequency. Another move to the right yields two half-cycles. A move down yields two half-cycles horizontally and a half-cycle vertically. The source data (8x8) is transformed to a linear combination of these 64 frequency squares.

MD-DCT-IV

The M-D DCT-IV is just an extension of 1-D DCT-IV on to M dimensional domain. The 2-D DCT-IV of a matirx or an image is given by

$$X_{k,l} = \sum_{n=0}^{N-1}\sum_{m=0}^{M-1} x_{n,m} cos(\frac{(2n+1)(2k+1)\pi}{4N}) cos(\frac{(2n+1)(2l+1)\pi}{4M})$$ where k=0,1,2...,N-1 and l=0,1,2...,M-1.

We can compute the MD DCT-IV using the regular row-column method or we can use the polynomial transform method[19] for the fast and efficient computation. The main idea of this algorithm is to use the Polynomial Transform to convert the multidimensional DCT into a series of 1-D DCTs directly. MD DCT-IV also has several applications in various fiields.

Computation

Although the direct application of these formulas would require O(N^2) operations, it is possible to compute the same thing with only O(N log N) complexity by factorizing the computation similarly to the fast Fourier transform (FFT). One can also compute DCTs via FFTs combined with O(N) pre- and post-processing steps. In general, O(N log N) methods to compute DCTs are known as fast cosine transform (FCT) algorithms.

The most efficient algorithms, in principle, are usually those that are specialized directly for the DCT, as opposed to using an ordinary FFT plus O(N) extra operations. However, even "specialized" DCT algorithms (including all of those that achieve the lowest known arithmetic counts, at least for power-of-two sizes) are typically closely related to FFT algorithms—since DCTs are essentially DFTs of real-even data, one can design a fast DCT algorithm by taking an FFT and eliminating the redundant operations due to this symmetry. This can even be done automatically (Frigo & Johnson, 2005). Algorithms based on the Cooley–Tukey FFT algorithm are most common, but any other FFT algorithm is also applicable. For example, the Winograd FFT algorithm leads to minimal-multiplication algorithms for the DFT, albeit generally at the cost of more additions, and a similar algorithm was proposed by Feig & Winograd (1992) for the DCT. Because the algorithms for DFTs, DCTs, and similar transforms are all so closely related, any improvement in algorithms for one transform will theoretically lead to immediate gains for the other transforms as well (Duhamel & Vetterli 1990).

While DCT algorithms that employ an unmodified FFT often have some theoretical overhead compared to the best specialized DCT algorithms, the former also have a distinct advantage: highly optimized FFT programs are widely available. Thus, in practice, it is often easier to obtain high performance for general lengths N with FFT-based algorithms. (Performance on modern hardware is typically not dominated simply by arithmetic counts, and optimization requires substantial engineering effort.) Specialized DCT algorithms, on the other hand, see widespread use for transforms of small, fixed sizes such as the 8×8 DCT-II used in JPEG compression, or the small DCTs (or MDCTs) typically used in audio compression. (Reduced code size may also be a reason to use a specialized DCT for embedded-device applications.)

In fact, even the DCT algorithms using an ordinary FFT are sometimes equivalent to pruning the redundant operations from a larger FFT of real-symmetric data, and they can even be optimal from the perspective of arithmetic counts. For example, a type-II DCT is equivalent to a DFT of size $4N$ with real-even symmetry whose even-indexed elements are zero. One of the most common methods for computing this via an FFT (e.g. the method used in FFTPACK and FFTW) was described by Narasimha & Peterson (1978) and Makhoul (1980), and this method in hindsight can be seen as one step of a radix-4 decimation-in-time Cooley–Tukey algorithm applied to the "logical" real-even DFT corresponding to the DCT II. (The radix-4 step reduces the size $4N$ DFT to four size-N DFTs of real data, two of which are zero and two of which are equal to one another by the even symmetry, hence giving a single size-N FFT of real data plus $O(N)$ butterflies.) Because the even-indexed elements are zero, this radix-4 step is exactly the same as a split-radix step; if the subsequent size-N real-data FFT is also performed by a real-data split-radix algorithm (as in Sorensen et al. 1987), then the resulting algorithm actually matches what was long the lowest published arithmetic count for the power-of-two DCT-II ($2N\log_2 N - N + 2$ real-arithmetic operations[a]). So, there is nothing intrinsically bad about computing the DCT via an FFT from an arithmetic perspective—it is sometimes merely a question of whether the corresponding FFT algorithm is optimal. (As a practical matter, the function-call overhead in invoking a separate FFT routine might be significant for small N, but this is an implementation rather than an algorithmic question since it can be solved by unrolling/inlining.)

Example of IDCT

Consider this 8x8 grayscale image of capital letter A.

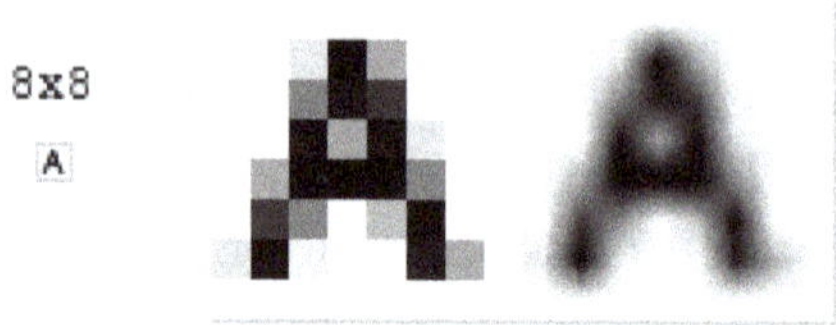

Original size, scaled 10x (nearest neighbor), scaled 10x (bilinear).

DCT of the image.

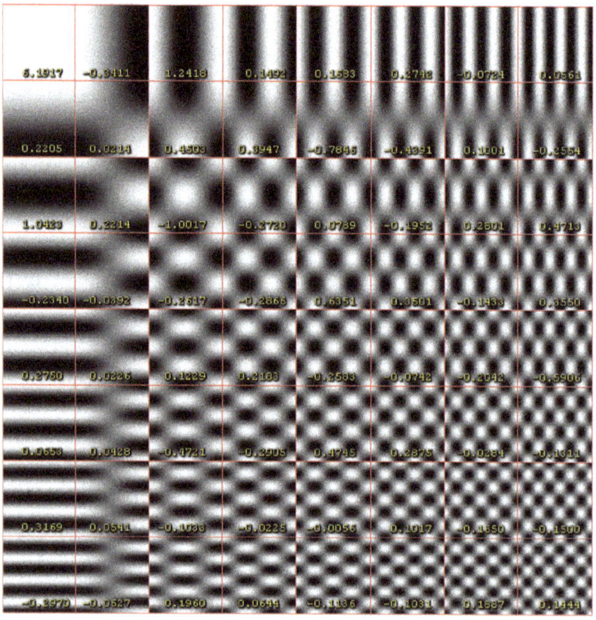

Each basis function is multiplied by its coefficient and then this product is added to the final image.

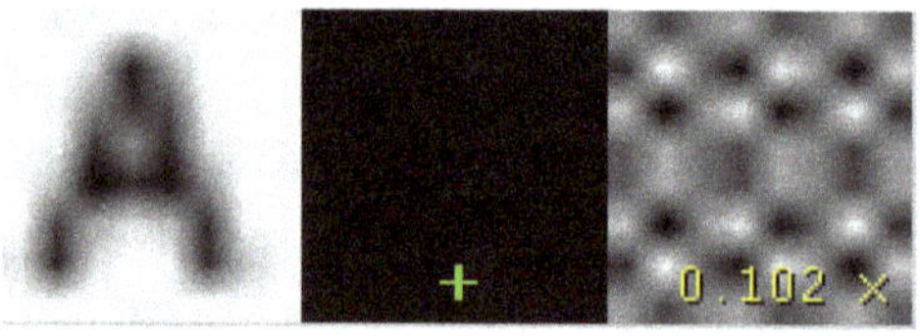

On the left is the final image. In the middle is the weighted function (multiplied by a coefficient) which is added to the final image. On the right is the current function and corresponding coefficient. Images are scaled (using bilinear interpolation) by factor 10×.

Parallel Processing (DSP Implementation)

In digital signal processing (DSP), parallel processing is a technique duplicating function units to operate different tasks (signals) simultaneously. Accordingly, we can perform the same processing for different signals on the corresponding duplicated function units. Further, due to the features of parallel processing, the parallel DSP design often contains multiple outputs, resulting in higher throughput than not parallel.

Conceptual Example

Consider a function unit (F_0) and three tasks (T_0, T_1 and T_2). The required time for the function unit F_0 to process those tasks is t_0,t_1 and t_2 respectively. Then, if we operate these three tasks in a sequential order, the required time to complete them is $t_0 + t_1 + t_2$.

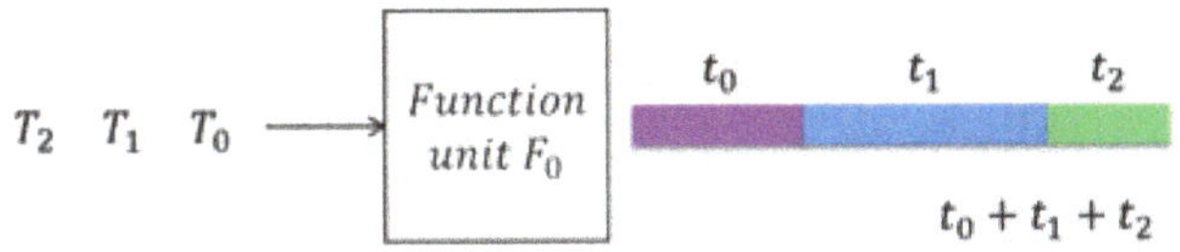

However, if we duplicate the function unit to another two copies (F), the aggregate time is reduced to max(t_0,t_1,t_2), which is smaller than in a sequential order.

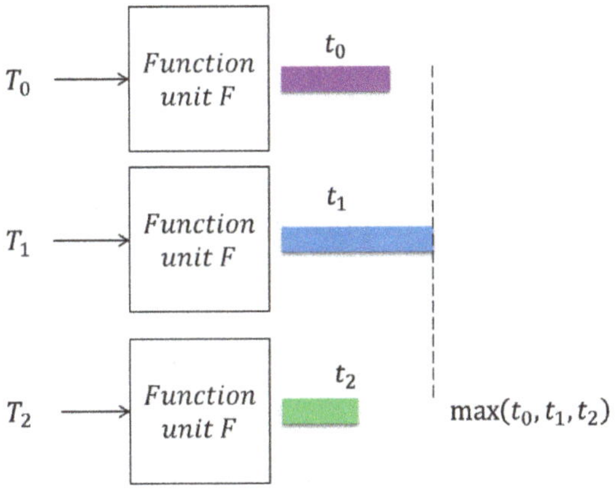

Versus Pipelining

Mechanism:

- Parallel: duplicated function units working in parallel
    - Each task is processed entirely by a different function unit.
- Pipelining: different function units working in parallel
    - Each task is split into a sequence of sub-tasks, which are handled by specialized and different function units.

Objective:

- Pipelining leads to a reduction in the critical path, which can increase the sample speed or reduce power consumption at the same speed.

- Parallel processing techniques require multiple outputs, which are computed in parallel in a clock period. Therefore, the effective sample speed is increased by the level of parallelism.

Consider a condition that we are able to apply both parallel processing and pipelining techniques, it is better to choose parallel processing techniques with the following reasons

- Pipelining usually causes I/O bottlenecks
- Parallel processing is also utilized for reduction of power consumption while using slow clocks
- The hybrid method of pipelining and parallel processing further increase the speed of the architecture

Parallel FIR Filters

Consider a 3-tap FIR filter:

$$y(n) = ax(n) + bx(n-1) + cx(n-2)$$

which is shown in the following figure.

Assume the calculation time for multiplication units is T_m and T_a for add units. The sample period is given by $T_{\text{sample}} \geq T_m + 2T_a$

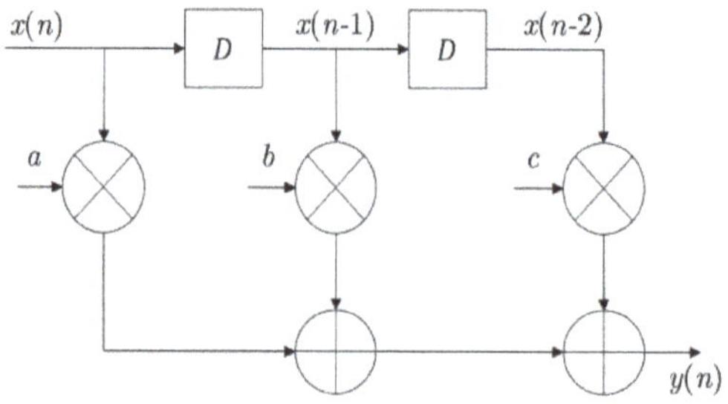

By parallelizing it, the resultant architecture is shown as follows. The sample rate now becomes

$$T_{\text{sample}} \geq \frac{T_{\text{clock}}}{N} = \frac{T_m + 2T_a}{3}$$

where N represents the number of copies.

Please note that, in a parallel system, $T_{\text{sample}} \neq T_{\text{clock}}$ while $T_{\text{sample}} = T_{\text{clock}}$ holds in a pipelined system.

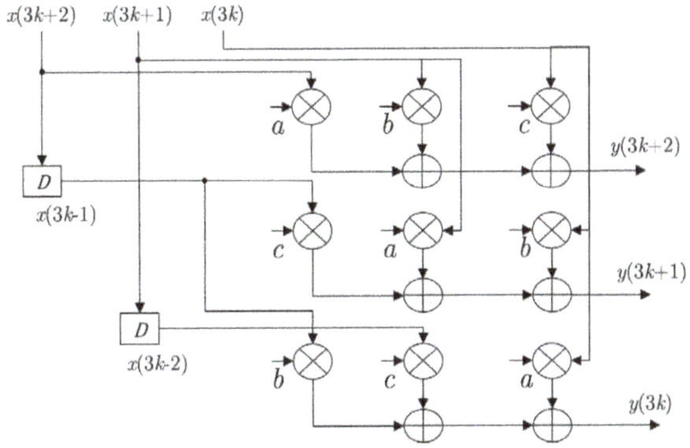

Parallel 1st-order IIR Filters

Consider the transfer function of a 1st-order IIR filter formulated as

$$H(z)=\frac{z^{-1}}{1-az^{-1}}$$

where $|a| \leq 1$ for stability, and such filter has only one pole located at $z = a$;

The corresponding recursive representation is

$$y(n+1)=ay(n)+u(n)$$

Consider the design of a 4-parallel architecture ($N = 4$). In such parallel system, each delay element means a block delay and the clock period is four times the sample period.

Therefore, by iterating the recursion with $n = 4k$, we have

$$y(n+4)=a^4y(n)+a^3u(n)+a^2u(n+1)+au(n+2)+u(n+3)$$

$$\rightarrow y(4k+4)=a^4y(4k)+a^3u(4k)+a^2u(4k+1)+au(4k+2)+u(4k+3)$$

The corresponding architecture is shown as follows.

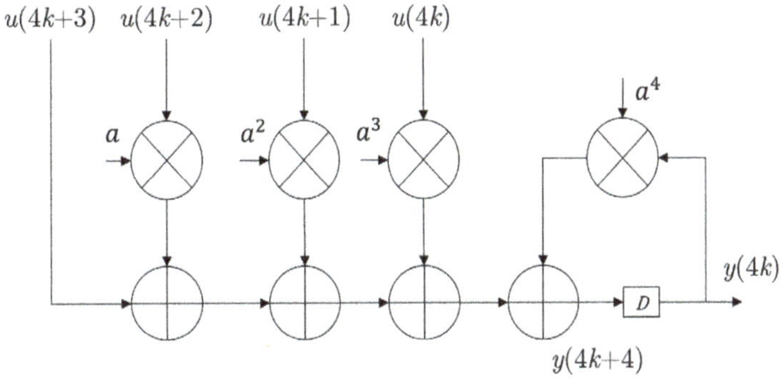

The resultant parallel design has the following properties.

- The pole of the original filter is at $z = a$ while the pole for the parallel system is at $z = a^4$ which is closer to the origin.
- The pole movement improves the robustness of the system to the round-off noise.
- Hardware complexity of this architecture: $N \times N$ multiply-add operations.

The square increase in hardware complexity can be reduced by exploiting the concurrency and the incremental computation to avoid repeated computing.

Parallel Processing for Low Power

Another advantage for the parallel processing techniques is that it can reduce the power consumption of a system by reducing the supply voltage.

Consider the following power consumption in a normal CMOS circuit.

$$P_{\text{seq}} = C_{\text{total}} \cdot V_0^2 \cdot f$$

where the C_{total} represents the total capacitance of the CMOS circuit.

For a parallel version, the charging capacitance remains the same but the total capacitance increases by N times.

In order to maintain the same sample rate, the clock period of the N-parallel circuit increases to N times the propagation delay of the original circuit.

It makes the charging time prolongs N times. The supply voltage can be reduced to βV_0.

Therefore, the power consumption of the N-parallel system can be formulated as

$$P_{\text{para}} = (NC_{\text{total}}) \cdot (\beta V_0^2) \cdot \frac{f}{N} = \beta^2 P_{\text{seq}}$$

where β can be computed by

$$N(\beta V_0 - V_t)^2 = \beta(V_0 - V_t)^2.$$

Pipelining (DSP Implementation)

Pipelining is an important technique used in several applications such as digital signal processing (DSP) systems, microprocessors, etc. It originates from the idea of a water pipe with continuous water sent in without waiting for the water in the pipe to come out. Accordingly, it results in speed enhancement for the critical path in most DSP systems. For example, it can either increase the clock speed or reduce the power consumption at the same speed in a DSP system.

Concept

Pipelining allows different functional units of a system to run concurrently. Consider an informal example in the following figure. A system includes three sub-function units (F_0, F_1 and F_2). Assume that there are three independent tasks (T_0, T_1 and T_2) being performed by these three function units. The time for each function unit to complete a task is the same and will occupy a slot in the schedule.

If we put these three units and tasks in a sequential order, the required time to complete them is five slots.

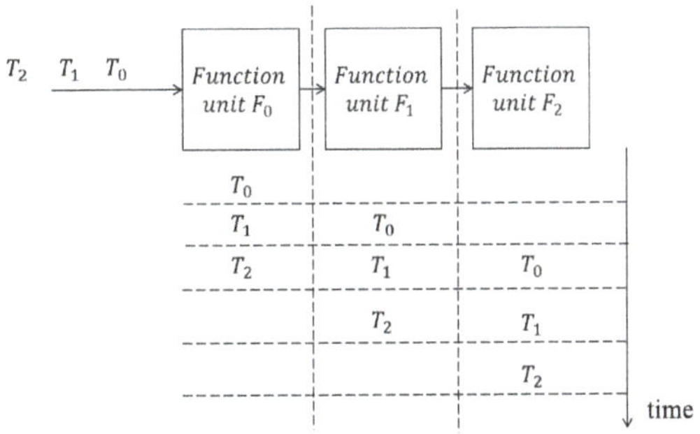

However, if we pipeline T_0 to T_2 concurrently, the aggregate time is reduced to three slots.

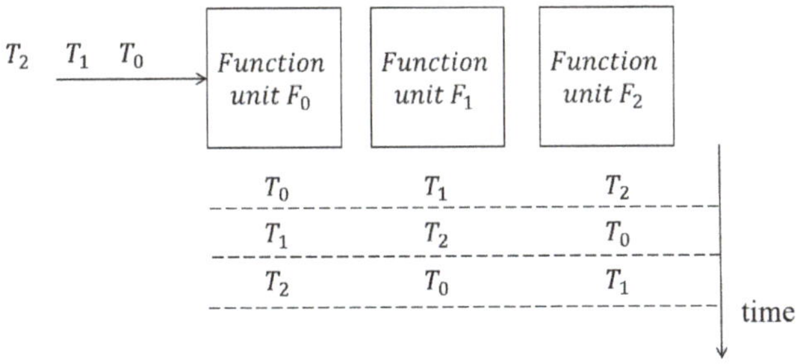

Therefore, it is possible for an adequate pipelined design to achieve significant enhancement on speed.

Costs and Disadvantages

Pipelining cannot decrease the processing time required for a single task. The advantage of pipelining is that it increases the throughput of the system when processing a stream of tasks.

Applying too many pipelined functions can lead to increased latency - that is, the time required for a single task to propagate through the full pipe is prolonged. A pipelined system may also require more resources (buffers, circuits, processing units, memory etc.), if the reuse of resources across different stages is restricted.

Comparison with Parallel Approaches

Another technique to enhance the efficiency through concurrency is parallel processing. The core difference is that parallel techniques usually duplicate function units and distribute multiple input tasks at once amongst them. Therefore, it can complete more tasks per unit time but may suffer more expensive resource costs.

For the previous example, the parallel technique duplicates each function units into another two. Accordingly, all the tasks can be operated upon by the duplicated function units with the same function simultaneously. The time to complete these three tasks is reduced to three slots.

Pipelining in FIR Filters

Consider a 3-tap FIR filter:

$$y(n) = ax(n) + bx(n-1) + cx(n-2)$$

which is as shown in the following figure.

Assume the calculation time for multiplication units is T_m and T_a for add units. The critical path, representing the minimum time required for processing a new sample, is limited by 1 multiplication and 2 add function units. Therefore, the sample period is given by

$$T_{\text{sample}} \geq T_m + 2T_a$$

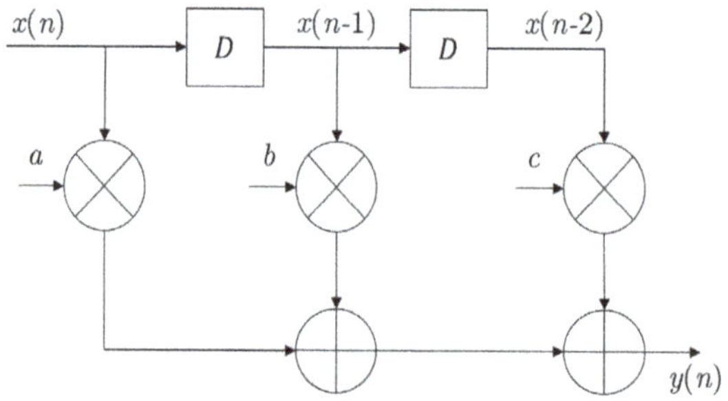

However, such structure may not be suitable for the design with the requirement of high speed. To reduce the sampling period, we can introduce extra pipelining registers along the critical data path. Then the structure is partitioned into two stages and the data produced in the first stage will be stored in the introduced registers, delaying one clock to the second stage. The data in first three clocks is recorded in the following table. Under such pipelined structure, the sample period is reduced to

$$T_{\text{sample}} \geq T_m + T_a.$$

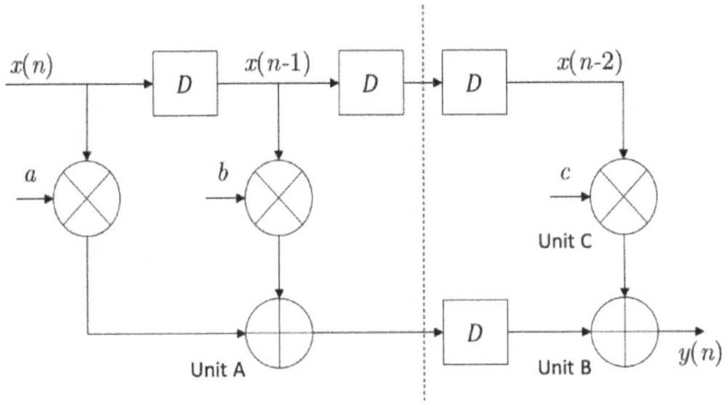

Clock	Input	Unit A	Unit B	Unit C	Output
0	$x(0)$	$ax(0)+bx(-1)$	–	–	–
1	$x(1)$	$ax(1)+bx(0)$	$ax(0)+bx(-1)$	$cx(-2)$	$y(0)$
2	$x(2)$	$ax(2)+bx(1)$	$ax(1)+bx(0)$	$cx(-1)$	$y(1)$
3	$x(3)$	$ax(3)+bx(2)$	$ax(2)+bx(1$	$cx(0)$	$y(2)$

Pipelining in 1st-order IIR Filters

By combining look-ahead techniques and pipelining, we are able to enhance the sample rate of target design. Look-ahead pipelining will add canceling poles and zeroes to the transfer function such that the coefficients of the following terms in the denominator of the transfer function are zero.

$$\{z^{-1}, \ldots, z^{-(M-1)}\}$$

Then, the output sample $y(n)$ can be computed in terms of the inputs and the output sample $y(n - M)$ such that there are M delay elements in the critical loop. These elements are then used to pipeline the critical loop by M stages so that the sample rate can be increased by a factor M.

Consider the 1st-order IIR filter transfer function

$$H(z) = \frac{1}{1 - az^{-1}}$$

The output $y(n)$ can be computed in terms of the input $u(n)$ and the previous output.

$$y(n) = ay(n-1) + u(n)$$

In a straightforward structure to design such function, the sample rate of this recursive filter is restricted by the calculation time of one multiply-add operation.

To pipeline such design, we observe that H has a pole at

$$z = a, a \leq 1$$

Therefore, in a 3-stage pipelined equivalent stable filter, the transfer function can be derived by adding poles and zeros at

$$z = ae^{\pm(\frac{2j\pi}{3})}$$

and is given by

$$H(z) = \frac{1 + az^{-1} + a^2 z^{-2}}{1 - a^3 z^{-3}}$$

Therefore, the corresponding sample rate can be increased by a factor 3.

Impulse Invariance

Impulse invariance is a technique for designing discrete-time infinite-impulse-response (IIR) filters from continuous-time filters in which the impulse response of the continuous-time system is sampled to produce the impulse response of the discrete-time system. The frequency response of the discrete-time system will be a sum of shifted copies of the frequency response of the continuous-time system; if the continuous-time system is approximately band-limited to a frequency less than the Nyquist frequency of the sampling, then the frequency response of the discrete-time system will be approximately equal to it for frequencies below the Nyquist frequency.

Discussion

The continuous-time system's impulse response, $h_c(t)$, , is sampled with sampling period T to produce the discrete-time system's impulse response, $h[n]$.

$$h[n] = Th_c(nT)$$

Thus, the frequency responses of the two systems are related by

$$H(e^{j\omega}) = \sum_{k=-\infty}^{\infty} H_c\left(j\frac{\omega}{T} + j\frac{2\pi}{T}k\right)$$

If the continuous time filter is approximately band-limited (i.e. $H_c(j\Omega) < \delta$ when $|\Omega| \geq \pi/T$), then the frequency response of the discrete-time system will be approximately the continuous-time system's frequency response for frequencies below π radians per sample (below the Nyquist frequency 1/(2*T*) Hz):

$$H(e^{j\omega}) = H_c(j\omega / T) \text{ for } |\omega| \leq \pi$$

Comparison to the Bilinear Transform

Note that aliasing will occur, including aliasing below the Nyquist frequency to the extent that the continuous-time filter's response is nonzero above that frequency. The bilinear transform is an alternative to impulse invariance that uses a different mapping that maps the continuous-time system's frequency response, out to infinite frequency, into the range of frequencies up to the Nyquist frequency in the discrete-time case, as opposed to mapping frequencies linearly with circular overlap as impulse invariance does.

Effect on Poles in System Function

If the continuous poles at $s = s_k$, the system function can be written in partial fraction expansion as

$$H_c(s) = \sum_{k=1}^{N} \frac{A_k}{s - s_k}$$

Thus, using the inverse Laplace transform, the impulse response is

$$h_c(t) = \begin{cases} \sum_{k=1}^{N} A_k e^{s_k t}, & t \geq 0 \\ 0, & \textbf{otherwise} \end{cases}$$

The corresponding discrete-time system's impulse response is then defined as the following

$$h[n] = Th_c(nT)$$

$$h[n] = T \sum_{k=1}^{N} A_k e^{s_k n T} u[n]$$

Performing a z-transform on the discrete-time impulse response produces the following discrete-time system function

$$H(z) = T \sum_{k=1}^{N} \frac{A_k}{1 - e^{s_k T} z^{-1}}$$

Thus the poles from the continuous-time system function are translated to poles at $z = e^{s_k T}$. The zeros, if any, are not so simply mapped.

Poles and Zeros

If the system function has zeros as well as poles, they can be mapped the same way, but the result is no longer an impulse invariance result: the discrete-time impulse response is not equal simply to samples of the continuous-time impulse response. This method is known as the matched Z-transform method, or pole–zero mapping. In the case of all-pole filters, the methods are equivalent.

Stability and Causality

Since poles in the continuous-time system at $s = s_k$ transform to poles in the discrete-time system at $z = \exp(s_k T)$, poles in the left half of the s-plane map to inside the unit circle in the z-plane; so if the continuous-time filter is causal and stable, then the discrete-time filter will be causal and stable as well.

Corrected Formula

When a causal continuous-time impulse response has a discontinuity at $t = 0$, the expressions above are not consistent. This is because $h_c(0)$ should really only contribute half its value to $h[0]$..

Making this correction gives

$$h[n] = T\left(h_c(nT) - \frac{1}{2}h_c(0)\delta[n]\right)$$

$$h[n] = T\sum_{k=1}^{N} A_k e^{s_k nT}\left(u[n] - \frac{1}{2}\delta[n]\right)$$

Performing a z-transform on the discrete-time impulse response produces the following discrete-time system function $H(z) = T\sum_{k=1}^{N}\frac{A_k}{1-e^{s_k T}z^{-1}} - \frac{T}{2}\sum_{k=1}^{N}A_k$.

References

- E. R. Kanasewich (1981). Time sequence analysis in geophysics (3rd ed.). University of Alberta. pp. 185–186. ISBN 978-0-88864-074-1.
- Cornelius T. Leondes (1996). Digital control systems implementation and computational techniques. Academic Press. p. 123. ISBN 978-0-12-012779-5.
- E. R. Kanasewich (1981). Time Sequence Analysis in Geophysics. University of Alberta. pp. 186, 249. ISBN 9780888640741.

- S. V. Narasimhan and S. Veena (2005). Signal processing: principles and implementation. Alpha Science Int'l Ltd. p. 260. ISBN 978-1-84265-199-5.
- Alexander D. Poularikas, ed. (2010). Transforms and Applications Handbook (3rd ed.). CRC Press. pp. 16.1–16.21. ISBN 978-1-4200-6652-4.
- Oppenheim, Alan (2010). Discrete Time Signal Processing Third Edition. Upper Saddle River, NJ: Pearson Higher Education, Inc. p. 504. ISBN 978-0-13-198842-2.
- Astrom, Karl J. (1990). Computer Controlled Systems, Theory and Design (Second ed.). Prentice-Hall. p. 212. ISBN 0-13-168600-3.
- Brigham, E. Oran (1988). The fast Fourier transform and its applications. Englewood Cliffs, N.J.: Prentice Hall. ISBN 0-13-307505-2.
- Oppenheim, Alan V.; Schafer, R. W.; and Buck, J. R. (1999). Discrete-time signal processing. Upper Saddle River, N.J.: Prentice Hall. ISBN 0-13-754920-2.
- Slides for VLSI Digital Signal Processing Systems: Design and Implementation John Wiley & Sons, 1999 (ISBN Number: 0-471-24186-5):

5

Digital Signal Processor

A digital signal processor is a microprocessor that is used in the operational needs of digital signal processing. The aim of digital signal processors is to measure and compress analog signals. Image processors, media processors and video scalers are also explained in the section.

Digital Signal Processor

A digital signal processor (DSP) is a specialized microprocessor (or a SIP block), with its architecture optimized for the operational needs of digital signal processing.

A digital signal processor chip found in a guitar effects unit.

The goal of DSPs is usually to measure, filter and/or compress continuous real-world analog signals. Most general-purpose microprocessors can also execute digital signal processing algorithms successfully, but dedicated DSPs usually have better power efficiency thus they are more suitable in portable devices such as mobile phones because of power consumption constraints. DSPs often use special memory architectures that are able to fetch multiple data and/or instructions at the same time.

Overview

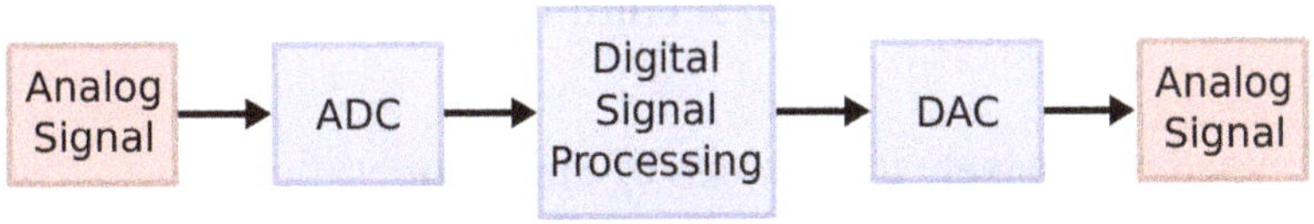

A typical digital processing system

Digital signal processing algorithms typically require a large number of mathematical operations to be performed quickly and repeatedly on a series of data samples. Signals (perhaps from audio or video sensors) are constantly converted from analog to digital, manipulated digitally, and then converted back to analog form. Many DSP applications have constraints on latency; that is, for the system to work, the DSP operation must be completed within some fixed time, and deferred (or batch) processing is not viable.

Most general-purpose microprocessors and operating systems can execute DSP algorithms successfully, but are not suitable for use in portable devices such as mobile phones and PDAs because of power efficiency constraints. A specialized digital signal processor, however, will tend to provide a lower-cost solution, with better performance, lower latency, and no requirements for specialized cooling or large batteries.

The architecture of a digital signal processor is optimized specifically for digital signal processing. Most also support some of the features as an applications processor or microcontroller, since signal processing is rarely the only task of a system. Some useful features for optimizing DSP algorithms are outlined below.

Architecture

Software Architecture

By the standards of general-purpose processors, DSP instruction sets are often highly irregular; while traditional instruction sets are made up of more general instructions that allow them to perform a wider variety of operations, instruction sets optimized for digital signal processing contain instructions for common mathematical operations that occur frequently in DSP calculations. Both traditional and DSP-optimized instruction sets are able to compute any arbitrary operation but an operation that might require multiple ARM or x86 instructions to compute might require only one instruction in a DSP optimized instruction set.

One implication for software architecture is that hand-optimized assembly-code routines are commonly packaged into libraries for re-use, instead of relying on advanced compiler technologies to handle essential algorithms. Even with modern compiler optimizations hand-optimized assembly code is more efficient and many common algorithms involved in DSP calculations are hand-written in order to take full advantage of the architectural optimizations.

Instruction Sets

- multiply–accumulates (MACs, including fused multiply–add, FMA) operations
    - used extensively in all kinds of matrix operations
        - convolution for filtering
        - dot product
        - polynomial evaluation
    - Fundamental DSP algorithms depend heavily on multiply–accumulate performance
        - FIR filters
        - Fast Fourier transform (FFT)
- Instructions to increase parallelism:
    - SIMD
    - VLIW
    - superscalar architecture
- Specialized instructions for modulo addressing in ring buffers and bit-reversed addressing mode for FFT cross-referencing
- Digital signal processors sometimes use time-stationary encoding to simplify hardware and increase coding efficiency.
- Multiple arithmetic units may require memory architectures to support several accesses per instruction cycle
- Special loop controls, such as architectural support for executing a few instruction words in a very tight loop without overhead for instruction fetches or exit testing

Data Instructions

- Saturation arithmetic, in which operations that produce overflows will accumulate at the maximum (or minimum) values that the register can hold rather than wrapping around (maximum+1 doesn't overflow to minimum as in many general-purpose CPUs, instead it stays at maximum). Sometimes various sticky bits operation modes are available.
- Fixed-point arithmetic is often used to speed up arithmetic processing
- Single-cycle operations to increase the benefits of pipelining

Program Flow

- Floating-point unit integrated directly into the datapath
- Pipelined architecture
- Highly parallel multiplier–accumulators (MAC units)
- Hardware-controlled looping, to reduce or eliminate the overhead required for looping operations

Hardware Architecture

Memory Architecture

DSPs are usually optimized for streaming data and use special memory architectures that are able to fetch multiple data and/or instructions at the same time, such as the Harvard architecture or Modified von Neumann architecture, which use separate program and data memories (sometimes even concurrent access on multiple data buses).

DSPs can sometimes rely on supporting code to know about cache hierarchies and the associated delays. This is a tradeoff that allows for better performance. In addition, extensive use of DMA is employed.

Addressing and Virtual Memory

DSPs frequently use multi-tasking operating systems, but have no support for virtual memory or memory protection. Operating systems that use virtual memory require more time for context switching among processes, which increases latency.

- Hardware modulo addressing
    - Allows circular buffers to be implemented without having to test for wrapping
- Bit-reversed addressing, a special addressing mode
    - useful for calculating FFTs
- Exclusion of a memory management unit
- Memory-address calculation unit

History

Prior to the advent of stand-alone DSP chips discussed below, most DSP applications were implemented using bit-slice processors. The AMD 2901 bit-slice chip with its family of components was a very popular choice. There were reference designs from AMD, but very often the specifics of a particular design were application specific. These bit slice architectures would sometimes include a peripheral multiplier chip. Examples of these multipliers were a series from TRW including the TDC1008 and TDC1010, some of which included an accumulator, providing the requisite multiply–accumulate (MAC) function.

In 1976, Richard Wiggins proposed the Speak & Spell concept to Paul Breedlove, Larry Brantingham, and Gene Frantz at Texas Instrument's Dallas research facility. Two years later in 1978 they produced the first Speak & Spell, with the technological centerpiece being the TMS5100, the industry's first digital signal processor. It also set other milestones, being the first chip to use Linear predictive coding to perform speech synthesis.

In 1978, Intel released the 2920 as an "analog signal processor". It had an on-chip ADC/DAC with an internal signal processor, but it didn't have a hardware multiplier and was not successful in the market. In 1979, AMI released the S2811. It was designed as a microprocessor peripheral, and it had to be initialized by the host. The S2811 was likewise not successful in the market.

In 1980 the first stand-alone, complete DSPs – the NEC µPD7720 and AT&T DSP1 – were presented at the International Solid-State Circuits Conference '80. Both processors were inspired by the research in PSTN telecommunications.

The Altamira DX-1 was another early DSP, utilizing quad integer pipelines with delayed branches and branch prediction.

Another DSP produced by Texas Instruments (TI), the TMS32010 presented in 1983, proved to be an even bigger success. It was based on the Harvard architecture, and so had separate instruction and data memory. It already had a special instruction set, with instructions like load-and-accumulate or multiply-and-accumulate. It could work on 16-bit numbers and needed 390 ns for a multiply–add operation. TI is now the market leader in general-purpose DSPs.

About five years later, the second generation of DSPs began to spread. They had 3 memories for storing two operands simultaneously and included hardware to accelerate tight loops, they also had an addressing unit capable of loop-addressing. Some of them operated on 24-bit variables and a typical model only required about 21 ns for a MAC. Members of this generation were for example the AT&T DSP16A or the Motorola 56000.

The main improvement in the third generation was the appearance of application-specific units and instructions in the data path, or sometimes as coprocessors. These units allowed direct hardware acceleration of very specific but complex mathematical problems, like the Fourier-transform or matrix operations. Some chips, like the Motorola MC68356, even included more than one processor core to work in parallel. Other DSPs from 1995 are the TI TMS320C541 or the TMS 320C80.

The fourth generation is best characterized by the changes in the instruction set and the instruction encoding/decoding. SIMD extensions were added, VLIW and the superscalar architecture appeared. As always, the clock-speeds have increased, a 3 ns MAC now became possible.

Modern DSPs

Modern signal processors yield greater performance; this is due in part to both technological and architectural advancements like lower design rules, fast-access two-level cache, (E)DMA circuitry and a wider bus system. Not all DSPs provide the same speed and many kinds of signal processors exist, each one of them being better suited for a specific task, ranging in price from about US$1.50 to US$300

Texas Instruments produces the C6000 series DSPs, which have clock speeds of 1.2 GHz and implement separate instruction and data caches. They also have an 8 MiB 2nd level cache and 64 EDMA channels. The top models are capable of as many as 8000 MIPS (instructions per second), use VLIW (very long instruction word), perform eight operations per clock-cycle and are compatible with a broad range of external peripherals and various buses (PCI/serial/etc). TMS320C6474 chips each have three such DSPs, and the newest generation C6000 chips support floating point as well as fixed point processing.

Freescale produces a multi-core DSP family, the MSC81xx. The MSC81xx is based on StarCore Architecture processors and the latest MSC8144 DSP combines four programmable SC3400 StarCore DSP cores. Each SC3400 StarCore DSP core has a clock speed of 1 GHz.

XMOS produces a multi-core multi-threaded line of processor well suited to DSP operations, They come in various speeds ranging from 400 to 1600 MIPS. The processors have a multi-threaded architecture that allows up to 8 real-time threads per core, meaning that a 4 core device would support up to 32 real time threads. Threads communicate between each other with buffered channels that are capable of up to 80 Mbit/s. The devices are easily programmable in C and aim at bridging the gap between conventional micro-controllers and FPGAs

CEVA, Inc. produces and licenses three distinct families of DSPs. Perhaps the best known and most widely deployed is the CEVA-TeakLite DSP family, a classic memory-based architecture, with 16-bit or 32-bit word-widths and single or dual MACs. The CEVA-X DSP family offers a combination of VLIW and SIMD architectures, with different members of the family offering dual or quad 16-bit MACs. The CEVA-XC DSP family targets Software-defined Radio (SDR) modem designs and leverages a unique combination of VLIW and Vector architectures with 32 16-bit MACs.

Analog Devices produce the SHARC-based DSP and range in performance from 66 MHz/198 MFLOPS (million floating-point operations per second) to 400 MHz/2400

MFLOPS. Some models support multiple multipliers and ALUs, SIMD instructions and audio processing-specific components and peripherals. The Blackfin family of embedded digital signal processors combine the features of a DSP with those of a general use processor. As a result, these processors can run simple operating systems like μCLinux, velOSity and Nucleus RTOS while operating on real-time data.

NXP Semiconductors produce DSPs based on TriMedia VLIW technology, optimized for audio and video processing. In some products the DSP core is hidden as a fixed-function block into a SoC, but NXP also provides a range of flexible single core media processors. The TriMedia media processors support both fixed-point arithmetic as well as floating-point arithmetic, and have specific instructions to deal with complex filters and entropy coding.

CSR produces the Quatro family of SoCs that contain one or more custom Imaging DSPs optimized for processing document image data for scanner and copier applications.

Most DSPs use fixed-point arithmetic, because in real world signal processing the additional range provided by floating point is not needed, and there is a large speed benefit and cost benefit due to reduced hardware complexity. Floating point DSPs may be invaluable in applications where a wide dynamic range is required. Product developers might also use floating point DSPs to reduce the cost and complexity of software development in exchange for more expensive hardware, since it is generally easier to implement algorithms in floating point.

Generally, DSPs are dedicated integrated circuits; however DSP functionality can also be produced by using field-programmable gate array chips (FPGAs).

Embedded general-purpose RISC processors are becoming increasingly DSP like in functionality. For example, the OMAP3 processors include a ARM Cortex-A8 and C6000 DSP.

In Communications a new breed of DSPs offering the fusion of both DSP functions and H/W acceleration function is making its way into the mainstream. Such Modem processors include ASOCS ModemX and CEVA's XC4000.

Digital Signal Controller

A digital signal controller (DSC) is a hybrid of microcontrollers and digital signal processors (DSPs). Like microcontrollers, DSCs have fast interrupt responses, offer control-oriented peripherals like PWMs and watchdog timers, and are usually programmed using the C programming language, although they can be programmed using the device's native assembly language. On the DSP side, they incorporate features found on most DSPs such as single-cycle multiply–accumulate (MAC) units, barrel shifters, and large accumulators. Not all vendors have adopted the term DSC. The term was first introduced by Microchip Technology in 2002 with the launch of their 6000 series DSCs and subsequently adopted by most, but not all DSC vendors. For example, Infineon and Renesas refer to their DSCs as microcontrollers.)

DSCs are used in a wide range of applications, but the majority go into motor control, power conversion, and sensor processing applications. Currently DSCs are being marketed as green technologies for their potential to reduce power consumption in electric motors and power supplies.

In order of market share, the top three DSC vendors are Texas Instruments, Freescale, and Microchip Technology, according to market research firm Forward Concepts (2007). These three companies dominate the DSC market, with other vendors such as Infineon and Renesas taking a smaller slice of the pie.

DSC Chips

NOTE: Data is from 2012 (Microchip and TI) and table currently only includes offering from the top 3 DSC vendors.

Vendor	Device	Clock Speed (MHz)	Flash (kB)	PWM channels, resolution, duty cycle
Microchip	dsPIC30F	30	6–144	4–8 (16 bits, 1 or 16.5 ns depending on part)
	dsPIC33F	40	12–256	up 18 PWM (16 bits, 12.5 ns)
	dsPIC33E	70	64-512	up 16 PWM (16 bits, 8.32 ns)
Texas Instruments	TMS320F28x	60–150	32–512	16 PWM (13 bits, 150 ps)
	TMS320LF240x	40	16–64	7–16 PWM (11 bits, 150 ps)
Freescale	MC56F83x	60	48–280	12 PWM (15 bits, 10 ns)
	MC56F80x	32	12–64	5–6 PWM (15 bits, 10 ns)
	MC56F81x	40	40–572	12 PWM (15 bits, 10 ns)

DSC Software

DSCs, like microcontrollers and DSPs, require software support. There are a growing number of software packages that offer the features required by both DSP applications and microcontroller applications. With a broader set of requirements, software solutions are more rare. They require: development tools, DSP libraries, optimization for DSP processing, fast interrupt handling, multi-threading and a tiny footprint.

Image Processor

Nikon EXPEED, a system on a chip including an image processor, video processor, digital signal processor (DSP) and a 32-bit microcontroller controlling the chip

An image processor, image processing engine, also called media processor, is a specialized digital signal processor (DSP) used for image processing in digital cameras, mobile phones or other devices. Image processors often employ parallel computing even with SIMD or MIMD technologies to increase speed and efficiency. The digital image processing engine can perform a range of tasks. To increase the system integration on embedded devices, often it is a system on a chip with multi-core processor architecture.

Function

Bayer Transformation

The photodiodes employed in an image sensor are color-blind by nature: they can only record shades of grey. To get color into the picture, they are covered with different color filters: red, green and blue (RGB) according to the pattern designated by the Bayer filter - named after its inventor. As each photodiode records the color information for exactly one pixel of the image, without an image processor there would be a green pixel next to each red and blue pixel. (Actually, with most sensors there are two green for each blue and red diodes.)

This process, however, is quite complex and involves a number of different operations. Its quality depends largely on the effectiveness of the algorithms applied to the raw data coming from the sensor. The mathematically manipulated data becomes the photo file recorded.

Demosaicing

As stated above, the image processor evaluates the color and brightness data of a given pixel, compares them with the data from neighboring pixels and then uses a demosaicing algorithm to produce an appropriate colour and brightness value for the pixel. The image processor also assesses the whole picture to guess at the correct distribution of contrast. By adjusting the gamma value (heightening or lowering the contrast range of an image's mid-tones) subtle tonal gradations, such as in human skin or the blue of the sky, become much more realistic.

Noise Reduction

Noise is a phenomenon found in any electronic circuitry. In digital photography its effect is often visible as random spots of obviously wrong colour in an otherwise smoothly-coloured area. Noise increases with temperature and exposure times. When higher ISO settings are chosen the electronic signal in the image sensor is amplified, which at the same time increases the noise level, leading to a lower signal-to-noise ratio. The image processor attempts to separate the noise from the image information and to remove it. This can be quite a challenge, as the image may contain areas with fine textures which, if treated as noise, may lose some of their definition.

Image Sharpening

As the color and brightness values for each pixel are interpolated some image softening is applied to even out any fuzziness that has occurred. To preserve the impression of depth, clarity and fine details, the image processor must sharpen edges and contours. It therefore must detect edges correctly and reproduce them smoothly and without over-sharpening.

Models

Image processor users are using industry standard products, application-specific standard products (ASSP) or even application-specific integrated circuits (ASIC) with trade names: Canon's is called DIGIC, Nikon's EXPEED, Olympus' TruePic, Panasonic's VENUS Engine and Sony's BIONZ. Some are known to be based on the Fujitsu Milbeaut, the Texas Instruments OMAP, Panasonic MN103, Zoran Coach, Altek Sunny or Sanyo image/video processors.

ARM architecture processors with its NEON SIMD Media Processing Engines (MPE) are often used in mobile phones.

Processor Brand Names

- Canon - DIGIC (based on Texas Instruments OMAP)
- Casio - EXILIM engine
- Epson - EDiART
- Fujifilm - EXR III or X Processor Pro

- Minolta / Konica Minolta - SUPHEED with CxProcess
- Leica - MAESTRO (based on Fujitsu Milbeaut)
- Nikon - EXPEED (based on Fujitsu Milbeaut)
- Olympus - TruePic (based on Panasonic MN103/MN103S)
- Panasonic - Venus engine (based on Panasonic MN103/MN103S)
- Pentax - PRIME (Pentax Real IMage Engine) (newer variants based on Fujitsu Milbeaut)
- Ricoh - GR engine (GR digital), Smooth Imaging Engine
- Samsung - DRIMe (based on Samsung Exynos)
- Sanyo - Platinum engine
- Sigma - True (newer variants based on Fujitsu Milbeaut)
- Sharp - ProPix
- Sony - BIONZ
- HTC - ImageSense

Speed

With the ever higher pixel count in image sensors, the image processor's speed becomes more critical: photographers don't want to wait for the camera's image processor to complete its job before they can carry on shooting - they don't even want to notice some processing is going on inside the camera. Therefore, image processors must be optimised to cope with more data in the same or even a shorter period of time.

Media Processor

A media processor, mostly used as an image / video processor, is a microprocessor-based system-on-a-chip which is designed to deal with digital streaming data in real-time (e.g. display refresh) rates. These devices can also be considered a class of digital signal processors (DSPs).

Unlike graphics processing units (GPUs), which are used for computer displays, media processors are targeted at digital televisions and set-top boxes.

The streaming digital media classes include:

- uncompressed video

- compressed digital video - e.g. MPEG-1, MPEG-2, MPEG-4
- digital audio- e.g. PCM, AAC

Such SOCs are composed of:

- a microprocessor optimized to deal with these media datatypes
- a memory interface
- streaming media interfaces
- specialized functional units to help deal with the various digital media codecs

The microprocessor might have these optimizations:

- vector processing or SIMD functional units to efficiently deal with these media datatypes
- DSP-like features

Previous to media processors, these streaming media datatypes were processed using fixed-function, hardwired ASICs, which could not be updated in the field. This was a big disadvantage when any of the media standards were changed. Since media processors are software programmed devices, the processing done on them could be updated with new software releases. This allowed new generations of systems to be created without hardware redesign. For set-top boxes this even allows for the possibility of in-the-field upgrade by downloading of new software through cable or satellite networks.

Companies that pioneered the idea of media processors (and created the marketing term of media processor) included:

- MicroUnity MediaProcessor - Cancelled in 1996 before introduction
- IBM Mfast - Described at the Microprocessor Forum in 1995, planned to ship in mid-1997 but was cancelled before introduction
- Equator Semiconductor BSP line - their processors are used in Hitachi televisions, company acquired by Pixelworks
- Chromatic Research MPact line - their products were used on some PC graphics cards in the mid-1990s, company acquired by ATI Technologies
- Philips TriMedia line - used in Philips, Dell, Sony, etc. consumer electronics, Philips Semiconductors split off from Philips and became NXP Semiconductors in 2006

Consumer electronics companies have successfully dominated this market by designing their own media processors and integrating them into their video products. Companies such as Philips, Samsung, Matsushita, Fujitsu, Mitsubishi have their own in-house media processor devices.

Newer generations of such devices now use various forms of multiprocessing—multiple CPUs or DSPs, in order to deal with the vastly increased computational needs when dealing with high definition television signals.

Video Scaler

A blow-up of a small section of a 1024x768 (VESA XGA) resolution image; the individual pixels are more visible in its scaled form than its normal resolution.

A video scaler is a system which converts video signals from one display resolution to another; typically, scalers are used to convert a signal from a lower resolution (such as 480p standard definition) to a higher resolution (such as 1080i high definition), a process known as "upconversion" or "upscaling" (by contrast, converting from high to low resolution is known as "downconversion" or "downscaling").

Video scalers are typically found inside consumer electronics devices such as televisions, video game consoles, and DVD or Blu-ray disc players, but can also be found in other AV equipment (such as video editing and television broadcasting equipment). Video scalers can also be a completely separate devices, often providing simple video switching capabilities. These units are commonly found as part of home theatre or projected presentation systems. They are often combined with other video processing devices or algorithms to create a video processor that improves the apparent definition of video signals.

Video scalers are primarily a digital device; however, they can be combined with an analog-to-digital converter (ADC, or digitizer) and a digital-to-analog converter (DAC) to support analog inputs and outputs.

Process

The "native resolution" of a display is how many physical pixels make up each row and column of the visible area on the display's output surface. There are many different video signals in use which are not the same resolution (neither are all of the displays), thus some form of resolution adaptation is required to properly frame a video signal to a display device. For example, within the United States, there are NTSC, ATSC, and VESA video standards each with several different resolution video formats. Multiple common resolutions are also used for high-definition television; 720p, 1080i, and 1080p.

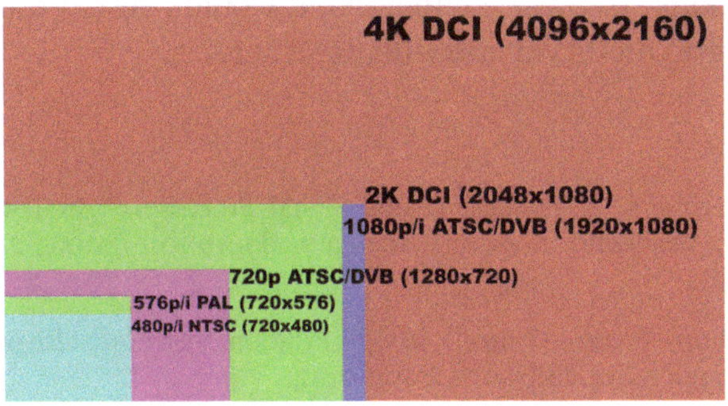

This is a comparison of several common video resolutions. The more pixels in an image the greater the possibility for finer detail and fidelity.

While scaling a video signal does allow it to match the size of a particular display, the process can result in an increased number of visual artifacts in the signal, such as ringing and posterization.

Scaling by Television Channels

Television channels which air a mixture of 16:9 (or high definition) programming and 4:3 (or standard definition) programming may employ scaling and/or cropping in order to make the programming fill the entire screen, as opposed to pillarboxing the feed instead, in order to maintain consistency in format. Likewise, as opposed to "center-cropping", channels may downscale programming produced in 16:9 for broadcast on their 4:3 feeds through letterboxing—either as a full 16:9 letterbox, or a partial 14:9 letterbox—a technique used primarily by European broadcasters during the transition to digital terrestrial television. The Active Format Description standard is a system of variables defining various scaling, letterboxing, and pillarboxing states; broadcasting equipment and televisions can be configured to automatically switch to the appropriate state based on the AFD flag encoded in the content and the aspect ratio of the display.

When the U.S. cable network TNT introduced an HD feed in 2004, it controversially employed a stretching system known as FlexView (which was also offered to other broadcasters). FlexView used a nonlinear method to stretch more near the edges of the screen than in the center of it. The practice was imposed by the senior vice president of broadcast engineering at TNT, Clyde D. Smith, who argued that pillarboxing could cause burn-in on plasma televisions, some older HDTVs could not stretch 4:3 content automatically, the quality of stretching on some displays was poor, and also desired a more consistent viewing experience with no "jarring" transitions to 4:3 programming. Despite TNT's intentions, the system was frequently criticized by viewers of high definition channels, with some nicknaming the effect "Stretch-O-Vision".

In 2014, FXX faced similar criticism for its use of cropping and scaling on reruns of *The Simpsons* (which only started producing episodes in HD beginning in its 20th season), as its cropping method caused various visual gags to be lost. In February 2015, FXX announced that in response to these complaints, it would present these episodes in their original 4:3 aspect ratio on its video-on-demand service.

References

- Dyer, S. A.; Harms, B. K. (1993). "Digital Signal Processing". In Yovits, M. C. Advances in Computers. 37. Academic Press. pp. 104–107. doi:10.1016/S0065-2458(08)60403-9. ISBN 9780120121373.
- Liptak, B. G. (2006). Process Control and Optimization. Instrument Engineers' Handbook. 2 (4th ed.). CRC Press. pp. 11–12. ISBN 9780849310812.
- "FXX ruins the punchline by inexplicably cropping old standard definition 'Simpsons' episodes". The Verge. Retrieved 13 February 2015.
- "Speak & Spell, the First Use of a Digital Signal Processing IC for Speech Generation, 1978". IEEE Milestones. IEEE. Retrieved 2012-03-02.

6

Sampling: An Integrated Study

Sampling is the reduction of continuous signal to a discrete signal. Sampling can be classified into undersampling, oversampling, upsampling and decimation. This chapter has been carefully written to provide an easy understanding of the varied facets of sampling.

Sampling (Signal Processing)

In signal processing, sampling is the reduction of a continuous signal to a discrete signal. A common example is the conversion of a sound wave (a continuous signal) to a sequence of samples (a discrete-time signal).

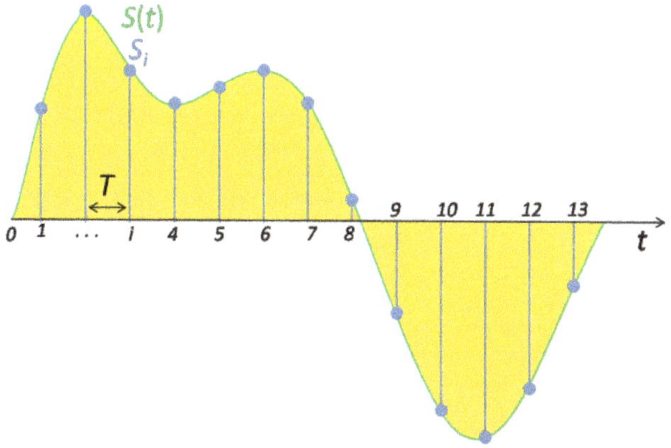

Signal sampling representation. The continuous signal is represented with a green colored line while the discrete samples are indicated by the blue vertical lines.

A sample is a value or set of values at a point in time and/or space.

A sampler is a subsystem or operation that extracts samples from a continuous signal.

A theoretical ideal sampler produces samples equivalent to the instantaneous value of the continuous signal at the desired points.

Theory

Sampling can be done for functions varying in space, time, or any other dimension, and similar results are obtained in two or more dimensions.

For functions that vary with time, let $s(t)$ be a continuous function (or "signal") to be sampled, and let sampling be performed by measuring the value of the continuous function every T seconds, which is called the sampling interval or the sampling period. Then the sampled function is given by the sequence:

$s(nT)$, for integer values of n.

The sampling frequency or sampling rate, f_s, is the average number of samples obtained in one second (*samples per second*), thus $f_s = 1/T$.

Reconstructing a continuous function from samples is done by interpolation algorithms. The Whittaker–Shannon interpolation formula is mathematically equivalent to an ideal lowpass filter whose input is a sequence of Dirac delta functions that are modulated (multiplied) by the sample values. When the time interval between adjacent samples is a constant (T), the sequence of delta functions is called a Dirac comb. Mathematically, the modulated Dirac comb is equivalent to the product of the comb function with $s(t)$. That purely mathematical abstraction is sometimes referred to as *impulse sampling*.

Most sampled signals are not simply stored and reconstructed. But the fidelity of a theoretical reconstruction is a customary measure of the effectiveness of sampling. That fidelity is reduced when $s(t)$ contains frequency components whose periodicity is smaller than 2 samples; or equivalently the ratio of cycles to samples exceeds ½ (see Aliasing). The quantity ½ *cycles/sample* × f_s *samples/sec* = $f_s/2$ *cycles/sec* (hertz) is known as the Nyquist frequency of the sampler. Therefore, $s(t)$ is usually the output of a lowpass filter, functionally known as an *anti-aliasing filter*. Without an anti-aliasing filter, frequencies higher than the Nyquist frequency will influence the samples in a way that is misinterpreted by the interpolation process.

Practical Considerations

In practice, the continuous signal is sampled using an analog-to-digital converter (ADC), a device with various physical limitations. This results in deviations from the theoretically perfect reconstruction, collectively referred to as distortion.

Various types of distortion can occur, including:

- Aliasing. Some amount of aliasing is inevitable because only theoretical, infinitely long, functions can have no frequency content above the Nyquist frequency. Aliasing can be made arbitrarily small by using a sufficiently large order of the anti-aliasing filter.
- Aperture error results from the fact that the sample is obtained as a time average within a sampling region, rather than just being equal to the signal value at the sampling instant. In a capacitor-based sample and hold circuit, aperture error is introduced because the capacitor cannot instantly change voltage thus requiring the sample to have non-zero width.

- Jitter or deviation from the precise sample timing intervals.
- Noise, including thermal sensor noise, analog circuit noise, etc.
- Slew rate limit error, caused by the inability of the ADC input value to change sufficiently rapidly.
- Quantization as a consequence of the finite precision of words that represent the converted values.
- Error due to other non-linear effects of the mapping of input voltage to converted output value (in addition to the effects of quantization).

Although the use of oversampling can completely eliminate aperture error and aliasing by shifting them out of the pass band, this technique cannot be practically used above a few GHz, and may be prohibitively expensive at much lower frequencies. Furthermore, while oversampling can reduce quantization error and non-linearity, it cannot eliminate these entirely. Consequently, practical ADCs at audio frequencies typically do not exhibit aliasing, aperture error, and are not limited by quantization error. Instead, analog noise dominates. At RF and microwave frequencies where oversampling is impractical and filters are expensive, aperture error, quantization error and aliasing can be significant limitations.

Jitter, noise, and quantization are often analyzed by modeling them as random errors added to the sample values. Integration and zero-order hold effects can be analyzed as a form of low-pass filtering. The non-linearities of either ADC or DAC are analyzed by replacing the ideal linear function mapping with a proposed nonlinear function.

Applications

Audio Sampling

Digital audio uses pulse-code modulation and digital signals for sound reproduction. This includes analog-to-digital conversion (ADC), digital-to-analog conversion (DAC), storage, and transmission. In effect, the system commonly referred to as digital is in fact a discrete-time, discrete-level analog of a previous electrical analog. While modern systems can be quite subtle in their methods, the primary usefulness of a digital system is the ability to store, retrieve and transmit signals without any loss of quality.

Sampling Rate

A commonly seen measure of sampling is S/s, which stands for "Samples per second." As an example, 1 MS/s is one million samples per second.

When it is necessary to capture audio covering the entire 20–20,000 Hz range of human hearing, such as when recording music or many types of acoustic events, audio waveforms are typically sampled at 44.1 kHz (CD), 48 kHz, 88.2 kHz, or 96 kHz. The approximately double-rate requirement is a consequence of the Nyquist theorem. Sampling rates higher than about 50 kHz to 60 kHz cannot supply more usable information for human listeners. Early professional audio equipment manufacturers chose sampling rates in the region of 50 kHz for this reason.

There has been an industry trend towards sampling rates well beyond the basic requirements: such as 96 kHz and even 192 kHz This is in contrast with laboratory experiments, which have failed to show that ultrasonic frequencies are audible to human observers; however in some cases ultrasonic sounds do interact with and modulate the audible part of the frequency spectrum (intermodulation distortion). It is noteworthy that intermodulation distortion is not present in the live audio and so it represents an artificial coloration to the live sound. One advantage of higher sampling rates is that they can relax the low-pass filter design requirements for ADCs and DACs, but with modern oversampling sigma-delta converters this advantage is less important.

The Audio Engineering Society recommends 48 kHz sampling rate for most applications but gives recognition to 44.1 kHz for Compact Disc and other consumer uses, 32 kHz for transmission-related applications, and 96 kHz for higher bandwidth or relaxed anti-aliasing filtering.

A more complete list of common audio sample rates is:

Sampling rate	Use
8,000 Hz	Telephone and encrypted walkie-talkie, wireless intercom and wireless microphone transmission; adequate for human speech but without sibilance; ess sounds like eff (/s/, /f/).
11,025 Hz	One quarter the sampling rate of audio CDs; used for lower-quality PCM, MPEG audio and for audio analysis of subwoofer bandpasses.
16,000 Hz	Wideband frequency extension over standard telephone narrowband 8,000 Hz. Used in most modern VoIP and VVoIP communication products.
22,050 Hz	One half the sampling rate of audio CDs; used for lower-quality PCM and MPEG audio and for audio analysis of low frequency energy. Suitable for digitizing early 20th century audio formats such as 78s.
32,000 Hz	miniDV digital video camcorder, video tapes with extra channels of audio (e.g. DVCAM with 4 Channels of audio), DAT (LP mode), Germany's Digitales Satellitenradio, NICAM digital audio, used alongside analogue television sound in some countries. High-quality digital wireless microphones. Suitable for digitizing FM radio.
37,800 Hz	CD-XA audio
44,056 Hz	Used by digital audio locked to NTSC color video signals (3 samples per line, 245 lines per field, 59.94 fields per second = 29.97 frames per second).
44,100 Hz	Audio CD, also most commonly used with MPEG-1 audio (VCD, SVCD, MP3). Originally chosen by Sony because it could be recorded on modified video equipment running at either 25 frames per second (PAL) or 30 frame/s (using an NTSC monochrome video recorder) and cover the 20 kHz bandwidth thought necessary to match professional analog recording equipment of the time. A PCM adaptor would fit digital audio samples into the analog video channel of, for example, PAL video tapes using 3 samples per line, 588 lines per frame, 25 frames per second.

47,250 Hz	world's first commercial PCM sound recorder by Nippon Columbia (Denon)
48,000 Hz	The standard audio sampling rate used by professional digital video equipment such as tape recorders, video servers, vision mixers and so on. This rate was chosen because it could reconstruct frequencies up to 22 kHz and work with 29.97 frames per second NTSC video - as well as 25 frame/s, 30 frame/s and 24 frame/s systems. With 29.97 frame/s systems it is necessary to handle 1601.6 audio samples per frame delivering an integer number of audio samples only every fifth video frame. Also used for sound with consumer video formats like DV, digital TV, DVD, and films. The professional Serial Digital Interface (SDI) and High-definition Serial Digital Interface (HD-SDI) used to connect broadcast television equipment together uses this audio sampling frequency. Most professional audio gear uses 48 kHz sampling, including mixing consoles, and digital recording devices.
50,000 Hz	First commercial digital audio recorders from the late 70s from 3M and Soundstream.
50,400 Hz	Sampling rate used by the Mitsubishi X-80 digital audio recorder.
88,200 Hz	Sampling rate used by some professional recording equipment when the destination is CD (multiples of 44,100 Hz). Some pro audio gear uses (or is able to select) 88.2 kHz sampling, including mixers, EQs, compressors, reverb, crossovers and recording devices.
96,000 Hz	DVD-Audio, some LPCM DVD tracks, BD-ROM (Blu-ray Disc) audio tracks, HD DVD (High-Definition DVD) audio tracks. Some professional recording and production equipment is able to select 96 kHz sampling. This sampling frequency is twice the 48 kHz standard commonly used with audio on professional equipment.
176,400 Hz	Sampling rate used by HDCD recorders and other professional applications for CD production.
192,000 Hz	DVD-Audio, some LPCM DVD tracks, BD-ROM (Blu-ray Disc) audio tracks, and HD DVD (High-Definition DVD) audio tracks, High-Definition audio recording devices and audio editing software. This sampling frequency is four times the 48 kHz standard commonly used with audio on professional video equipment.
352,800 Hz	Digital eXtreme Definition, used for recording and editing Super Audio CDs, as 1-bit DSD is not suited for editing. Eight times the frequency of 44.1 kHz.
2,822,400 Hz	SACD, 1-bit delta-sigma modulation process known as Direct Stream Digital, co-developed by Sony and Philips.
5,644,800 Hz	Double-Rate DSD, 1-bit Direct Stream Digital at 2x the rate of the SACD. Used in some professional DSD recorders.

Bit Depth

Audio is typically recorded at 8-, 16-, and 24-bit depth, which yield a theoretical maximum Signal-to-quantization-noise ratio (SQNR) for a pure sine wave of, approximately, 49.93 dB, 98.09 dB and 122.17 dB. CD quality audio uses 16-bit samples. Thermal noise limits the true number of bits that can be used in quantization. Few analog systems have signal to noise ratios (SNR) exceeding 120 dB. However, digital signal processing operations can have very high dynamic range, consequently it is common to perform mixing and mastering operations at 32-bit precision and then convert to 16 or 24 bit for distribution.

Speech Sampling

Speech signals, i.e., signals intended to carry only human speech, can usually be sampled at a much lower rate. For most phonemes, almost all of the energy is contained in the 5 Hz-4 kHz range, allowing a sampling rate of 8 kHz. This is the sampling rate used by nearly all telephony systems, which use the G.711 sampling and quantization specifications.

Video Sampling

Standard-definition television (SDTV) uses either 720 by 480 pixels (US NTSC 525-line) or 704 by 576 pixels (UK PAL 625-line) for the visible picture area.

High-definition television (HDTV) uses 720p (progressive), 1080i (interlaced), and 1080p (progressive, also known as Full-HD).

In digital video, the temporal sampling rate is defined the frame rate – or rather the field rate – rather than the notional pixel clock. The image sampling frequency is the repetition rate of the sensor integration period. Since the integration period may be significantly shorter than the time between repetitions, the sampling frequency can be different from the inverse of the sample time:

- 50 Hz – PAL video
- 60 / 1.001 Hz ~= 59.94 Hz – NTSC video

Video digital-to-analog converters operate in the megahertz range (from ~3 MHz for low quality composite video scalers in early games consoles, to 250 MHz or more for the highest-resolution VGA output).

When analog video is converted to digital video, a different sampling process occurs, this time at the pixel frequency, corresponding to a spatial sampling rate along scan lines. A common pixel sampling rate is:

- 13.5 MHz – CCIR 601, D1 video

Spatial sampling in the other direction is determined by the spacing of scan lines in the raster. The sampling rates and resolutions in both spatial directions can be measured in units of lines per picture height.

Spatial aliasing of high-frequency luma or chroma video components shows up as a moiré pattern.

3D Sampling

The process of volume rendering samples a 3D grid of voxels to produce 3D renderings of sliced (tomographic) data. The 3D grid is assumed to represent a continuous region of 3D space. Volume rendering is common in medial imaging, X-ray computed tomography (CT/CAT), Magnetic resonance imaging (MRI), Positron Emission Tomography (PET) are some examples. It is also used for Seismic tomography and other applications.

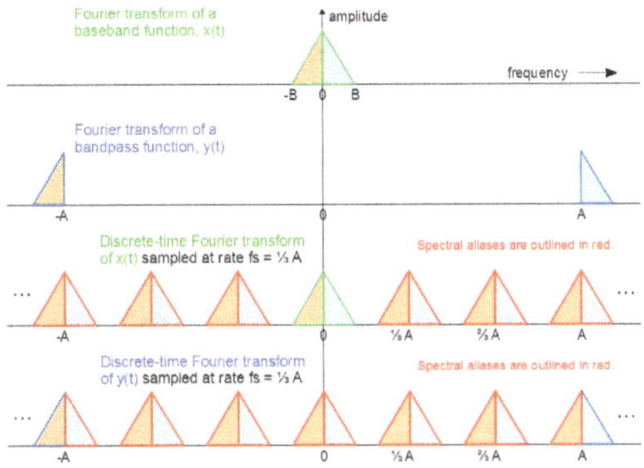

The top 2 graphs depict Fourier transforms of 2 different functions that produce the same results when sampled at a particular rate. The baseband function is sampled faster than its Nyquist rate, and the bandpass function is undersampled, effectively converting it to baseband. The lower graphs indicate how identical spectral results are created by the aliases of the sampling process.

Undersampling

When a bandpass signal is sampled slower than its Nyquist rate, the samples are indistinguishable from samples of a low-frequency alias of the high-frequency signal. That is often done purposefully in such a way that the lowest-frequency alias satisfies the Nyquist criterion, because the bandpass signal is still uniquely represented and recoverable. Such undersampling is also known as *bandpass sampling*, *harmonic sampling*, *IF sampling*, and *direct IF to digital conversion*.

Oversampling

Oversampling is used in most modern analog-to-digital converters to reduce the distortion introduced by practical digital-to-analog converters, such as a zero-order hold instead of idealizations like the Whittaker–Shannon interpolation formula.

Complex Sampling

Complex sampling (I/Q sampling) is the simultaneous sampling of two different, but related, waveforms, resulting in pairs of samples that are subsequently treated as complex numbers. When one waveform, $\hat{s}(t)$, is the Hilbert transform of the other waveform, $s(t)$, the complex-valued function, $s_a(t) \overset{\text{def}}{=} s(t) + j \cdot \hat{s}(t)$, is called an analytic signal, whose Fourier transform is zero for all negative values of frequency. In that case, the Nyquist rate for a waveform with no frequencies ≥ ***B*** can be reduced to just *B* (complex samples/sec), instead of 2*B* (real samples/sec). More apparently, the equivalent baseband waveform, $s_a(t)\cdot e^{-j2\pi\frac{B}{2}t}$, also has a Nyquist rate of *B*, because all of its non-zero frequency content is shifted into the interval [-B/2, B/2).

Although complex-valued samples can be obtained as described above, they are also created by manipulating samples of a real-valued waveform. For instance, the equivalent baseband waveform can be created without explicitly computing $\hat{s}(t)$, by processing the product sequence, $\left[s(nT)\cdot e^{-j2\pi\frac{B}{2}Tn}\right]$, through a digital lowpass filter whose cutoff frequency is B/2. Computing only every other sample of the output sequence reduces the sample-rate commensurate with the reduced Nyquist rate. The result is half as many complex-val-ued samples as the original number of real samples. No information is lost, and the original s(t) waveform can be recovered, if necessary.

Nyquist–Shannon Sampling Theorem

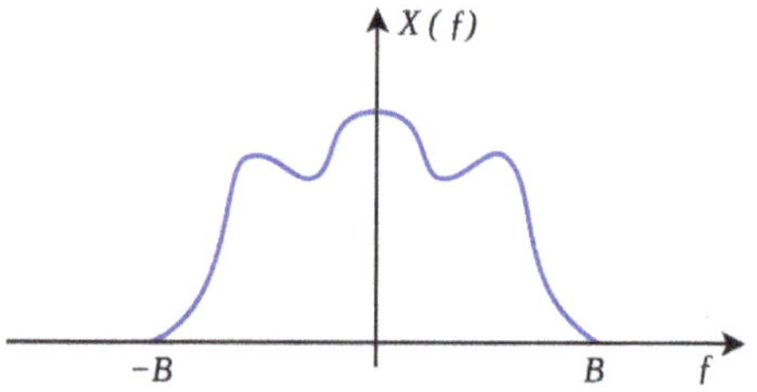

Example of magnitude of the Fourier transform of a bandlimited function

In the field of digital signal processing, the sampling theorem is a fundamental bridge between continuous-time signals (often called "analog signals") and discrete-time signals (often called "digital signals"). It establishes a sufficient condition for a sample rate that permits a discrete sequence of *samples* to capture all the information from a continuous-time signal of finite bandwidth.

Strictly speaking, the theorem only applies to a class of mathematical functions having a Fourier transform that is zero outside of a finite region of frequencies. Intuitively we expect that when one reduces a continuous function to a discrete sequence and interpolates back to a continuous function, the fidelity of the result depends on the density (or sample rate) of the original samples. The sampling theorem introduces the concept of a sample rate that is sufficient for perfect fidelity for the class of functions that are bandlimited to a given bandwidth, such that no actual information is lost in the sampling process. It expresses the sufficient sample rate in terms of the bandwidth for the class of functions. The theorem also leads to a formula for perfectly reconstructing the original continuous-time function from the samples.

Perfect reconstruction may still be possible when the sample-rate criterion is not satis-fied, provided other constraints on the signal are known. In some cases (when the sample-rate criterion is not satisfied), utilizing additional constraints allows for approximate reconstructions. The fidelity of these reconstructions can be verified and quantified utilizing Bochner's theorem.

The name *Nyquist–Shannon sampling theorem* honors Harry Nyquist and Claude Shannon. The theorem was also discovered independently by E. T. Whittaker, by Vladimir Kotelnikov, and by others. It is thus also known by the names *Nyquist–Shannon–Kotelnikov, Whittaker–Shannon–Kotelnikov, Whittaker–Nyquist–Kotelnikov–Shannon*, and *cardinal theorem of interpolation.*

Introduction

Sampling is a process of converting a signal (for example, a function of continuous time and/or space) into a numeric sequence (a function of discrete time and/or space). Shannon's version of the theorem states:

If a function x(t) contains no frequencies higher than B hertz, it is completely determined by giving its ordinates at a series of points spaced $1/(2B)$ seconds apart.

A sufficient sample-rate is therefore $2B$ samples/second, or anything larger. Equivalently, for a given sample rate f_s, perfect reconstruction is guaranteed possible for a bandlimit $B < f_s/2$.

When the bandlimit is too high (or there is no bandlimit), the reconstruction exhibits imperfections known as aliasing. Modern statements of the theorem are sometimes careful to explicitly state that $x(t)$ must contain no sinusoidal component at exactly frequency B, or that B must be strictly less than ½ the sample rate. The two thresholds, $2B$ and $f_s/2$ are respectively called the Nyquist rate and Nyquist frequency. And respectively, they are attributes of $x(t)$ and of the sampling equipment. The condition described by these inequalities is called the Nyquist criterion, or sometimes the *Raabe condition*. The theorem is also applicable to functions of other domains, such as *space,* in the case of a digitized image. The only change, in the case of other domains, is the units of measure applied to t, f_s, and B.

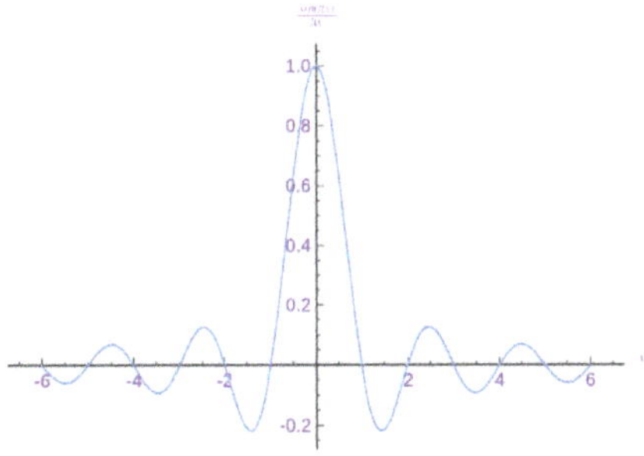

The normalized sinc function: sin(πx) / (πx) ... showing the central peak at x= 0, and zero-crossings at the other integer values of x.

The symbol $T = 1/f_s$ is customarily used to represent the interval between samples and is called the sample period or sampling interval. And the samples of function $x(t)$ are commonly denoted by $x[n] = x(nT)$ (alternatively "x_n" in older signal processing literature), for all integer values of n. A mathematically ideal way to interpolate the sequence involves the use of sinc functions. Each sample in the sequence is replaced by a sinc function, centered on the time axis at the original location of the sample, nT, with the amplitude of the sinc function scaled to the sample value, $x[n]$. Subsequently, the sinc functions are summed into a continuous function. A mathematically equivalent method is to convolve one sinc function with a series of Dirac delta pulses, weighted by the sample values. Neither method is numerically practical. Instead, some type of approximation of the sinc functions, finite in length, is used. The imperfections attributable to the approximation are known as *interpolation error*.

Practical digital-to-analog converters produce neither scaled and delayed sinc functions, nor ideal Dirac pulses. Instead they produce a piecewise-constant sequence of scaled and delayed rectangular pulses (the zero-order hold), usually followed by an "anti-imaging filter" to clean up spurious high-frequency content.

Aliasing

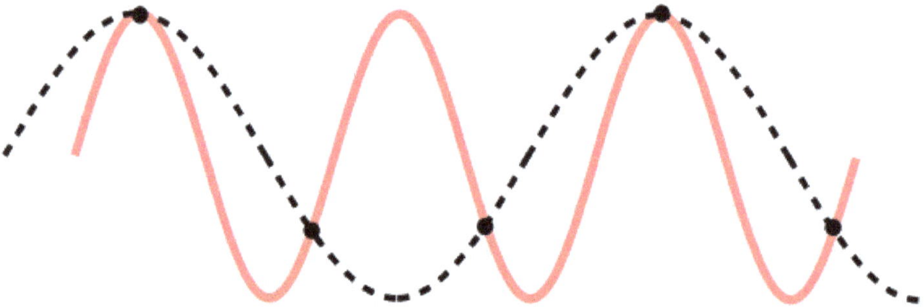

The samples of two sine waves can be identical when at least one of them is at a frequency above half the sample rate.

When $x(t)$ is a function with a Fourier transform, $X(f)$:

$$X(f) \stackrel{\text{def}}{=} \int_{-\infty}^{\infty} x(t)\, e^{-i2\pi ft}\, dt,$$

the Poisson summation formula indicates that the samples, $x(nT)$, of $x(t)$ are sufficient to create a periodic summation of $X(f)$. The result is:

$$X_s(f) \stackrel{\text{def}}{=} \sum_{k=-\infty}^{\infty} X(f - kf_s) = \sum_{n=-\infty}^{\infty} T \cdot x(nT)\, e^{-i2\pi nTf}, \qquad \textbf{(Eq.1)}$$

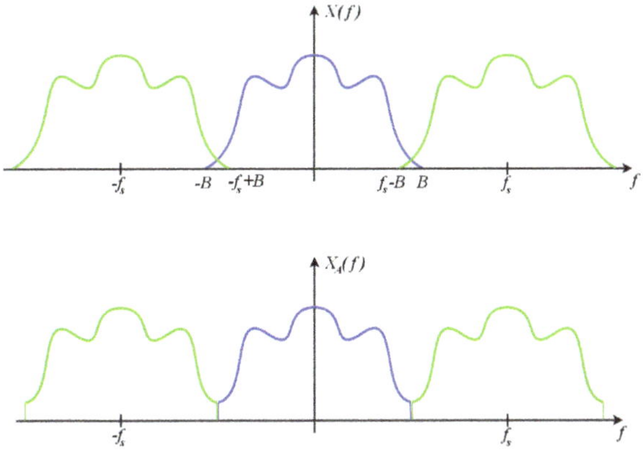

$X(f)$ (top blue) and $X_A(f)$ (bottom blue) are continuous Fourier transforms of two *different* functions, $x(t)$ and $x_A(t)$ (not shown). When the functions are sampled at rate f_s, the images (green) are added to the original transforms (blue) when one examines the discrete-time Fourier transforms (DTFT) of the sequences. In this hypothetical example, the DTFTs are identical, which means *the sampled sequences are identical*, even though the original continuous pre-sampled functions are not. If these were audio signals, $x(t)$ and $x_A(t)$ might not sound the same. But their samples (taken at rate f_s) are identical and would lead to identical reproduced sounds; thus $x_A(t)$ is an alias of $x(t)$ at this sample rate.

which is a periodic function and its equivalent representation as a Fourier series, whose coefficients are $T{\bullet}x(nT)$. This function is also known as the discrete-time Fourier transform (DTFT) of the sequence $T{\bullet}x(nT)$, for integers n.

As depicted, copies of $X(f)$ are shifted by multiples of f_s and combined by addition. For a band-limited function ($X(f) = 0$ for all $|f| \geq B$), and sufficiently large f_s, it is possible for the copies to remain distinct from each other. But if the Nyquist criterion is not satisfied, adjacent copies overlap, and it is not possible in general to discern an unambiguous $X(f)$. Any frequency component above $f_s/2$ is indistinguishable from a lower-frequency component, called an *alias*, associated with one of the copies. In such cases, the customary interpolation techniques produce the alias, rather than the original component. When the sample-rate is pre-determined by other considerations (such as an industry standard), $x(t)$ is usually filtered to reduce its high frequencies to acceptable levels before it is sampled. The type of filter required is a lowpass filter, and in this application it is called an anti-aliasing filter.

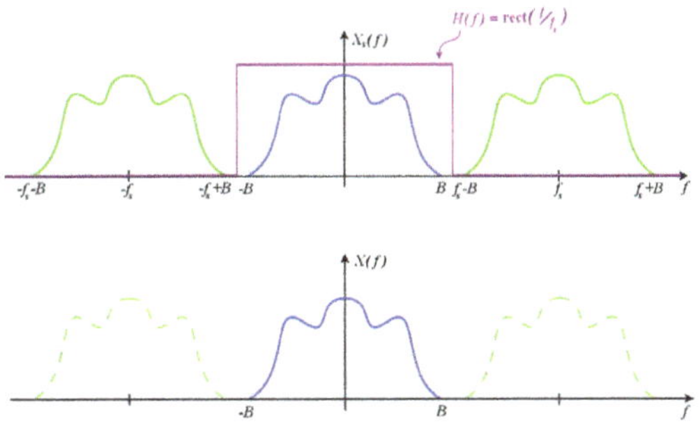

Spectrum, $X_s(f)$, of a properly sampled bandlimited signal (blue) and the adjacent DTFT images (green) that do not overlap. A *brick-wall* low-pass filter, $H(f)$, removes the images, leaves the original spectrum, $X(f)$, and recovers the original signal from its samples.

Derivation as a Special Case of Poisson Summation

When there is no overlap of the copies (aka "images") of $X(f)$, the $k = 0$ term of $X_s(f)$ can be recovered by the product:

$$X(f) = H(f) \cdot X_s(f), \quad \text{where:}$$

$$H(f) \stackrel{\text{def}}{=} \begin{cases} 1 & |f| < B \\ 0 & |f| > f_s - B. \end{cases}$$

At this point, the sampling theorem is proved, since $X(f)$ uniquely determines $x(t)$.

All that remains is to derive the formula for reconstruction. $H(f)$ need not be precisely defined in the region $[B, f_s - B]$ because $X_s(f)$ is zero in that region. However, the worst case is when $B = f_s/2$, the Nyquist frequency. A function that is sufficient for that and all less severe cases is:

$$H(f) = \text{rect}\left(\frac{f}{f_s}\right) = \begin{cases} 1 & |f| < \frac{f_s}{2} \\ 0 & |f| > \frac{f_s}{2}, \end{cases}$$

where rect(•) is the rectangular function. Therefore:

$$X(f) = \text{rect}\left(\frac{f}{f_s}\right) \cdot X_s(f)$$

$$= \text{rect}(Tf) \cdot \sum_{n=-\infty}^{\infty} T \cdot x(nT)\, e^{-i2\pi nTf} \qquad \text{(from Eq.1, above).}$$

$$= \sum_{n=-\infty}^{\infty} x(nT) \cdot \underbrace{T \cdot \text{rect}(Tf) \cdot e^{-i2\pi nTf}}_{\mathcal{F}\left\{\text{sinc}\left(\frac{t-nT}{T}\right)\right\}}.$$

The inverse transform of both sides produces the Whittaker–Shannon interpolation formula:

$$x(t) = \sum_{n=-\infty}^{\infty} x(nT) \cdot \text{sinc}\left(\frac{t-nT}{T}\right),$$

which shows how the samples, $x(nT)$, can be combined to reconstruct $x(t)$.

- Larger-than-necessary values of f_s (smaller values of T), called *oversampling*, have no effect on the outcome of the reconstruction and have the benefit of leaving room for a *transition band* in which $H(f)$ is free to take intermediate values. Undersampling, which causes aliasing, is not in general a reversible operation.
- Theoretically, the interpolation formula can be implemented as a low pass filter, whose impulse response is sinc(t/T) and whose input is $\sum_{n=-\infty}^{\infty} x(nT)\cdot\delta(t-nT)$, which is a Dirac comb function modulated by the signal samples. Practical digital-to-analog converters (DAC) implement an approximation like the zero-order hold. In that case, oversampling can reduce the approximation error.

Shannon's Original Proof

Poisson shows that the Fourier series in Eq.1 produces the periodic summation of $X(f)$, regardless of f_s and B. Shannon, however, only derives the series coefficients for the case $f_s = 2B$. Virtually quoting Shannon's original paper:

Let $X(\omega)$ be the spectrum of $x(t)$. Then

$$x(t)=\frac{1}{2\pi}\int_{-\infty}^{\infty} X(\omega)e^{i\omega t}\,d\omega$$

$$=\frac{1}{2\pi}\int_{-2\pi B}^{2\pi B} X(\omega)e^{i\omega t}\,d\omega$$

since $X(\omega)$ is assumed to be zero outside the band $|\frac{\omega}{2\pi}|< B$. If we let

$$t=\frac{n}{2B}$$

where n is any positive or negative integer, we obtain

$$x\left(\frac{n}{2B}\right)=\frac{1}{2\pi}\int_{-2\pi B}^{2\pi B} X(\omega)e^{i\omega\frac{n}{2B}}\,d\omega.$$

On the left are values of $x(t)$ at the sampling points. The integral on the right will be recognized as essentially the nth coefficient in a Fourier-series expansion of the function $X(\omega)$, taking the interval $-B$ to B as a fundamental period. This means that the values of the samples $x(n/2B)$ determine the Fourier coefficients in the series expansion of $X(\omega)$. Thus they determine $X(\omega)$, since $X(\omega)$ is zero for frequencies greater than B, and for lower frequencies $X(\omega)$ is determined if its Fourier coefficients are determined. But $X(\omega)$ determines the original function $x(t)$ completely, since a function is determined if its spectrum is known. Therefore the original samples determine the function $x(t)$ completely.

Shannon's proof of the theorem is complete at that point, but he goes on to discuss reconstruction via sinc functions, what we now call the Whittaker–Shannon interpolation formula as discussed above. He does not derive or prove the properties of the sinc function, but these would have been familiar to engineers reading his works at the time, since the Fourier pair relationship between rect (the rectangular function) and sinc was well known.

Let x_n be the nth sample. Then the function $x(t)$ is represented by:

$$x(t) = \sum_{n=-\infty}^{\infty} x_n \frac{\sin \pi(2Bt - n)}{\pi(2Bt - n)}.$$

As in the other proof, the existence of the Fourier transform of the original signal is assumed, so the proof does not say whether the sampling theorem extends to bandlimited stationary random processes.

Notes

The actual coefficient formula contains an additional factor of $1/2B = T$. So Shannon's coefficients are $T \cdot x(nT)$, which agrees with Eq.1.

Application to Multivariable Signals and Images

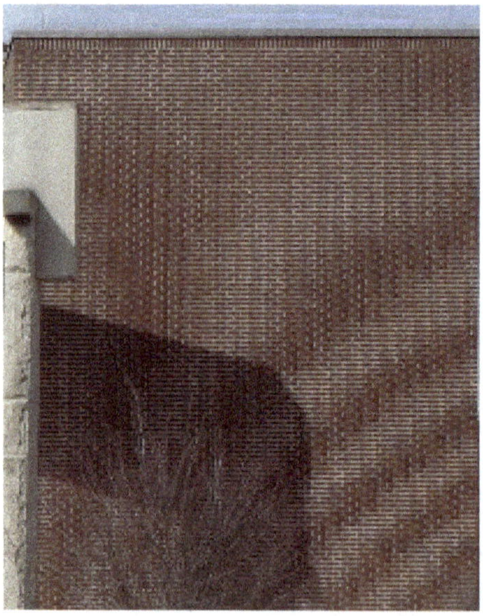

Subsampled image showing a Moiré pattern

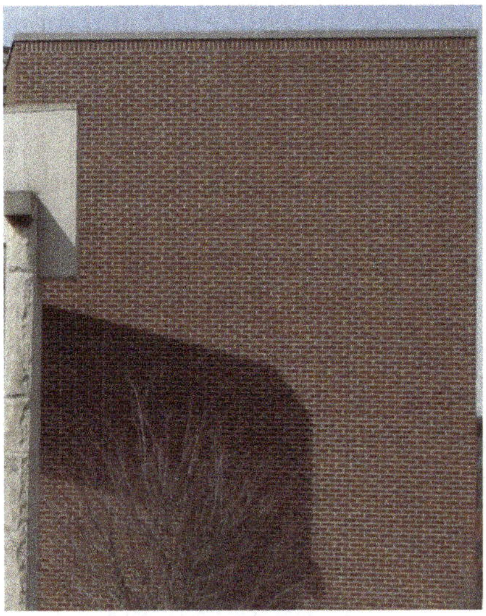

Properly sampled image

The sampling theorem is usually formulated for functions of a single variable. Consequently, the theorem is directly applicable to time-dependent signals and is normally formulated in that context. However, the sampling theorem can be extended in a straightforward way to functions of arbitrarily many variables. Grayscale images, for example, are often represented as two-dimensional arrays (or matrices) of real numbers representing the relative intensities of pixels (picture elements) located at the intersections of row and column sample locations. As a result, images require two independent variables, or indices, to specify each pixel uniquely—one for the row, and one for the column.

Color images typically consist of a composite of three separate grayscale images, one to represent each of the three primary colors—red, green, and blue, or *RGB* for short. Other colorspaces using 3-vectors for colors include HSV, CIELAB, XYZ, etc. Some colorspaces such as cyan, magenta, yellow, and black (CMYK) may represent color by four dimensions. All of these are treated as vector-valued functions over a two-dimensional sampled domain.

Similar to one-dimensional discrete-time signals, images can also suffer from aliasing if the sampling resolution, or pixel density, is inadequate. For example, a digital photograph of a striped shirt with high frequencies (in other words, the distance between the stripes is small), can cause aliasing of the shirt when it is sampled by the camera's image sensor. The aliasing appears as a moiré pattern. The "solution" to higher sampling in the spatial domain for this case would be to move closer to the shirt, use a higher resolution sensor, or to optically blur the image before acquiring it with the sensor.

Another example is shown to the right in the brick patterns. The top image shows the effects when the sampling theorem's condition is not satisfied. When software rescales an image (the same process that creates the thumbnail shown in the lower image) it, in effect, runs the image through a low-pass filter first and then downsamples the image to result in a smaller image that does not exhibit the moiré pattern. The top image is what happens when the image is downsampled without low-pass filtering: aliasing results.

The application of the sampling theorem to images should be made with care. For example, the sampling process in any standard image sensor (CCD or CMOS camera) is relatively far from the ideal sampling which would measure the image intensity at a single point. Instead these devices have a relatively large sensor area at each sample point in order to obtain sufficient amount of light. In other words, any detector has a finite-width point spread function. The analog optical image intensity function which is sampled by the sensor device is not in general bandlimited, and the non-ideal sampling is itself a useful type of low-pass filter, though not always sufficient to remove enough high frequencies to sufficiently reduce aliasing. When the area of the sampling spot (the size of the pixel sensor) is not large enough to provide sufficient spatial anti-aliasing, a separate anti-aliasing filter (optical low-pass filter) is typically included in a camera system to further blur the optical image. Despite images having these problems in relation to the sampling theorem, the theorem can be used to describe the basics of down and up sampling of images.

Critical Frequency

To illustrate the necessity of $f_s > 2B$, consider the family of sinusoids generated by different values of θ in this formula:

$$x(t) = \frac{\cos(2\pi Bt + \theta)}{\cos(\theta)} = \cos(2\pi Bt) - \sin(2\pi Bt)\tan(\theta), \quad -\pi/2 < \theta < \pi/2.$$

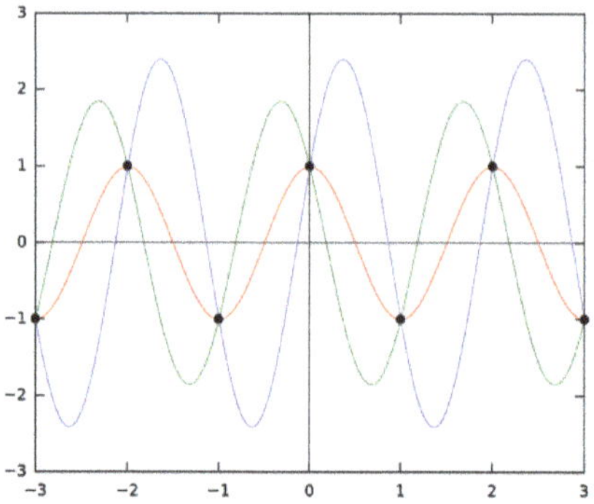

A family of sinusoids at the critical frequency, all having the same sample sequences of alternating +1 and −1. That is, they all are aliases of each other, even though their frequency is not above half the sample rate.

With $f_s = 2B$ or equivalently $T = 1/(2B)$, the samples are given by:

$$x(nT) = \cos(\pi n) - \underbrace{\sin(\pi n)}_{0}\tan(\theta) = (-1)^n$$

regardless of the value of θ. That sort of ambiguity is the reason for the *strict* inequality of the sampling theorem's condition.

Sampling of Non-baseband Signals

As discussed by Shannon:

> A similar result is true if the band does not start at zero frequency but at some higher value, and can be proved by a linear translation (corresponding physically to single-sideband modulation) of the zero-frequency case. In this case the elementary pulse is obtained from $\sin(x)/x$ by single-side-band modulation.

That is, a sufficient no-loss condition for sampling signals that do not have baseband components exists that involves the *width* of the non-zero frequency interval as opposed to its highest frequency component.

For example, in order to sample the FM radio signals in the frequency range of 100-102 MHz, it is not necessary to sample at 204 MHz (twice the upper frequency), but rather it is sufficient to sample at 4 MHz (twice the width of the frequency interval).

A bandpass condition is that $X(f) = 0$, for all nonnegative f outside the open band of frequencies:

$$\left(\frac{N}{2}f_s, \frac{N+1}{2}f_s\right),$$

for some nonnegative integer N. This formulation includes the normal baseband condition as the case N=0.

The corresponding interpolation function is the impulse response of an ideal brick-wall bandpass filter (as opposed to the ideal brick-wall lowpass filter used above) with cutoffs at the upper and lower edges of the specified band, which is the difference between a pair of lowpass impulse responses:

$$(N+1)sinc\left(\frac{(N+1)t}{T}\right) - N sinc\left(\frac{Nt}{T}\right).$$

Other generalizations, for example to signals occupying multiple non-contiguous bands, are possible as well. Even the most generalized form of the sampling theorem does not have a provably true converse. That is, one cannot conclude that information is necessarily lost just because the conditions of the sampling theorem are not satisfied; from an engineering perspective, however, it is generally safe to assume that if the sampling theorem is not satisfied then information will most likely be lost.

Nonuniform Sampling

The sampling theory of Shannon can be generalized for the case of nonuniform sampling, that is, samples not taken equally spaced in time. The Shannon sampling theory for non-uniform sampling states that a band-limited signal can be perfectly reconstructed from its samples if the average sampling rate satisfies the Nyquist condition. Therefore, although uniformly spaced samples may result in easier reconstruction algorithms, it is not a necessary condition for perfect reconstruction.

The general theory for non-baseband and nonuniform samples was developed in 1967 by Landau. He proved that the average sampling rate (uniform or otherwise) must be twice the *occupied* bandwidth of the signal, assuming it is *a priori* known what portion of the spectrum was occupied. In the late 1990s, this work was partially extended to cover signals of when the amount of occupied bandwidth was known, but the actual occupied portion of the spectrum was unknown. In the 2000s, a complete theory was developed (see the section Beyond Nyquist below) using compressed sensing. In particular, the theory, using signal processing language, is described in this 2009 paper. They show, among other things, that if the frequency locations are unknown, then it is necessary to sample at least at twice the Nyquist criteria; in other words, you must pay at least a factor of 2 for not knowing the location of the spectrum. Note that minimum sampling requirements do not necessarily guarantee stability.

Sampling Below the Nyquist Rate Under Additional Restrictions

The Nyquist–Shannon sampling theorem provides a sufficient condition for the sampling and reconstruction of a band-limited signal. When reconstruction is done via the Whittaker–Shannon interpolation formula, the Nyquist criterion is also a necessary condition to avoid aliasing, in the sense that if samples are taken at a slower rate than twice the band limit, then there are some signals that will not be correctly reconstructed. However, if further restrictions are imposed on the signal, then the Nyquist criterion may no longer be a necessary condition.

A non-trivial example of exploiting extra assumptions about the signal is given by the recent field of compressed sensing, which allows for full reconstruction with a sub-Nyquist sampling rate. Specifically, this applies to signals that are sparse (or compressible) in some domain. As an example, compressed sensing deals with signals that may have a low overall bandwidth (say, the *effective* bandwidth EB), but the frequency locations are unknown, rather than all together in a single band, so that the passband technique doesn't apply. In other words, the frequency spectrum is sparse. Traditionally, the necessary sampling rate is thus $2B$. Using compressed sensing techniques, the signal could be perfectly reconstructed if it is sampled at a rate slightly lower than $2EB$. The downside of this approach is that reconstruction is no longer given by a formula, but instead by the solution to a convex optimization program which requires well-studied but nonlinear methods.

Historical Background

The sampling theorem was implied by the work of Harry Nyquist in 1928, in which he showed that up to $2B$ independent pulse samples could be sent through a system of bandwidth B; but he did not explicitly consider the problem of sampling and reconstruction of continuous signals. About the same time, Karl Küpfmüller showed a similar result, and discussed the sinc-function impulse response of a band-limiting filter, via its integral, the step response *Integralsinus*; this bandlimiting and reconstruction filter that is so central to the sampling theorem is sometimes referred to as a *Küpfmüller filter* (but seldom so in English).

The sampling theorem, essentially a dual of Nyquist's result, was proved by Claude E. Shannon. V. A. Kotelnikov published similar results in 1933, as did the mathematician E. T. Whittaker in 1915, J. M. Whittaker in 1935, and Gabor in 1946 ("Theory of communication"). In 1999, the Eduard Rhein Foundation awarded Kotelnikov their Basic Research Award "for the first theoretically exact formulation of the sampling theorem."

In 1948 and 1949, Claude E. Shannon published the two revolutionary papers in which he founded the information theory. In Shannon 1948 the sampling theorem is formulated as "Theorem 13": Let f(t) contain no frequencies over W. Then

$$f(t) = \sum_{n=-\infty}^{\infty} X_n \frac{\sin \pi(2Wt - n)}{\pi(2Wt - n)},$$

where $X_n = f(n/2W)$. It was not until these papers were published that the theorem known as "Shannon's sampling theorem" became common property among communication engineers, although Shannon himself writes that this is a fact which is common knowledge in the communication art. A few lines further on, however, he adds: ... "but in spite of its evident importance [it] seems not to have appeared explicitly in the literature of communication theory".

Other Discoverers

Others who have independently discovered or played roles in the development of the sampling theorem have been discussed in several historical articles, for example by Jerri and by Lüke. For example, Lüke points out that H. Raabe, an assistant to Küpfmüller, proved the theorem in his 1939 Ph.D. dissertation; the term *Raabe condition* came to be associated with the criterion for unambiguous representation (sampling rate greater than twice the bandwidth). Meijering mentions several other discoverers and names in a paragraph and pair of footnotes:

As pointed out by Higgins , the sampling theorem should really be considered in two parts, as done above: the first stating the fact that a bandlimited function is completely determined by its samples, the second describing how to reconstruct the function using its samples. Both parts of the sampling theorem were given in a somewhat different form by J. M. Whittaker [350, 351, 353] and before him also by Ogura [241, 242]. They were probably not aware of the fact that the first part of the theorem had been stated as early as 1897 by Borel . As we have seen, Borel also used around that time what became known as the cardinal series. However, he appears not to have made the link . In later years it became known that the sampling theorem had been presented before Shannon to the Russian communication community by Kotel'nikov . In more implicit, verbal form, it had also been described in the German literature by Raabe . Several authors [33, 205] have mentioned that Someya introduced the theorem in the Japanese literature parallel to Shannon. In the English literature, Weston introduced it independently of Shannon around the same time.

Several authors, following Black , have claimed that this first part of the sampling theorem was stated even earlier by Cauchy, in a paper published in 1841. However, the paper of Cauchy does not contain such a statement, as has been pointed out by Higgins .

As a consequence of the discovery of the several independent introductions of the sampling theorem, people started to refer to the theorem by including the names of the aforementioned authors, resulting in such catchphrases as "the Whittaker–Kotel'nikov–Shannon (WKS) sampling theorem" or even "the Whittaker–Kotel'nikov–Raabe–Shannon–Someya sampling theorem" . To avoid confusion, perhaps the best thing to do is to refer to it as the sampling theorem, "rather than trying to find a title that does justice to all claimants" .

Why Nyquist?

Exactly how, when, or why Harry Nyquist had his name attached to the sampling theorem remains obscure. The term *Nyquist Sampling Theorem* (capitalized thus) appeared as early as 1959 in a book from his former employer, Bell Labs, and appeared again in 1963, and not capitalized in 1965. It had been called the *Shannon Sampling Theorem* as early as 1954, but also just *the sampling theorem* by several other books in the early 1950s.

In 1958, Blackman and Tukey cited Nyquist's 1928 paper as a reference for *the sampling theorem of information theory*, even though that paper does not treat sampling and reconstruction of continuous signals as others did. Their glossary of terms includes these entries:

> *Sampling theorem (of information theory)*
>
> Nyquist's result that equi-spaced data, with two or more points per cycle of highest frequency, allows reconstruction of band-limited functions. (See *Cardinal theorem.*)
>
> *Cardinal theorem (of interpolation theory)*

> A precise statement of the conditions under which values given at a doubly infinite set of equally spaced points can be interpolated to yield a continuous band-limited function with the aid of the function
>
> $$\frac{\sin(x-x_i)}{x-x_i}.$$

Exactly what "Nyquist's result" they are referring to remains mysterious.

When Shannon stated and proved the sampling theorem in his 1949 paper, according to Meijering "he referred to the critical sampling interval $T = 1/(2W)$ as the *Nyquist interval* corresponding to the band W, in recognition of Nyquist's discovery of the fundamental importance of this interval in connection with telegraphy." This explains Nyquist's name on the critical interval, but not on the theorem.

Similarly, Nyquist's name was attached to *Nyquist rate* in 1953 by Harold S. Black:

> "If the essential frequency range is limited to B cycles per second, $2B$ was given by Nyquist as the maximum number of code elements per second that could be unambiguously resolved, assuming the peak interference is less half a quantum step. This rate is generally referred to as signaling at the Nyquist rate and $1/(2B)$ has been termed a *Nyquist interval.*" (bold added for emphasis; italics as in the original)

According to the OED, this may be the origin of the term *Nyquist rate*. In Black's usage, it is not a sampling rate, but a signaling rate.

Undersampling

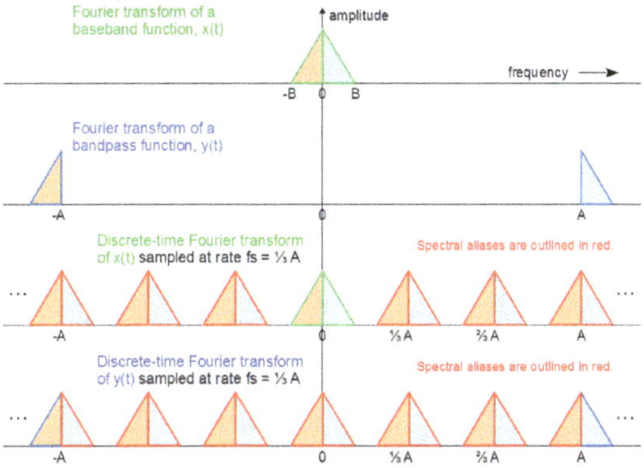

Fig 1: The top 2 graphs depict Fourier transforms of 2 different functions that produce the same results when sampled at a particular rate. The baseband function is sampled faster than its Nyquist rate, and the bandpass function is undersampled, effectively converting it to baseband. The lower graphs indicate how identical spectral results are created by the aliases of the sampling process.

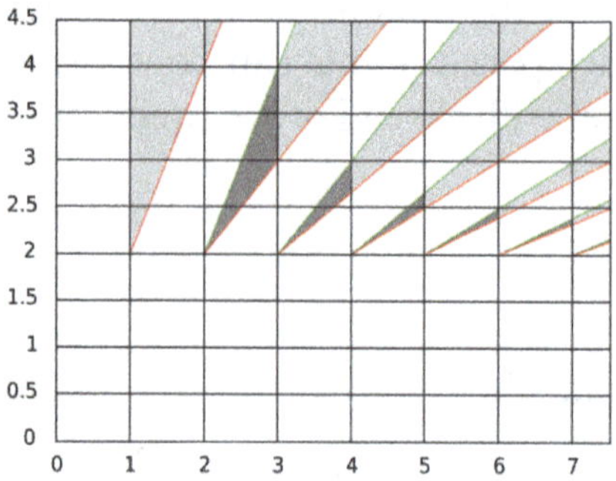

Plot of sample rates (y axis) versus the upper edge frequency (x axis) for a band of width 1; grays areas are combinations that are "allowed" in the sense that no two frequencies in the band alias to same frequency. The darker gray areas correspond to undersampling with the maximum value of n in the equations of this section.

In signal processing, undersampling or bandpass sampling is a technique where one samples a bandpass-filtered signal at a sample rate below its Nyquist rate (twice the upper cutoff frequency), but is still able to reconstruct the signal.

When one undersamples a bandpass signal, the samples are indistinguishable from the samples of a low-frequency alias of the high-frequency signal. Such sampling is also known as bandpass sampling, harmonic sampling, IF sampling, and direct IF-to-digital conversion.

Description

The Fourier transforms of real-valued functions are symmetrical around the 0 Hz axis. After sampling, only a periodic summation of the Fourier transform (called discrete-time Fourier transform) is still available. The individual, frequency-shifted copies of the original transform are called *aliases*. The frequency offset between adjacent aliases is the sampling-rate, denoted by f_s. When the aliases are mutually exclusive (spectrally), the original transform and the original continuous function, or a frequency-shifted version of it (if desired), can be recovered from the samples. The first and third graphs of Figure 1 depict a baseband spectrum before and after being sampled at a rate that completely separates the aliases.

The second graph of Figure 1 depicts the frequency profile of a bandpass function occupying the band (A, $A+B$) (shaded blue) and its mirror image (shaded beige). The condition for a non-destructive sample rate is that the aliases of both bands do not overlap when shifted by all integer multiples of f_s. The fourth graph depicts the spectral result of sampling at the same rate as the baseband function. The rate was chosen by finding the lowest rate that is an integer sub-multiple of A and also satisfies the baseband Nyquist criterion: $f_s > 2B$. Consequently, the bandpass function has effectively been converted to baseband. All the other rates that avoid overlap are given by these more general criteria, where A and $A+B$ are replaced by f_L and f_H, respectively:

$$\frac{2f_H}{n} \le f_s \le \frac{2f_L}{n-1}, \text{ for any integer } n \text{ satisfying: } 1 \le n \le \left\lfloor \frac{f_H}{f_H - f_L} \right\rfloor$$

The highest n for which the condition is satisfied leads to the lowest possible sampling rates.

Important signals of this sort include a radio's intermediate-frequency (IF), radio-frequency (RF) signal, and the individual *channels* of a filter bank.

If $n > 1$, then the conditions result in what is sometimes referred to as *undersampling*, *bandpass sampling*, or using a sampling rate less than the Nyquist rate ($2f_H$). For the case of a given sampling frequency, simpler formulae for the constraints on the signal's spectral band are given below.

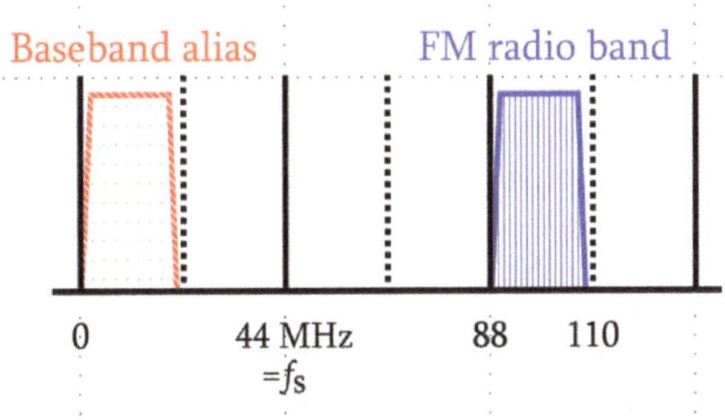

Spectrum of the FM radio band (88–108 MHz) and its baseband alias under 44 MHz ($n = 5$) sampling. An anti-alias filter quite tight to the FM radio band is required, and there's not room for stations at nearby expansion channels such as 87.9 without aliasing.

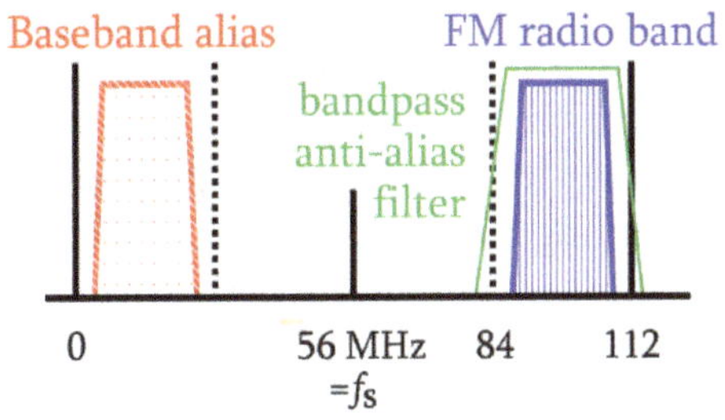

Spectrum of the FM radio band (88–108 MHz) and its baseband alias under 56 MHz ($n = 4$) sampling, showing plenty of room for bandpass anti-aliasing filter transition bands. The baseband image is frequency-reversed in this case (even n).

Example: Consider FM radio to illustrate the idea of undersampling.

In the US, FM radio operates on the frequency band from f_L = 88 MHz to f_H = 108 MHz. The bandwidth is given by

$$W = f_H - f_L = 108 \text{ MHz} - 88 \text{ MHz} = 20 \text{ MHz}$$

The sampling conditions are satisfied for

$$1 \le n \le \lfloor 5.4 \rfloor = \lfloor \frac{108\ \text{MHz}}{20\ \text{MHz}} \rfloor$$

Therefore, n can be 1, 2, 3, 4, or 5.

The value $n = 5$ gives the lowest sampling frequencies interval $43.2\ \text{MHz} < f_s < 44\ \text{MHz}$ and this is a scenario of undersampling. In this case, the signal spectrum fits between 2 and 2.5 times the sampling rate (higher than 86.4–88 MHz but lower than 108–110 MHz).

A lower value of n will also lead to a useful sampling rate. For example, using $n = 4$, the FM band spectrum fits easily between 1.5 and 2.0 times the sampling rate, for a sampling rate near 56 MHz (multiples of the Nyquist frequency being 28, 56, 84, 112, etc.).

When undersampling a real-world signal, the sampling circuit must be fast enough to capture the highest signal frequency of interest. Theoretically, each sample should be taken during an infinitesimally short interval, but this is not practically feasible. Instead, the sampling of the signal should be made in a short enough interval that it can represent the instantaneous value of the signal with the highest frequency. This means that in the FM radio example above, the sampling circuit must be able to capture a signal with a frequency of 108 MHz, not 43.2 MHz. Thus, the sampling frequency may be only a little bit greater than 43.2 MHz, but the input bandwidth of the system must be at least 108 MHz. Similarly, the accuracy of the sampling timing, or aperture uncertainty of the sampler, frequently the analog-to-digital converter, must be appropriate for the frequencies being sampled 108MHz, not the lower sample rate.

If the sampling theorem is interpreted as requiring twice the highest frequency, then the required sampling rate would be assumed to be greater than the *Nyquist rate* 216 MHz. While this does satisfy the last condition on the sampling rate, it is grossly oversampled.

Note that if a band is sampled with $n > 1$, then a band-pass filter is required for the anti-aliasing filter, instead of a lowpass filter.

As we have seen, the normal baseband condition for reversible sampling is that $X(f) = 0$ outside the interval: $\left(-\frac{1}{2} f_s, \frac{1}{2} f_s\right)$,

and the reconstructive interpolation function, or lowpass filter impulse response, is $sinc(t / T)$.

To accommodate undersampling, the bandpass condition is that $X(f) = 0$ outside the union of open positive and negative frequency bands

$$\left(-\frac{n}{2}f_s, -\frac{n-1}{2}f_s\right) \cup \left(\frac{n-1}{2}f_s, \frac{n}{2}f_s\right) \text{for some positive integer } n.$$

which includes the normal baseband condition as case $n = 1$ (except that where the intervals come together at 0 frequency, they can be closed).

The corresponding interpolation function is the bandpass filter given by this difference of lowpass impulse responses:

$$n\,sinc\left(\frac{nt}{T}\right) - (n-1)\,sinc\left(\frac{(n-1)t}{T}\right).$$

On the other hand, reconstruction is not usually the goal with sampled IF or RF signals. Rather, the sample sequence can be treated as ordinary samples of the signal frequency-shifted to near baseband, and digital demodulation can proceed on that basis, recognizing the spectrum mirroring when n is even.

Further generalizations of undersampling for the case of signals with multiple bands are possible, and signals over multidimensional domains (space or space-time) and have been worked out in detail by Igor Kluvánek.

Oversampling

In signal processing, oversampling is the process of sampling a signal with a sampling frequency significantly higher than the Nyquist rate. Theoretically a bandwidth-limited signal can be perfectly reconstructed if sampled above the Nyquist rate, which is twice the highest frequency in the signal. Oversampling improves resolution, reduces noise and helps avoid aliasing and phase distortion by relaxing anti-aliasing filter performance requirements.

A signal is said to be oversampled by a factor of N if it is sampled at N times the Nyquist rate.

Motivation

There are three main reasons for performing oversampling:

Anti-aliasing

Oversampling can make it easier to realize analog anti-aliasing filters. Without oversampling, it is very difficult to implement filters with the sharp cutoff necessary to maximize use of the available bandwidth without exceeding the Nyquist limit. By increasing the bandwidth of the sampled signal, design constraints for the anti-aliasing filter may be relaxed. Once sampled, the signal can be digitally filtered and downsampled to the desired sampling frequency. In modern integrated circuit technology, digital filters are easier to implement than comparable analog filters.

Resolution

In practice, oversampling is implemented in order to achieve cheaper higher-resolution A/D and D/A conversion. For instance, to implement a 24-bit converter, it is sufficient to use a 20-bit converter that can run at 256 times the target sampling rate. Combining 256 consecutive 20-bit samples can increase the signal-to-noise ratio at the voltage level by a factor of 16 (the square root of the number of samples averaged), effectively adding 4 bits to the resolution and producing a single sample with 24-bit resolution.

When oversampling by a factor of N, the dynamic range increases by log2(N) bits, because there are N times as many possible values for the sum.

However, the SNR increases by sqrt(N), (not by N as in the article). Summing up uncorrelated noise increases its amplitude by sqrt(N), while summing up a coherent signal increases its average by N. As a result, the SNR (or signal/noise) increases by sqrt(N). In the example, that means while with N=256 there is an increase in Dynamic range by 8 bits, and the content of "coherent signal" increases by N, but the noise changes by a factor of sqrt(N)=sqrt(256)=16 in the example (not to be confused with an increase of 16 bits), so the SNR changes by a factor of 16.

The number of samples required to get n bits of additional data precision is

$$\text{number of samples} = (2^n)^2 = 2^{2n}.$$

To get the mean sample scaled up to an integer with n additional bits, the sum of 2^{2n} samples is divided by 2^n:

$$\text{scaled mean} = \frac{\sum_{i=0}^{2^{2n}-1} 2^n data_i}{2^{2n}} = \frac{\sum_{i=0}^{2^{2n}-1} data_i}{2^n}.$$

This averaging is only possible if the signal contains equally distributed noise which is enough to be observed by the A/D converter. If not, in the case of a stationary input signal, all 2^n samples would have the same value and the resulting average would be identical to this value; so in this case, oversampling would have made no improvement. (In similar cases where the A/D converter sees no noise and the input signal is changing over time, oversampling still improves the result, but to an inconsistent/unpredictable extent.) This is an interesting counter-intuitive example where adding some dithering noise to the input signal can improve (rather than degrade) the final result because the dither noise allows oversampling to work to improve resolution (or dynamic range). In many practical applications, a small increase in noise is well worth a substantial increase in measurement resolution. In practice, the dithering noise can often be placed outside the frequency range of interest to the measurement, so that this noise can be subsequently filtered out in the digital domain—resulting in a final measurement (in the frequency range of interest) with both higher resolution and lower noise.

Noise

If multiple samples are taken of the same quantity with uncorrelated noise added to each sample, then averaging N samples reduces the noise power by a factor of $1/N$. If, for example, we oversample by a factor of 4, the signal-to-noise ratio in terms of power improves by factor of 4 which corresponds to a factor of 2 improvement in terms of voltage.

Certain kinds of A/D converters known as delta-sigma converters produce disproportionately more quantization noise in the upper portion of their output spectrum. By running these converters at some multiple of the target sampling rate, and low-pass filtering the oversampled signal down to half the target sampling rate, a final result with *less* noise (over the entire band of the converter) can be obtained. Delta-sigma converters use a technique called noise shaping to move the quantization noise to the higher frequencies.

Example

Consider a signal with a bandwidth or highest frequency of $B = 100$ Hz. The sampling theorem states that sampling frequency would have to be greater than 200 Hz. Sampling at four times that rate requires a sampling frequency of 800 Hz. This gives the anti-aliasing filter a transition band of 300 Hz ($(f_s/2) - B = (800\text{ Hz}/2) - 100\text{ Hz} = 300\text{ Hz}$) instead of 0 Hz if the sampling frequency was 200 Hz.

Achieving an anti-aliasing filter with 0 Hz transition band is unrealistic whereas an anti-aliasing filter with a transition band of 300 Hz is not difficult to create.

Oversampling in Reconstruction

The term oversampling is also used to denote a process used in the reconstruction phase of digital-to-analog conversion, in which an intermediate high sampling rate is used between the digital input and the analogue output. Here, samples are interpolated in the digital domain to add additional samples in between, thereby converting the data to a higher sample rate, which is a form of upsampling. When the resulting higher-rate samples are converted to analog, a less complex/expensive analog low pass filter is required to remove the high-frequency content, which will consist of reflected images of the real signal created by the zero-order hold of the digital-to-analog converter. Essentially, this is a way to shift some of the complexity of the filtering into the digital domain and achieves the same benefit as oversampling in analog-to-digital conversion.

Upsampling

Upsampling is interpolation, applied in the context of digital signal processing and sample rate conversion. When upsampling is performed on a sequence of samples of a continuous function or signal, it produces an approximation of the sequence that would have been obtained by sampling the signal at a higher rate (or density, as in the case of a photograph). For example, if compact disc audio is upsampled by a factor of 5/4, the resulting sample-rate increases from 44,100 Hz to 55,125 Hz.

Upsampling by an Integer Factor

Interpolation by an integer factor L can be explained as a 2-step process, with an equivalent implementation that is more efficient:

1. Create a sequence, $x_L[n]$, comprising the original samples, $x[n]$, separated by $L-1$ zeros.
2. Smooth out the discontinuities with a lowpass filter, which replaces the zeros.

In this application the filter is called an interpolation filter, and its design is discussed below. When the interpolation filter is an FIR type, its efficiency can be improved, because the zeros contribute nothing to its dot product calculations. It is an easy matter to omit them from both the data stream and the calculations. The calculation performed by an efficient interpolating FIR filter for each output sample is a dot product:

$$y[j+nL]=\sum_{k=0}^{K}x[n-k]\cdot h[j+kL],\quad j=0,1,\ldots,L-1,$$

where the $h[\bullet]$ sequence is the impulse response, and K is the largest value of k for which $h[j+kL]$ is non-zero. In the case $L=2$, $h[\bullet]$ can be designed as a half-band filter, where almost half of the coefficients are zero and need not be included in the dot products. Impulse response coefficients taken at intervals of L form a subsequence, and there are L such subsequences (called phases) multiplexed together. Each of L phases of the impulse response is filtering the same sequential values of the $x[\bullet]$ data stream and producing one of L sequential output values. In some multi-processor architectures, these dot products are performed simultaneously, in which case it is called a polyphase filter.

For completeness, we now mention that a possible, but unlikely, implementation of each phase is to replace the coefficients of the other phases with zeros in a copy of the $h[\bullet]$ array, and process the $x_L[n]$, sequence at L times faster than the original input rate. $L - 1$ of every L outputs are zero, and the real values are supplied by the other phases. Adding them all together produces the desired $y[\bullet]$ sequence. Adding a zero is equivalent to discarding it. The equivalence of computing and discarding $L - 1$ zeros vs computing just every L^{th} output is known as the *second Noble identity*.

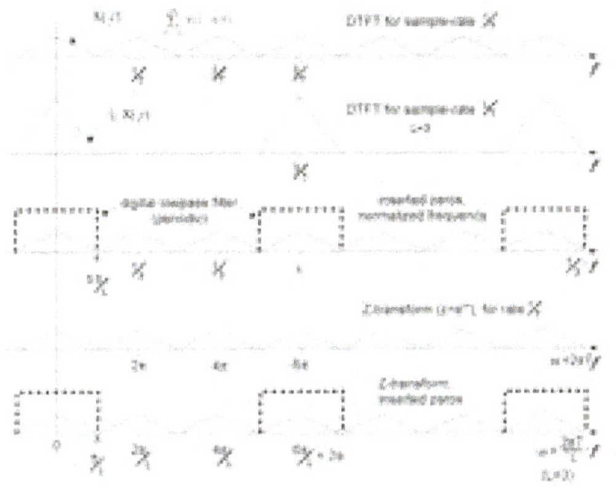

Spectral depictions of zero-fill and interpolation by lowpass filtering

Interpolation Filter Design

Let $X(f)$ be the Fourier transform of any function, $x(t)$, whose samples at some interval, T, equal the $x[n]$ sequence. Then the discrete-time Fourier transform (DTFT) of the $x[n]$ sequence is the Fourier series representation of a periodic summation of $X(f)$:

$$\underbrace{\sum_{n=-\infty}^{\infty} \overbrace{x(nT)}^{x[n]}\, e^{-i2\pi fnT}}_{\text{DTFT}} = \frac{1}{T}\sum_{k=-\infty}^{\infty} X(f - k/T). \qquad \textbf{(Eq.1)}$$

When T has units of seconds, f has units of hertz. Sampling L times faster (at interval T/L) increases the periodicity by a factor of L:

$$\frac{L}{T}\sum_{k=-\infty}^{\infty} X\left(f - k\cdot\frac{L}{T}\right), \qquad \textbf{(Eq.2)}$$

which is also the desired result of interpolation. An example of both these distributions is depicted in the top two graphs of Fig.1.

When the additional samples are inserted zeros, they increase the data rate, but they have no effect on the frequency distribution until the zeros are replaced by the interpolation filter. Many filter design programs use frequency units of *cycles/sample*, which is achieved by normalizing the frequency axis, based on the new data rate (L/T). The result is shown in the third graph of Fig.1. Also shown is the passband of the interpolation filter needed to make the third graph resemble the second one. Its cutoff frequency is $\frac{0.5}{L}$. In terms of actual frequency, the cutoff is $\frac{0.5}{T}$ Hz, which is the Nyquist frequency of the original x[n] sequence.

The same result can be obtained from Z-transforms, constrained to values of complex-variable, z, of the form $z = e^{i\omega}$. Then the transform is the same Fourier series with different frequency normalization. By comparison with Eq.1, we deduce:

$$\sum_{n=-\infty}^{\infty} x[n]\, z^{-n} = \sum_{n=-\infty}^{\infty} x[n]\, e^{-i\omega n} = \frac{1}{T}\sum_{k=-\infty}^{\infty} \underbrace{X\left(\tfrac{\omega}{2\pi T} - \tfrac{k}{T}\right)}_{X\left(\frac{\omega - 2\pi k}{2\pi T}\right)},$$

which is depicted by the fourth graph in Fig.1. When the zeros are inserted, the transform becomes:

$$\sum_{n=-\infty}^{\infty} x[n]\, z^{-nL} = \sum_{n=-\infty}^{\infty} x[n]\, e^{-i\omega Ln} = \frac{1}{T}\sum_{k=-\infty}^{\infty} \underbrace{X\left(\tfrac{\omega L}{2\pi T} - \tfrac{k}{T}\right)}_{X\left(\frac{\omega - 2\pi k/L}{2\pi T/L}\right)},$$

depicted by the bottom graph. In these normalizations, the effective data-rate is always represented by the constant 2π (*radians/sample*) instead of 1. In those units, the interpolation filter bandwidth is π/L, as show on the bottom graph. The corresponding physical frequency is $\frac{\pi}{L} \cdot \frac{L}{2\pi T} = \frac{0.5}{T}$ Hz, the original Nyquist frequency.

Upsampling by a Rational Fraction

Let L/M denote the upsampling factor, where $L > M$.

1. Upsample by a factor of L
2. Downsample by a factor of M

Upsampling requires a lowpass filter after increasing the data rate, and downsampling requires a lowpass filter before decimation. Therefore, both operations can be accomplished by a single filter with the lower of the two cutoff frequencies. For the $L > M$ case, the interpolation filter cutoff, $\frac{0.5}{L}$ *cycles per intermediate sample*, is the lower frequency.

Decimation (Signal Processing)

In digital signal processing, decimation is the process of reducing the sampling rate of a signal. Complementary to interpolation, which increases sampling rate, it is a specific case of sample rate conversion in a multi-rate digital signal processing system. Decimation utilises filtering to mitigate aliasing distortion, which can occur when simply downsampling a signal. A system component that performs decimation is called a decimator.

In General

Decimation reduces the data rate or the size of the data. The decimation factor is usually an integer or a rational fraction greater than one. This factor multiplies the sampling time or, equivalently, divides the sampling rate. For example, if 16-bit compact disc audio (sampled at 44,100 Hz) is decimated to 22,050 Hz, the audio is said to be decimated by a factor of 2. The bit rate is also reduced in half, from 1,411,200 bit/s to 705,600 bit/s, assuming that each sample retains its bit depth of 16 bits.

By an Integer Factor

Decimation by an integer factor, M, can be explained as a 2-step process, with an equivalent implementation that is more efficient:

1. Reduce high-frequency signal components with a digital lowpass filter.
2. Downsample the filtered signal by M; that is, keep only every M^{th} sample.

Downsampling alone causes high-frequency signal components to be misinterpreted by subsequent users of the data, which is a form of distortion called aliasing. The first step, if necessary, is to suppress aliasing to an acceptable level. In this application, the filter is called an anti-aliasing filter, and its design is discussed below. Also see undersampling for information about downsampling bandpass functions and signals.

When the anti-aliasing filter is an IIR design, it relies on feedback from output to input, prior to the downsampling step. With FIR filtering, it is an easy matter to compute only every M^{th} output. The calculation performed by a decimating FIR filter for the nth output sample is a dot product:

$$y[n]=\sum_{k=0}^{K-1} x[nM-k]\cdot h[k],$$

where the h[•] sequence is the impulse response, and K is its length. x[•] represents the input sequence being downsampled. In a general purpose processor, after computing y[n], the easiest way to compute y[n+1] is to advance the starting index in the x[•] array by M, and recompute the dot product. In the case M=2, h[•] can be designed as a half-band filter, where almost half of the coefficients are zero and need not be included in the dot products.

Impulse response coefficients taken at intervals of M form a subsequence, and there are M such subsequences (phases) multiplexed together. The dot product is the sum of the dot products of each subsequence with the corresponding samples of the x[•] sequence.

Furthermore, because of downsampling by M, the stream of x[•] samples involved in any one of the M dot products is never involved in the other dot products. Thus M low-order FIR filters are each filtering one of M multiplexed *phases* of the input stream, and the M outputs are being summed. This viewpoint offers a different implementation that might be advantageous in a multi-processor architecture. In other words, the input stream is demultiplexed and sent through a bank of M filters whose outputs are summed. When implemented that way, it is called a polyphase filter.

For completeness, we now mention that a possible, but unlikely, implementation of each phase is to replace the coefficients of the other phases with zeros in a copy of the h[•] array, process the original x[•] sequence at the input rate, and decimate the output by a factor of M. The equivalence of this inefficient method and the implementation described above is known as the *first Noble identity*.

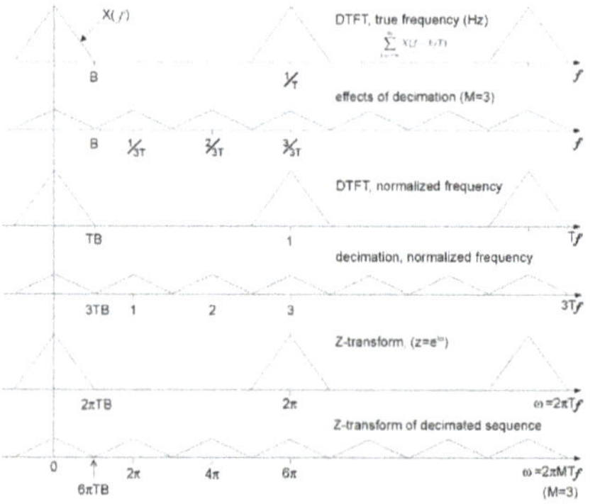

Spectral effects of decimation compared on 3 popular frequency scale conventions

Anti-aliasing Filter

The requirements of the anti-aliasing filter can be deduced from any of the 3 pairs of graphs in Fig. 1. Note that all 3 pairs are identical, except for the units of the abscissa variables. The upper graph of each pair is an example of the periodic frequency distribution of a sampled function, x(t), with Fourier transform, X(f). The lower graph is the new distribution that results when x(t) is sampled 3 times slower, or (equivalently) when the original sample sequence is decimated by a factor of M=3. In all 3 cases, the condition that ensures the copies of X(f) don't overlap each other is the same: $B < \frac{1}{M} \cdot \frac{1}{2T}$, where T is the interval between samples, 1/T is the sample-rate, and 1/2T is the Nyquist frequency. The anti-aliasing filter that can ensure the condition is met has a cutoff frequency less than $\frac{1}{M}$ times the Nyquist frequency.

The abscissa of the top pair of graphs represents the discrete-time Fourier transform (DTFT), which is a Fourier series representation of a periodic summation of X(f):

$$\underbrace{\sum_{n=-\infty}^{\infty} \overbrace{x(nT)}^{x[n]}\, e^{-i2\pi fnT}}_{\textbf{DTFT}} = \frac{1}{T}\sum_{k=-\infty}^{\infty} X(f - k/T). \qquad \textbf{(Eq.1)}$$

When T has units of seconds, f has units of hertz. Replacing T with MT in the formulas above gives the DTFT of the decimated sequence, x[nM]:

$$\sum_{n=-\infty}^{\infty} x(n \cdot MT)\, e^{-i2\pi fn(MT)} = \frac{1}{MT}\sum_{k=-\infty}^{\infty} X\left(f - \frac{k}{(MT)}\right).$$

The periodic summation has been reduced in amplitude and periodicity by a factor of M, as depicted in the second graph of Fig. 1. Aliasing occurs when adjacent copies of X(f) overlap. The purpose of the anti-aliasing filter is to ensure that the reduced periodicity does not create overlap.

In the middle pair of graphs, the frequency variable f, has been replaced by normalized frequency, which creates a periodicity of 1 and a Nyquist frequency of ½. A common practice in filter design programs is to assume those values and request only the corresponding cutoff frequency in the same units. In other words, the cutoff frequency $B_{max} = \frac{1}{M} \cdot \frac{1}{2T}$, is normalized to $TB_{max} = \frac{1}{M} \cdot \frac{1}{2} = \frac{0.5}{M}$. The units of this quantity are $(seconds/sample) \times (cycles/second) = cycles/sample$.

The bottom pair of graphs represent the Z-transforms of the original sequence and the decimated sequence, constrained to values of complex-variable, z, of the form $z = e^{i\omega}$. Then the transform of the x[n] sequence has the form of a Fourier series. By comparison with Eq.1, we deduce:

$$\sum_{n=-\infty}^{\infty} x[n]\, z^{-n} = \sum_{n=-\infty}^{\infty} x(nT)\, e^{-i\omega n} = \frac{1}{T}\sum_{k=-\infty}^{\infty} \underbrace{X\left(\tfrac{\omega}{2\pi T} - \tfrac{k}{T}\right)}_{X\left(\frac{\omega - 2\pi k}{2\pi T}\right)},$$

which is depicted by the fifth graph in Fig. 1. Similarly, the sixth graph depicts:

$$\sum_{n=-\infty}^{\infty} x[nM]\, z^{-n} = \sum_{n=-\infty}^{\infty} x(nMT)\, e^{-i\omega n} = \frac{1}{MT}\sum_{k=-\infty}^{\infty} \underbrace{X\left(\tfrac{\omega}{2\pi MT} - \tfrac{k}{MT}\right)}_{X\left(\frac{\omega - 2\pi k}{2\pi MT}\right)}.$$

By a Rational Factor

Let M/L denote the decimation factor, where: $M, L \in Z$; $M > L$.

1. Interpolate by a factor of L
2. Decimate by a factor of M

Interpolation requires a lowpass filter after increasing the data rate, and decimation requires a lowpass filter before decimation. Therefore, both operations can be accomplished by a single filter with the lower of the two cutoff frequencies. For the $M > L$ case, the anti-aliasing filter cutoff, $\frac{0.5}{M}$ *cycles per intermediate sample*, is the lower frequency.

By an Irrational Factor

Techniques for decimation (and sample-rate conversion in general) by factor $R \in R^+$ include polynomial interpolation and the Farrow structure.

Aliasing

Properly sampled image of a brick wall (requires screen of sufficient resolution to prevent moiré pattern)

Spatial aliasing in the form of a moiré pattern

In signal processing and related disciplines, aliasing is an effect that causes different signals to become indistinguishable (or *aliases* of one another) when sampled. It also refers to the distortion or artifact that results when the signal reconstructed from samples is different from the original continuous signal.

Aliasing can occur in signals sampled in time, for instance digital audio, and is referred to as temporal aliasing. Aliasing can also occur in spatially sampled signals, for instance moiré patterns in digital images. Aliasing in spatially sampled signals is called spatial aliasing.

Aliasing is generally avoided by applying low pass filters- anti-aliasing filters to the analog signal before sampling.

Description

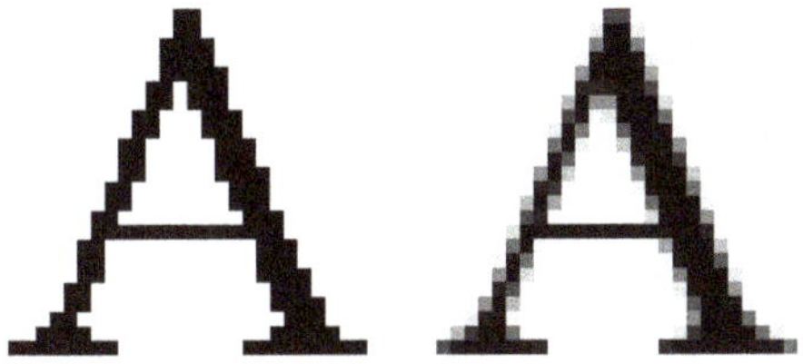

Left: An aliased image of the letter *A* in Times New Roman. Right: An *anti-aliased* image.

When a digital image is viewed, a reconstruction is performed by a display or printer device, and by the eyes and the brain. If the image data is processed in some ways during sampling or reconstruction, the reconstructed image will differ from the original image, and an alias is seen.

An example of spatial aliasing is the moiré pattern observed in a poorly pixelized image of a brick wall. Spatial anti-aliasing techniques avoid such poor pixelizations. Aliasing can be caused either by the sampling stage or the reconstruction stage; these may be distinguished by calling sampling aliasing *prealiasing* and reconstruction aliasing *postaliasing*.

Temporal aliasing is a major concern in the sampling of video and audio signals. Music, for instance, may contain high-frequency components that are inaudible to humans. If a piece of music is sampled at 32000 samples per second (Hz), any frequency components above 16000 Hz (the Nyquist frequency for this sampling rate) will cause aliasing when the music is reproduced by a digital to analog converter (DAC). To prevent this an anti-aliasing filter is used to remove components above the Nyquist frequency prior to sampling.

In video or cinematography, temporal aliasing results from the limited frame rate, and causes the wagon-wheel effect, whereby a spoked wheel appears to rotate too slowly or even backwards. Aliasing has changed its apparent frequency of rotation. A reversal of direction can be described as a negative frequency. Temporal aliasing frequencies in video and cinematography are determined by the frame rate of the camera, but the relative intensity of the aliased frequencies is determined by the shutter timing (exposure time) or the use of a temporal aliasing reduction filter during filming.

Like the video camera, most sampling schemes are periodic; that is, they have a characteristic sampling frequency in time or in space. Digital cameras provide a certain number of samples (pixels) per degree or per radian, or samples per mm in the focal plane of the camera. Audio signals are sampled (digitized) with an analog-to-digital converter, which produces a constant number of samples per second. Some of the most dramatic and subtle examples of aliasing occur when the signal being sampled also has periodic content.

Bandlimited Functions

Actual signals have finite duration and their frequency content, as defined by the Fourier transform, has no upper bound. Some amount of aliasing always occurs when such functions are sampled. Functions whose frequency content is bounded (*bandlimited*) have infinite duration. If sampled at a high enough rate, determined by the *bandwidth*, the original function can in theory be perfectly reconstructed from the infinite set of samples.

Bandpass Signals

Sometimes aliasing is used intentionally on signals with no low-frequency content, called *bandpass* signals. Undersampling, which creates low-frequency aliases, can produce the same result, with less effort, as frequency-shifting the signal to lower frequencies before sampling at the lower rate. Some digital channelizers exploit aliasing in this way for computational efficiency.

Sampling Sinusoidal Functions

Sinusoids are an important type of periodic function, because realistic signals are often modeled as the summation of many sinusoids of different frequencies and different amplitudes (for example, with a Fourier series or transform). Understanding what aliasing does to the individual sinusoids is useful in understanding what happens to their sum.

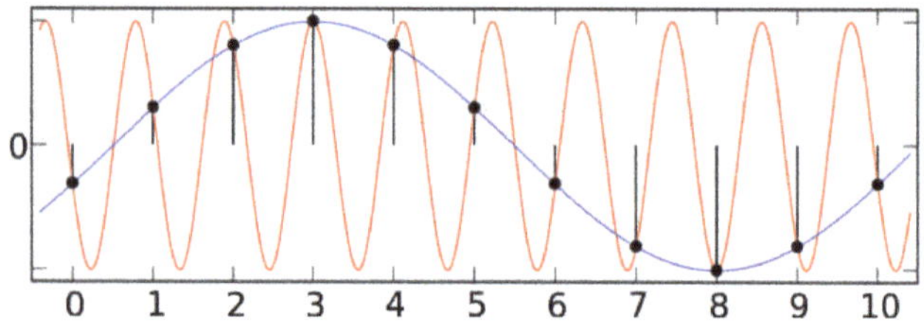

Two different sinusoids that fit the same set of samples.

Here, a plot depicts a set of samples whose sample-interval is 1, and two (of many) different sinusoids that could have produced the samples. The sample-rate in this case is $f_s = 1$. For instance, if the interval is 1 *second,* the rate is 1 *sample per second.* Nine cycles of the red sinusoid and 1 cycle of the blue sinusoid span an interval of 10 samples. The corresponding number of *cycles per sample* are $f_{\text{red}} = 0.9$ and $f_{\text{blue}} = 0.1.$. If these samples were produced by sampling functions $\cos(2\pi(0.9)x - \theta)$ and $\cos(2\pi(0.1)x - \phi)$, they could also have been produced by the trigonometrically identical functions $\cos(2\pi(-0.9)x + \theta)$ and $\cos(2\pi(-0.1)x + \phi)$ which introduces the useful concept of negative frequency.

In general, when a sinusoid of frequency f is sampled with frequency f_s, the resulting number of *cycles per sample* is f / f_s (known as normalized frequency), and the samples are indistinguishable from those of another sinusoid (called an *alias*) whose normalized frequency differs from f / f_s by any integer (positive or negative). Replacing negative frequency sinusoids by their equivalent positive frequency representations, we can express all the aliases of frequency f as $f_{\text{alias}}(N) \overset{\text{def}}{=} |f - Nf_s|$, for any integer N with $f_{\text{alias}}(0) = f$ being the true value, and N has units of *cycles per sample.* Then the $N = 1$ alias of f_{red} is f_{blue}, (and vice versa).

Aliasing matters when one attempts to reconstruct the original waveform from its samples. The most common reconstruction technique produces the smallest of the $f_{\text{alias}}(N)$ frequencies. So it is usually important that $f_{\text{alias}}(0)$ be the unique minimum. A necessary and sufficient condition for that is $f_s / 2 > |f|$, where $f_s / 2$ is commonly called the Nyquist frequency of a system that samples at rate f_s. In our example, the Nyquist condition is satisfied if the original signal is the blue sinusoid ($f = f_{\text{blue}}$). But if $f = f_{\text{red}} = 0.9$, the usual reconstruction method will produce the blue sinusoid instead of the red one.

Folding

In the example above, f_{red} and f_{blue} are symmetrical around the frequency $f_s / 2$. And in general, as f increases from 0 to $f_s / 2$, $f_{\text{alias}}(1)$ decreases from f_s to $f_s / 2$. Similarly, as f increases from $f / 2$ to f_s, $f_{\text{alias}}(1)$ continues decreasing from $f_s / 2$ to 0.

A graph of amplitude vs frequency for a single sinusoid at frequency $0.6 f_s$ and some of its aliases at $0.4 f_s$, $1.4 f_s$, and $1.6 f_s$ would look like the 4 black dots in the first figure below. The red lines depict the paths (loci) of the 4 dots if we were to adjust the frequency and amplitude of the sinusoid along the solid red segment (between $f_s/2$ and f_s). No matter what function we choose to change the amplitude vs frequency, the graph will exhibit symmetry between 0 and This symmetry is commonly referred to as folding, and another name for $f_s/2$ (the Nyquist frequency) is folding frequency. Folding is often observed in practice when viewing the frequency spectrum of real-valued samples, such as the second figure below.

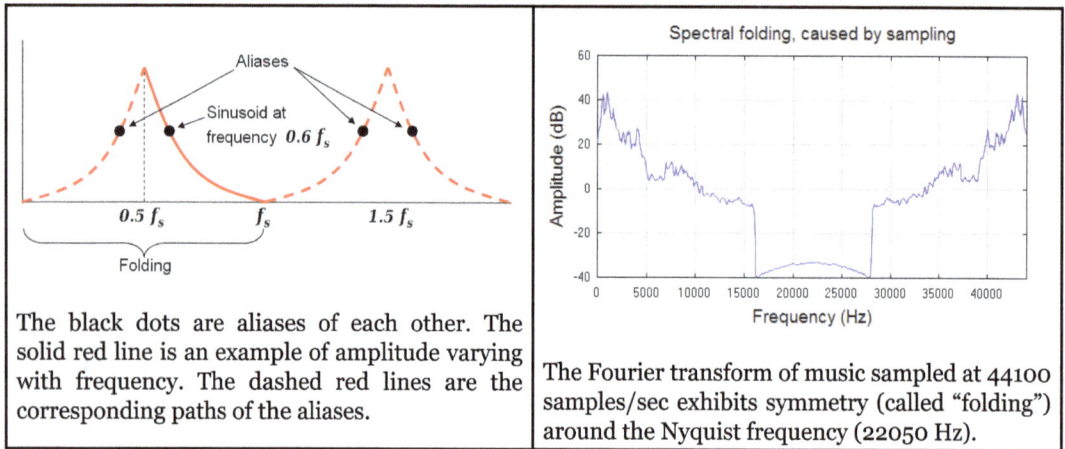

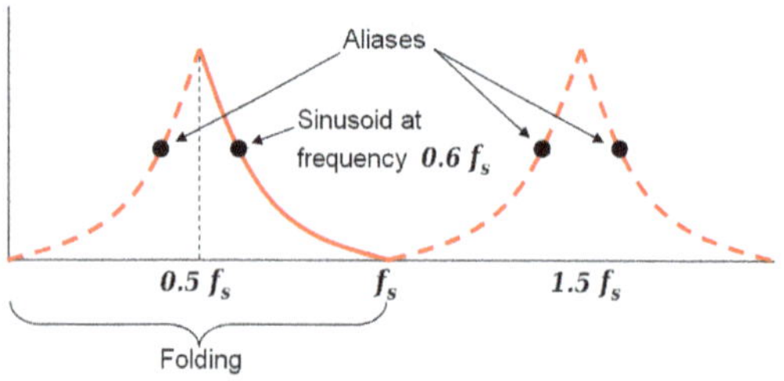

The black dots are aliases of each other. The solid red line is an example of amplitude varying with frequency. The dashed red lines are the corresponding paths of the aliases.

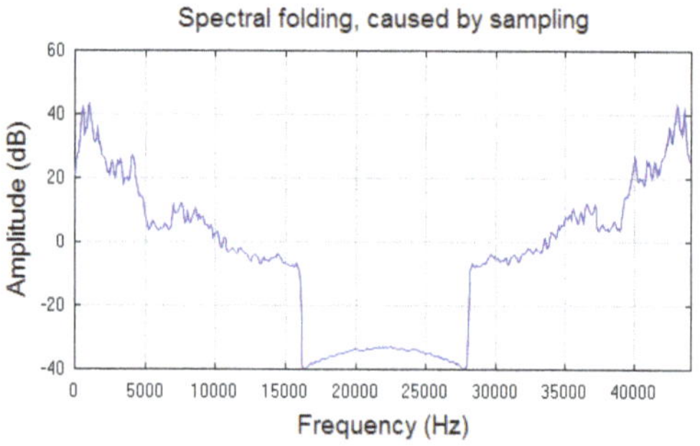

The Fourier transform of music sampled at 44100 samples/sec exhibits symmetry (called "folding") around the Nyquist frequency (22050 Hz).

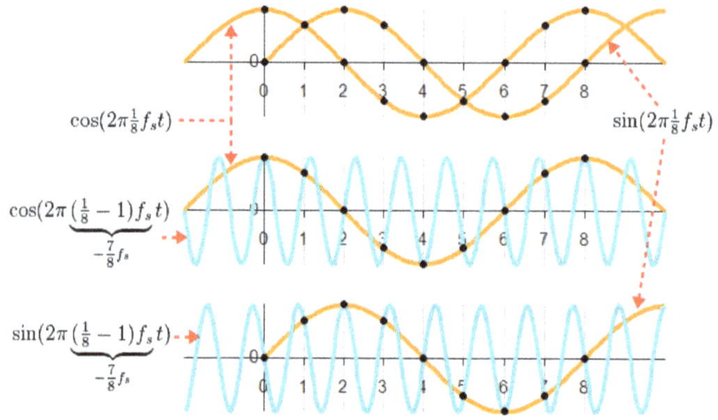

Two complex sinusoids, colored gold and cyan, that fit the same sets of real and imaginary sample points when sampled at the rate (f_s) indicated by the grid lines. The case shown here is:

$$f_{\text{cyan}} = f_{\text{alias}}(1) = f_{\text{gold}} - 1 \cdot f_s.$$

Complex Sinusoids

Complex sinusoids are waveforms whose samples are complex numbers, and the concept of negative frequency is necessary to distinguish them. In that case, the frequencies of the aliases are given by just: $f_{\text{alias}}(N) = f - Nf_s$. Therefore, as f increases from $f_s/2$ to f_s, $f_{\text{alias}}(1)$ goes from $-f_s/2$ up to 0. Consequently, complex sinusoids do not exhibit *folding*. Complex samples of real-valued sinusoids have zero-valued imaginary parts and do exhibit folding.

Sample Frequency

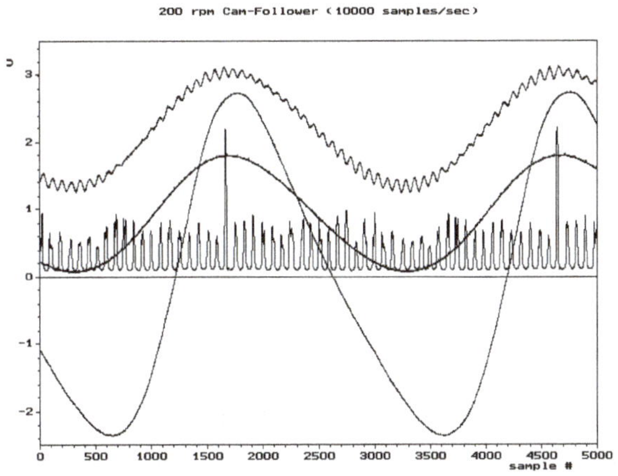

Illustration of 4 waveforms reconstructed from samples taken at six different rates. Two of the waveforms are sufficiently sampled to avoid aliasing at all six rates. The other two illustrate increasing distortion (aliasing) at the lower rates.

When the condition $f_s / 2 > f$ is met for the highest frequency component of the original signal, then it is met for all the frequency components, a condition called the Nyquist criterion. That is typically approximated by filtering the original signal to attenuate high frequency components before it is sampled. These attenuated high frequency components still generate low-frequency aliases, but typically at low enough amplitudes that they do not cause problems. A filter chosen in anticipation of a certain sample frequency is called an anti-aliasing filter.

The filtered signal can subsequently be reconstructed, by interpolation algorithms, without significant additional distortion. Most sampled signals are not simply stored and reconstructed. But the fidelity of a theoretical reconstruction (via the Whittaker–Shannon interpolation formula) is a customary measure of the effectiveness of sampling.

Historical Usage

Historically the term *aliasing* evolved from radio engineering because of the action of superheterodyne receivers. When the receiver shifts multiple signals down to lower frequencies, from RF to IF by heterodyning, an unwanted signal, from an RF frequency equally far from the local oscillator (LO) frequency as the desired signal, but on the wrong side of the LO, can end up at the same IF frequency as the wanted one. If it is strong enough it can interfere with reception of the desired signal. This unwanted signal is known as an *image* or *alias* of the desired signal.

Angular Aliasing

Aliasing occurs whenever the use of discrete elements to capture or produce a continuous signal causes frequency ambiguity.

Spatial aliasing, particular of angular frequency, can occur when reproducing a light field or sound field with discrete elements, as in 3D displays or wave field synthesis of sound.

This aliasing is visible in images such as posters with lenticular printing: if they have low angular resolution, then as one moves past them, say from left-to-right, the 2D image does not initially change (so it appears to move left), then as one moves to the next angular image, the image suddenly changes (so it jumps right) – and the frequency and amplitude of this side-to-side movement corresponds to the angular resolution of the image (and, for frequency, the speed of the viewer's lateral movement), which is the angular aliasing of the 4D light field.

The lack of parallax on viewer movement in 2D images and in 3-D film produced by stereoscopic glasses (in 3D films the effect is called "yawing", as the image appears to rotate on its axis) can similarly be seen as loss of angular resolution, all angular frequencies being aliased to 0 (constant).

More Examples

Online Audio Example

The qualitative effects of aliasing can be heard in the following audio demonstration. Six sawtooth waves are played in succession, with the first two sawtooths having a fundamental frequency of 440 Hz (A4), the second two having fundamental frequency of 880 Hz (A5), and the final two at 1760 Hz (A6). The sawtooths alternate between bandlimited (non-aliased) sawtooths and aliased sawtooths and the sampling rate is 22.05 kHz. The bandlimited sawtooths are synthesized from the sawtooth waveform's Fourier series such that no harmonics above the Nyquist frequency are present.

The aliasing distortion in the lower frequencies is increasingly obvious with higher fundamental frequencies, and while the bandlimited sawtooth is still clear at 1760 Hz, the aliased sawtooth is degraded and harsh with a buzzing audible at frequencies lower than the fundamental.

Direction Finding

A form of spatial aliasing can also occur in antenna arrays or microphone arrays used to estimate the direction of arrival of a wave signal, as in geophysical exploration by seismic waves. Waves must be sampled at more than two points per wavelength, or the wave arrival direction becomes ambiguous.

References

- Hiroshi Harada, Ramjee Prasad (2002). Simulation and Software Radio for Mobile Communications. Artech House. ISBN 1-58053-044-3.

- Lyons, Richard (2001). Understanding Digital Signal Processing. Prentice Hall. p. 304. ISBN 0-201-63467-8.
- Strang, Gilbert; Nguyen, Truong (1996-10-01). Wavelets and Filter Banks (2 ed.). Wellesley,MA: Wellesley-Cambridge Press. pp. 100–101. ISBN 0961408871.
- "Improving ADC Resolution by Oversampling and Averaging" (PDF). Silicon Laboratories Inc. Retrieved 17 January 2015.

7

Filter: A Comprehensive Study

A filter is a device that is used in the process of removing unwanted components from a signal. Filters can be linear, casual, analog or digital, discrete time or passive or active type etc. The topics discussed in this text are adaptive filters, digital filters, recursive least squares filters, finite impulsive responses, infinite impulse responses etc. The major components of filters are discussed in this chapter.

Filter (Signal Processing)

In signal processing, a filter is a device or process that removes some unwanted components or features from a signal. Filtering is a class of signal processing, the defining feature of filters being the complete or partial suppression of some aspect of the signal. Most often, this means removing some frequencies and not others in order to suppress interfering signals and reduce background noise. However, filters do not exclusively act in the frequency domain; especially in the field of image processing many other targets for filtering exist. Correlations can be removed for certain frequency components and not for others without having to act in the frequency domain.

There are many different bases of classifying filters and these overlap in many different ways; there is no simple hierarchical classification. Filters may be:

- linear or non-linear
- time-invariant or time-variant, also known as shift invariance. If the filter operates in a spatial domain then the characterization is space invariance.
- causal or not-causal: A filter is non-causal if its present output depends on future input. Filters processing time-domain signals in real-time must be causal, but not filters acting on spacial domain signals or deferred-time processing of time-domain signals.
- analog or digital
- discrete-time (sampled) or continuous-time
- passive or active type of continuous-time filter
- infinite impulse response (IIR) or finite impulse response (FIR) type of discrete-time or digital filter.

Linear Continuous-time Filters

Linear continuous-time circuit is perhaps the most common meaning for filter in the signal processing world, and simply "filter" is often taken to be synonymous. These circuits are generally designed to remove certain frequencies and allow others to pass. Circuits that perform this function are generally linear in their response, or at least approximately so. Any nonlinearity would potentially result in the output signal containing frequency components not present in the input signal.

The modern design methodology for linear continuous-time filters is called network synthesis. Some important filter families designed in this way are:

- Chebyshev filter, has the best approximation to the ideal response of any filter for a specified order and ripple.
- Butterworth filter, has a maximally flat frequency response.
- Bessel filter, has a maximally flat phase delay.
- Elliptic filter, has the steepest cutoff of any filter for a specified order and ripple.

The difference between these filter families is that they all use a different polynomial function to approximate to the ideal filter response. This results in each having a different transfer function.

Another older, less-used methodology is the image parameter method. Filters designed by this methodology are archaically called "wave filters". Some important filters designed by this method are:

- Constant k filter, the original and simplest form of wave filter.
- m-derived filter, a modification of the constant k with improved cutoff steepness and impedance matching.

Terminology

Some terms used to describe and classify linear filters:

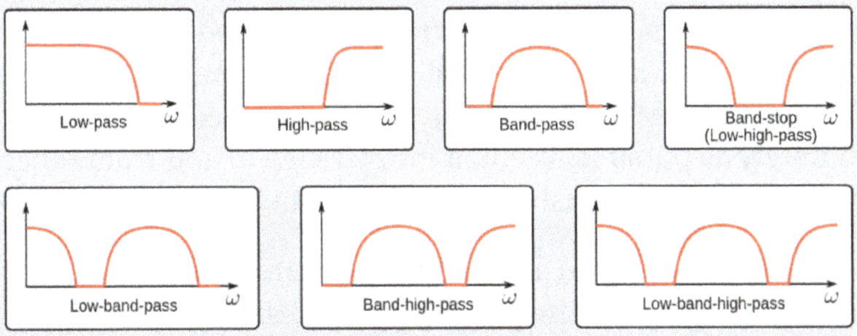

- The frequency response can be classified into a number of different bandforms describing which frequency bands the filter passes (the passband) and which it rejects (the stopband):
    - Low-pass filter – low frequencies are passed, high frequencies are attenuated.
    - High-pass filter – high frequencies are passed, low frequencies are attenuated.
    - Band-pass filter – only frequencies in a frequency band are passed.
    - Band-stop filter or band-reject filter – only frequencies in a frequency band are attenuated.
    - Notch filter – rejects just one specific frequency - an extreme band-stop filter.
    - Comb filter – has multiple regularly spaced narrow passbands giving the bandform the appearance of a comb.
    - All-pass filter – all frequencies are passed, but the phase of the output is modified.
- Cutoff frequency is the frequency beyond which the filter will not pass signals. It is usually measured at a specific attenuation such as 3 dB.
- Roll-off is the rate at which attenuation increases beyond the cut-off frequency.
- Transition band, the (usually narrow) band of frequencies between a passband and stopband.
- Ripple is the variation of the filter's insertion loss in the passband.
- The order of a filter is the degree of the approximating polynomial and in passive filters corresponds to the number of elements required to build it. Increasing order increases roll-off and brings the filter closer to the ideal response.

One important application of filters is in telecommunication. Many telecommunication systems use frequency-division multiplexing, where the system designers divide a wide frequency band into many narrower frequency bands called "slots" or "channels", and each stream of information is allocated one of those channels. The people who design the filters at each transmitter and each receiver try to balance passing the desired signal through as accurately as possible, keeping interference to and from other cooperating transmitters and noise sources outside the system as low as possible, at reasonable cost.

Multilevel and multiphase digital modulation systems require filters that have flat phase delay—are linear phase in the passband—to preserve pulse integrity in the time domain, giving less intersymbol interference than other kinds of filters.

On the other hand, analog audio systems using analog transmission can tolerate much larger ripples in phase delay, and so designers of such systems often deliberately sacrifice linear phase to get filters that are better in other ways—better stop-band rejection, lower passband amplitude ripple, lower cost, etc.

Technologies

Filters can be built in a number of different technologies. The same transfer function can be realised in several different ways, that is the mathematical properties of the filter are the same but the physical properties are quite different. Often the components in different technologies are directly analogous to each other and fulfill the same role in their respective filters. For instance, the resistors, inductors and capacitors of electronics correspond respectively to dampers, masses and springs in mechanics. Likewise, there are corresponding components in distributed element filters.

- Electronic filters were originally entirely passive consisting of resistance, inductance and capacitance. Active technology makes design easier and opens up new possibilities in filter specifications.
- Digital filters operate on signals represented in digital form. The essence of a digital filter is that it directly implements a mathematical algorithm, corresponding to the desired filter transfer function, in its programming or microcode.
- Mechanical filters are built out of mechanical components. In the vast majority of cases they are used to process an electronic signal and transducers are provided to convert this to and from a mechanical vibration. However, examples do exist of filters that have been designed for operation entirely in the mechanical domain.
- Distributed element filters are constructed out of components made from small pieces of transmission line or other distributed elements. There are structures in distributed element filters that directly correspond to the lumped elements of electronic filters, and others that are unique to this class of technology.
- Waveguide filters consist of waveguide components or components inserted in the waveguide. Waveguides are a class of transmission line and many structures of distributed element filters, for instance the stub (electronics), can also be implemented in waveguides.
- Crystal filters use quartz crystals as resonators, or some other piezoelectric material.
- Acoustic filters
- Optical filters were originally developed for purposes other than signal process-

ing such as lighting and photography. With the rise of optical fiber technology, however, optical filters increasingly find signal processing applications and signal processing filter terminology, such as longpass and shortpass, are entering the field.

The Transfer Function

The transfer function of a filter is most often defined in the domain of the complex frequencies. The back and forth passage to/from this domain is operated by the Laplace transform and its inverse (therefore, here below, the term "input signal" shall be understood as "the Laplace transform of" (the time representation of) the input signal, and so on).

The transfer function $H(s)$ of a filter is the ratio of the output signal $Y(s)$ to that of the input signal $X(s)$ as a function of the complex frequency s:

$$H(s) = \frac{Y(s)}{X(s)}$$

with $s = \sigma + j\omega$.

The transfer function of all linear time-invariant filters generally share certain characteristics:

- For filters which are constructed of discrete components, their transfer function must be the ratio of two polynomials in s, i.e. a rational function of s. The order of the transfer function will be the highest power of s encountered in either the numerator or the denominator.
- The polynomials of the transfer function will all have real coefficients. Therefore, the poles and zeroes of the transfer function will either be real or occur in complex conjugate pairs.
- Since the filters are assumed to be stable, the real part of all poles (i.e. zeroes of the denominator) will be negative, i.e. they will lie in the left half-plane in complex frequency space.

Distributed element filters do not, in general, produce rational functions but can often approximate to them.

The proper construction of a transfer function involves the Laplace transform, and therefore it is needed to assume null initial conditions, because

$$\mathcal{L}\left\{\frac{df}{dt}\right\} = s \cdot \mathcal{L}\{f(t)\} - f(0),$$

And when f(0)=0 we can get rid of the constants and use the usual expression

$$\mathcal{L}\left\{\frac{df}{dt}\right\} = s \cdot \mathcal{L}\{f(t)\}$$

An alternative to transfer functions is to give the behavior of the filter as a convolution. The convolution theorem, which holds for Laplace transforms, guarantees equivalence with transfer functions.

Classification

Filters may be specified by family and bandform. A filter's family is specified by the approximating polynomial used and each leads to certain characteristics of the transfer function of the filter. Some common filter families and their particular characteristics are:

- Butterworth filter – no gain ripple in pass band and stop band, slow cutoff
- Chebyshev filter (Type I) – no gain ripple in stop band, moderate cutoff
- Chebyshev filter (Type II) – no gain ripple in pass band, moderate cutoff
- Bessel filter – no group delay ripple, no gain ripple in both bands, slow gain cutoff
- Elliptic filter – gain ripple in pass and stop band, fast cutoff
- Optimum "L" filter
- Gaussian filter – no ripple in response to step function
- Hourglass filter
- Raised-cosine filter

Each family of filters can be specified to a particular order. The higher the order, the more the filter will approach the "ideal" filter; but also the longer the impulse response is and the longer the latency will be. An ideal filter has full transmission in the pass band, complete attenuation in the stop band, and an abrupt transition between the two bands, but this filter has infinite order (i.e., the response cannot be expressed as a linear differential equation with a finite sum) and infinite latency (i.e., its compact support in the Fourier transform forces its time response to be ever lasting).

Here is an image comparing Butterworth, Chebyshev, and elliptic filters. The filters in this illustration are all fifth-order low-pass filters. The particular implementation – analog or digital, passive or active – makes no difference; their output would be the same.

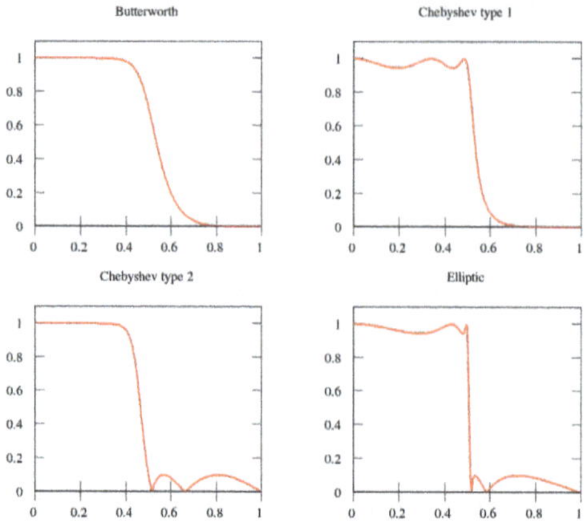

As is clear from the image, elliptic filters are sharper than all the others, but they show ripples on the whole bandwidth.

Any family can be used to implement a particular bandform of which frequencies are transmitted, and which, outside the passband, are more or less attenuated. The transfer function completely specifies the behavior of a linear filter, but not the particular technology used to implement it. In other words, there are a number of different ways of achieving a particular transfer function when designing a circuit. A particular bandform of filter can be obtained by transformation of a prototype filter of that family.

Impedance Matching

Impedance matching structures invariably take on the form of a filter, that is, a network of non-dissipative elements. For instance, in a passive electronics implementation, it would likely take the form of a ladder topology of inductors and capacitors. The design of matching networks shares much in common with filters and the design invariably will have a filtering action as an incidental consequence. Although the prime purpose of a matching network is not to filter, it is often the case that both functions are combined in the same circuit. The need for impedance matching does not arise while signals are in the digital domain.

Similar comments can be made regarding power dividers and directional couplers. When implemented in a distributed element format, these devices can take the form of a distributed element filter. There are four ports to be matched and widening the bandwidth requires filter-like structures to achieve this. The inverse is also true: distributed element filters can take the form of coupled lines.

Filter Design

Filter design is the process of designing a signal processing filter that satisfies a set of requirements, some of which are contradictory. The purpose is to find a realization of the filter that meets each of the requirements to a sufficient degree to make it useful.

The filter design process can be described as an optimization problem where each requirement contributes to an error function which should be minimized. Certain parts of the design process can be automated, but normally an experienced electrical engineer is needed to get a good result.

Typical Design Requirements

Typical requirements which are considered in the design process are:

- The filter should have a specific frequency response
- The filter should have a specific phase shift or group delay
- The filter should have a specific impulse response
- The filter should be causal
- The filter should be stable
- The filter should be localized (pulse or step inputs should result in finite time outputs)
- The computational complexity of the filter should be low
- The filter should be implemented in particular hardware or software

The Frequency Function

An important parameter is the required frequency response. In particular, the steepness and complexity of the response curve is a deciding factor for the filter order and feasibility.

A first order recursive filter will only have a single frequency-dependent component. This means that the slope of the frequency response is limited to 6 dB per octave. For many purposes, this is not sufficient. To achieve steeper slopes, higher order filters are required.

In relation to the desired frequency function, there may also be an accompanying *weighting* function which describes, for each frequency, how important it is that the resulting frequency function approximates the desired one. The larger weight, the more important is a close approximation.

Typical examples of frequency function are:

- A low-pass filter is used to cut unwanted high-frequency signals.
- A high-pass filter passes high frequencies fairly well; it is helpful as a filter to cut any unwanted low frequency components.

- A band-pass filter passes a limited range of frequencies.
- A band-stop filter passes frequencies above and below a certain range. A very narrow band-stop filter is known as a notch filter.
- A differentiator has an amplitude response proportional to the frequency.
- A low-shelf filter passes all frequencies, but increases or reduces frequencies below the shelf frequency by specified amount.
- A high-shelf filter passes all frequencies, but increases or reduces frequencies above the shelf frequency by specified amount.
- A peak EQ filter makes a peak or a dip in the frequency response, commonly used in parametric equalizers.

Phase and Group Delay

- An all-pass filter passes through all frequencies unchanged, but changes the phase of the signal. Filters of this type can be used to equalize the group delay of recursive filters. This filter is also used in phaser effects.
- A Hilbert transformer is a specific all-pass filter that passes sinusoids with unchanged amplitude but shifts each sinusoid phase by ±90°.
- A fractional delay filter is an all-pass that has a specified and constant group or phase delay for all frequencies.

The Impulse Response

There is a direct correspondence between the filter's frequency function and its impulse response: the former is the Fourier transform of the latter. That means that any requirement on the frequency function is a requirement on the impulse response, and vice versa.

However, in certain applications it may be the filter's impulse response that is explicit and the design process then aims at producing as close an approximation as possible to the requested impulse response given all other requirements.

In some cases it may even be relevant to consider a frequency function and impulse response of the filter which are chosen independently from each other. For example, we may want both a specific frequency function of the filter *and* that the resulting filter have a small effective width in the signal domain as possible. The latter condition can be realized by considering a very narrow function as the wanted impulse response of the filter even though this function has no relation to the desired frequency function. The goal of the design process is then to realize a filter which tries to meet both these contradicting design goals as much as possible.

Causality

In order to be implementable, any time-dependent filter (operating in real time) must be causal: the filter response only depends on the current and past inputs. A standard approach is to leave this requirement until the final step. If the resulting filter is not causal, it can be made causal by introducing an appropriate time-shift (or delay). If the filter is a part of a larger system (which it normally is) these types of delays have to be introduced with care since they affect the operation of the entire system.

Filters that do not operate in real time (e.g. for image processing) can be non-causal. This e.g. allows the design of zero delay recursive filters, where the group delay of a causal filter is canceled by its Hermitian non-causal filter.

Stability

A stable filter assures that every limited input signal produces a limited filter response. A filter which does not meet this requirement may in some situations prove useless or even harmful. Certain design approaches can guarantee stability, for example by using only feed-forward circuits such as an FIR filter. On the other hand, filters based on feedback circuits have other advantages and may therefore be preferred, even if this class of filters includes unstable filters. In this case, the filters must be carefully designed in order to avoid instability.

Locality

In certain applications we have to deal with signals which contain components which can be described as local phenomena, for example pulses or steps, which have certain time duration. A consequence of applying a filter to a signal is, in intuitive terms, that the duration of the local phenomena is extended by the width of the filter. This implies that it is sometimes important to keep the width of the filter's impulse response function as short as possible.

According to the uncertainty relation of the Fourier transform, the product of the width of the filter's impulse response function and the width of its frequency function must exceed a certain constant. This means that any requirement on the filter's locality also implies a bound on its frequency function's width. Consequently, it may not be possible to simultaneously meet requirements on the locality of the filter's impulse response function as well as on its frequency function. This is a typical example of contradicting requirements.

Computational Complexity

A general desire in any design is that the number of operations (additions and multiplications) needed to compute the filter response is as low as possible. In certain applications, this desire is a strict requirement, for example due to limited computational resources, limited power resources, or limited time. The last limitation is typical in real-time applications.

There are several ways in which a filter can have different computational complexity. For example, the order of a filter is more or less proportional to the number of operations. This means that by choosing a low order filter, the computation time can be reduced.

For discrete filters the computational complexity is more or less proportional to the number of filter coefficients. If the filter has many coefficients, for example in the case of multidimensional signals such as tomography data, it may be relevant to reduce the number of coefficients by removing those which are sufficiently close to zero. In multirate filters, the number of coefficients by taking advantage of its bandwidth limits, where the input signal is downsampled (e.g. to its critical frequency), and upsampled after filtering.

Another issue related to computational complexity is separability, that is, if and how a filter can be written as a convolution of two or more simpler filters. In particular, this issue is of importance for multidimensional filters, e.g., 2D filter which are used in image processing. In this case, a significant reduction in computational complexity can be obtained if the filter can be separated as the convolution of one 1D filter in the horizontal direction and one 1D filter in the vertical direction. A result of the filter design process may, e.g., be to approximate some desired filter as a separable filter or as a sum of separable filters.

Other Considerations

It must also be decided how the filter is going to be implemented:

- Analog filter
- Analog sampled filter
- Digital filter
- Mechanical filter

Analog Filters

The design of linear analog filters is for the most part covered in the linear filter section.

Digital Filters

Digital filters are classified into one of two basic forms, according to how they respond to a unit impulse:

- Finite impulse response, or FIR, filters express each output sample as a weighted sum of the last N input samples, where N is the order of the filter. FIR filters are normally non-recursive, meaning they do not use feedback and as such are inherently stable. A moving average filter or CIC filter are examples of FIR filters that are normally recursive (that use feedback). If the FIR coefficients are symmetrical (often the case), then such a filter is linear phase, so it delays signals of all frequencies equally which is important in many applications. It is also straightforward to avoid overflow in an FIR filter. The main disadvantage is that they may require significantly more processing and memory resources than cleverly designed IIR variants. FIR filters are generally easier to design than IIR filters - the Parks-McClellan filter design algorithm (based on the Remez algorithm) is one suitable method for designing quite good filters semi-automatically.

- Infinite impulse response, or IIR, filters are the digital counterpart to analog filters. Such a filter contains internal state, and the output and the next internal state are determined by a linear combination of the previous inputs and outputs (in other words, they use feedback, which FIR filters normally do not). In theory, the impulse response of such a filter never dies out completely, hence the name IIR, though in practice, this is not true given the finite resolution of computer arithmetic. IIR filters normally require less computing resources than an FIR filter of similar performance. However, due to the feedback, high order IIR filters may have problems with instability, arithmetic overflow, and limit cycles, and require careful design to avoid such pitfalls. Additionally, since the phase shift is inherently a non-linear function of frequency, the time delay through such a filter is frequency-dependent, which can be a problem in many situations. 2nd order IIR filters are often called 'biquads' and a common implementation of higher order filters is to cascade biquads. A useful reference for computing biquad coefficients is the RBJ Audio EQ Cookbook.

Sample Rate

Unless the sample rate is fixed by some outside constraint, selecting a suitable sample rate is an important design decision. A high rate will require more in terms of computational resources, but less in terms of anti-aliasing filters. Interference and beating with other signals in the system may also be an issue.

Anti-aliasing

For any digital filter design, it is crucial to analyze and avoid aliasing effects. Often, this is done by adding analog anti-aliasing filters at the input and output, thus avoiding any frequency component above the Nyquist frequency. The complexity (i.e., steepness) of such filters depends on the required signal to noise ratio and the ratio between the sampling rate and the highest frequency of the signal.

Theoretical Basis

Parts of the design problem relate to the fact that certain requirements are described in the frequency domain while others are expressed in the signal domain and that these may contradict. For example, it is not possible to obtain a filter which has both an arbitrary impulse response and arbitrary frequency function. Other effects which refer to relations between the signal and frequency domain are

- The uncertainty principle between the signal and frequency domains
- The variance extension theorem
- The asymptotic behaviour of one domain versus discontinuities in the other

The Uncertainty Principle

As stated in the uncertainty principle, the product of the width of the frequency function and the width of the impulse response cannot be smaller than a specific constant. This implies that if a specific frequency function is requested, corresponding to a specific frequency width, the minimum width of the filter in the signal domain is set. Vice versa, if the maximum width of the response is given, this determines the smallest possible width in the frequency. This is a typical example of contradictory requirements where the filter design process may try to find a useful compromise.

The Variance Extension Theorem

Let σ_s^2 be the variance of the input signal and let σ_f^2 be the variance of the filter. The variance of the filter response, σ_r^2, is then given by

$$\sigma_r^2 = \sigma_s^2 + \sigma_f^2$$

This means that $\sigma_r > \sigma_f$ and implies that the localization of various features such as pulses or steps in the filter response is limited by the filter width in the signal domain. If a precise localization is requested, we need a filter of small width in the signal domain and, via the uncertainty principle, its width in the frequency domain cannot be arbitrary small.

Discontinuities Versus Asymptotic Behaviour

Let $f(t)$ be a function and let $F(\omega)$ be its Fourier transform. There is a theorem which states that if the first derivative of F which is discontinuous has order $n \geq 0$, then f has an asymptotic decay like t^{-n-1}.

A consequence of this theorem is that the frequency function of a filter should be as smooth as possible to allow its impulse response to have a fast decay, and thereby a short width.

Methodology

One common method for designing FIR filters is the Parks-McClellan filter design algorithm, based on the Remez exchange algorithm. Here the user specifies a desired frequency response, a weighting function for errors from this response, and a filter order N. The algorithm then finds the set of N coefficients that minimize the maximum deviation from the ideal. Intuitively, this finds the filter that is as close as you can get to the desired response given that you can use only N coefficients. This method is particularly easy in practice and at least one text includes a program that takes the desired filter and N and returns the optimum coefficients. One possible drawback to filters designed this way is that they contain many small ripples in the passband(s), since such a filter minimizes the peak error.

Another method to finding a discrete FIR filter is *filter optimization* described in Knutsson et al., which minimizes the integral of the square of the error, instead of its maximum value. In its basic form this approach requires that an ideal frequency function of the filter $F_I(\omega)$ is specified together with a frequency weighting function $W(\omega)$ and set of coordinates x_k in the signal domain where the filter coefficients are located.

An error function ε is defined as

$$\varepsilon = \| W \cdot (F_I - \mathcal{F}\{f\}) \|^2$$

where $f(x)$ is the discrete filter and $\mathcal{F}$ is the discrete-time Fourier transform defined on the specified set of coordinates. The norm used here is, formally, the usual norm on L^2 spaces. This means that ε measures the deviation between the requested frequency function of the filter, , and the actual frequency function of the realized filter, F_I. However, the deviation is also subject to the weighting function W before the error function is computed.

Once the error function is established, the optimal filter is given by the coefficients $f(x)$ which minimize . This can be done by solving the corresponding least squares problem. In practice, the L^2 norm has to be approximated by means of a suitable sum over discrete points in the frequency domain. In general, however, these points should be significantly more than the number of coefficients in the signal domain to obtain a useful approximation.

Simultaneous Optimization in Both Domains

The previous method can be extended to include an additional error term related to a desired filter impulse response in the signal domain, with a corresponding weighting function. The ideal impulse response can be chosen independently of the ideal frequency function and is in practice used to limit the effective width and to remove ringing effects of the resulting filter in the signal domain. This is done by choosing a narrow ideal filter impulse response function, e.g., an impulse, and a weighting function which grows fast with the distance from the origin, e.g., the distance squared. The optimal filter can still be calculated by solving a simple least squares problem and the resulting filter is then a "compromise" which has a total optimal fit to the ideal functions in both domains. An important parameter is the relative strength of the two weighting functions which determines in which domain it is more important to have a good fit relative to the ideal function.

Adaptive Filter

An adaptive filter is a system with a linear filter that has a transfer function controlled by variable parameters and a means to adjust those parameters according to an optimization algorithm. Because of the complexity of the optimization algorithms, almost all adaptive filters are digital filters. Adaptive filters are required for some applications because some parameters of the desired processing operation (for instance, the locations of reflective surfaces in a reverberant space) are not known in advance or are changing. The closed loop adaptive filter uses feedback in the form of an error signal to refine its transfer function.

Generally speaking, the closed loop adaptive process involves the use of a cost function, which is a criterion for optimum performance of the filter, to feed an algorithm, which determines how to modify filter transfer function to minimize the cost on the next iteration. The most common cost function is the mean square of the error signal.

As the power of digital signal processors has increased, adaptive filters have become much more common and are now routinely used in devices such as mobile phones and other communication devices, camcorders and digital cameras, and medical monitoring equipment.

Example Application

The recording of a heart beat (an ECG), may be corrupted by noise from the AC mains. The exact frequency of the power and its harmonics may vary from moment to moment.

One way to remove the noise is to filter the signal with a notch filter at the mains frequency and its vicinity, which could excessively degrade the quality of the ECG since the heart beat would also likely have frequency components in the rejected range.

To circumvent this potential loss of information, an adaptive filter could be used. The adaptive filter would take input both from the patient and from the mains and would thus be able to track the actual frequency of the noise as it fluctuates and subtract the noise from the recording. Such an adaptive technique generally allows for a filter with a smaller rejection range, which means, in this case, that the quality of the output signal is more accurate for medical purposes.

Block Diagram

The idea behind a closed loop adaptive filter is that a variable filter is adjusted until the error (the difference between the filter output and the desired signal) is minimized. The Least Mean Squares (LMS) filter and the Recursive Least Squares (RLS) filter are types of adaptive filter.

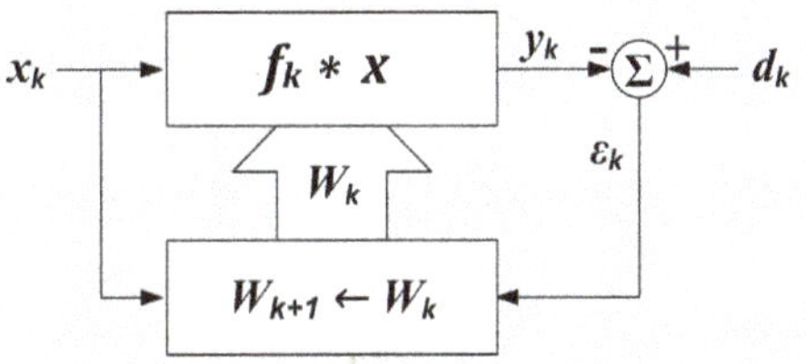

Adaptive Filter. k = sample number, x = reference input, X = set of recent values of x, d = desired input, W = set of filter coefficients, ε = error output, f = filter impulse response, * = convolution, Σ = summation, upper box=linear filter, lower box=adaption algorithm

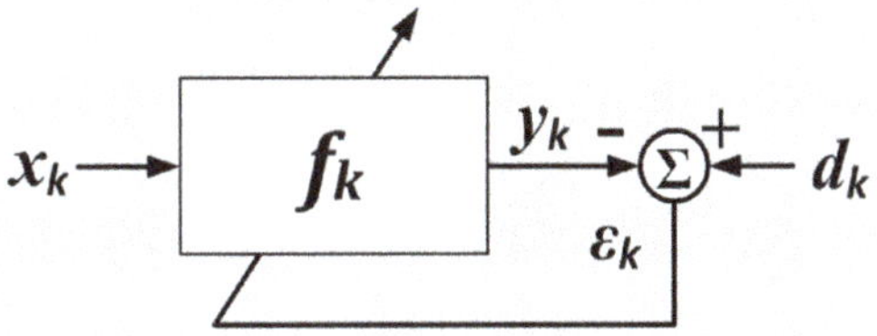

Adaptive Filter, compact representation. k = sample number, x = reference input, d = desired input, ε = error output, f = filter impulse response, Σ = summation, box=linear filter and adaption algorithm.

There are two input signals to the adaptive filter: d_k and x_k which are sometimes called the *primary input* and the *reference input* respectively.

d_k which includes the desired signal plus undesired interference and

x_k which includes the signals that are correlated to some of the undesired interference in d_k.

k represents the discrete sample number.

The filter is controlled by a set of L+1 coefficients or weights.

$\mathbf{W}_k = \left[w_{0k}, w_{1k}, ..., w_{Lk}\right]^T$ represents the set or vector of weights, which control the filter at sample time k.

where w_{lk} refers to the l'th weight at k'th time.

$\Delta\mathbf{W}_k$ represents the change in the weights that occurs as a result of adjustments computed at sample time k.

These changes will be applied after sample time k and before they are used at sample time k+1.

The output is usually ϵ_k but it could be y_k or it could even be the filter coefficients. (Widrow)

The input signals are defined as follows:

$$d_k = g_k + u_k + v_k$$

$$x_k = g_k' + u_k' + v_k'$$

where:

g = the desired signal,

g' = a signal that is correlated with the desired signal g ,

u = an undesired signal that is added to g , but not correlated with g or g'

u' = a signal that is correlated with the undesired signal u ,but not correlated with g or g' ,

v = an undesired signal (typically random noise) not correlated with g , g' , u, u' or v' ,

v' = an undesired signal (typically random noise) not correlated with g , g' , u, u' or v.

The output signals are defined as follows:

$$y_k = \hat{g}_k + \hat{u}_k + \hat{v}_k$$

$$\epsilon_k = d_k - y_k..$$

where:

$\hat{g}$ = the output of the filter if the input was only g' ,

$\hat{u}$ = the output of the filter if the input was only u' ,

$\hat{v}$ = the output of the filter if the input was only v' .

Tapped Delay Line FIR Filter

If the variable filter has a tapped delay line Finite Impulse Response (FIR) structure, then the impulse response is equal to the filter coefficients. The output of the filter is given by

$$y_k = \sum_{l=0}^{L} w_{lk}\, x_{(k-l)} = \hat{g}_k + \hat{u}_k + \hat{v}_k$$

where w_{lk} refers to the l'th weight at k'th time.

Ideal Case

In the ideal case $v \equiv 0, v' \equiv 0, g' \equiv 0$. All the undesired signals in d_k are represented by u_k. x_k consists entirely of a signal correlated with the undesired signal in u_k.

The output of the variable filter in the ideal case is

$$y_k = \hat{u}_k.$$

The error signal or cost function is the difference between d_k and y_k

$\epsilon_k = d_k - y_k = g_k + u_k - \hat{u}_k$.. The desired signal g_k passes through without being changed.

The error signal ϵ_k is minimized in the mean square sense when $[u_k - \hat{u}_k]$ is minimized. In other words, $\hat{u}_k$ is the best mean square estimate of u_k. In the ideal case, $u_k = \hat{u}_k$ and $\epsilon_k = g_k$, , and all that is left after the subtraction is g which is the unchanged desired signal with all undesired signals removed.

Signal Components in the Reference Input

In some situations, the reference input x_k includes components of the desired signal. This means g' ≠ 0.

Perfect cancelation of the undesired interference is not possible in the case, but improvement of the signal to interference ratio is possible. The output will be

$\epsilon_k = d_k - y_k = g_k - \hat{g}_k + u_k - \hat{u}_k$.. The desired signal will be modified (usually decreased).

The output signal to interference ratio has a simple formula referred to as *power inversion*.

$$\rho_{\text{out}}(z) = \frac{1}{\rho_{\text{ref}}(z)}.$$

where

$\rho_{\text{out}}(z)$ = output signal to interference ratio.

$\rho_{\text{ref}}(z)$ = reference signal to interference ratio.

z = frequency in the z-domain.

This formula means that the output signal to interference ratio at a particular frequency is the reciprocal of the reference signal to interference ratio.

Example: A fast food restaurant has a drive-up window. Before getting to the window, customers place their order by speaking into a microphone. The microphone also picks up noise from the engine and the environment. This microphone provides the primary signal. The signal power from the customer's voice and the noise power from the engine are equal. It is difficult for the employees in the restaurant to understand the customer. To reduce the amount of interference in the primary microphone, a second microphone is located where it is intended to pick up sounds from the engine. It also picks up the customer's voice. This microphone is the source of the reference signal. In this case, the engine noise is 50 times more powerful than the customer's voice. Once the canceler has converged, the primary signal to interference ratio will be improved from 1:1 to 50:1.

Adaptive Linear Combiner

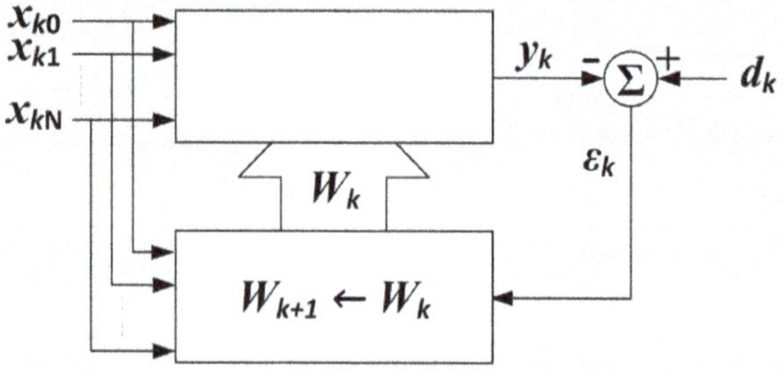

Adaptive linear combiner showing the combiner and the adaption process. k = sample number, n=input variable index, x = reference inputs, d = desired input, W = set of filter coefficients, ε = error output, Σ = summation, upper box=linear combiner, lower box=adaption algorithm.

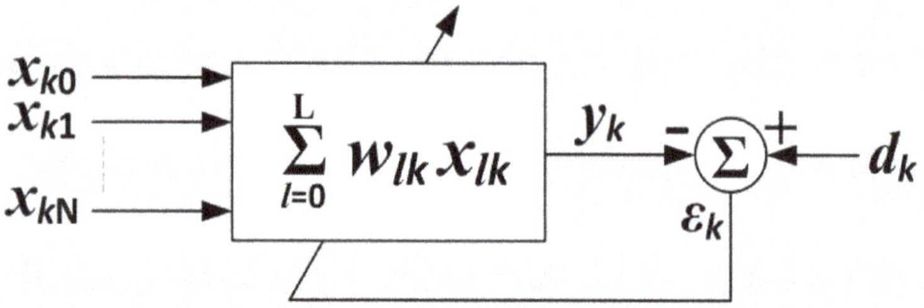

Adaptive linear combiner, compact representation. k = sample number, n=input variable index, x = reference inputs, d = desired input, ε = error output, Σ = summation.

The adaptive linear combiner (ALC) resembles the adaptive tapped delay line FIR filter except that there is no assumed relationship between the X values. If the X values were from the outputs of a tapped delay line, then the combination of tapped delay line and ALC would comprise an adaptive filter. However, the X values could be the values of an array of pixels. Or they could be the outputs of multiple tapped delay lines. The ALC finds use as an adaptive beam former for arrays of hydrophones or antennas.

$$y_k = \sum_{l=0}^{L} w_{lk}\, x_{lk} = \mathbf{W}_k^T \mathbf{x}_k$$

where w_{lk} refers to the l'th weight at k'th time.

LMS Algorithm

If the variable filter has a tapped delay line FIR structure, then the LMS update algorithm is especially simple. Typically, after each sample, the coefficients of the FIR filter are adjusted as follows:(Widrow)

for $l = 0 \ldots L$

$$w_{l,k+1} = w_{lk} + 2\mu \, \epsilon_k \, x_{k-l}$$

μ is called the *convergence factor.*

The LMS algorithm does not require that the X values have any particular relationship; therefor it can be used to adapt a linear combiner as well as an FIR filter. In this case the update formula is written as:

$$w_{l,k+1} = w_{lk} + 2\mu \, \epsilon_k \, x_{lk}$$

The effect of the LMS algorithm is at each time, k, to make a small change in each weight. The direction of the change is such that it would decrease the error if it had been applied at time k. The magnitude of the change in each weight depends on μ, the associated X value and the error at time k. The weights making the largest contribution to the output, y_k, are changed the most. If the error is zero, then there should be no change in the weights. If the associated value of X is zero, then changing the weight makes no difference, so it is not changed.

Convergence

μ controls how fast and how well the algorithm converges to the optimum filter coefficients. If μ is too large, the algorithm will not converge. If μ is too small the algorithm converges slowly and may not be able to track changing conditions. If μ is large but not too large to prevent convergence, the algorithm reaches steady state rapidly but continuously overshoots the optimum weight vector. Sometimes, μ is made large at first for rapid convergence and then decreased to minimize overshoot.

Widrow and Stearns state in 1985 that they have no knowledge of a proof that the LMS algorithm will converge in all cases.

However under certain assumptions about stationarity and independence it can be shown that the algorithm will converge if

$$0 < \mu < \frac{1}{\sigma^2}$$

where

$$\sigma^2 = \sum_{l=0}^{L} \sigma_l^2 = \text{sum of all input power}$$

σ_l is the RMS value of the l'th input

In the case of the tapped delay line filter, each input has the same RMS value because they are simply the same values delayed. In this case the total power is

$$\sigma^2 = (L+1)\sigma_0^2$$

where

σ_0 is the RMS value of x_k, the input stream.

This leads to a normalized LMS algorithm:

$w_{l,k+1} = w_{lk} + \left(\frac{2\mu_\sigma}{\sigma^2}\right)\epsilon_k\, x_{k-l}$ in which case the convergence criteria becomes: $0 < \mu_\sigma < 1$.

All-pass Filter

An all-pass filter is a signal processing filter that passes all frequencies equally in gain, but changes the phase relationship among various frequencies. It does this by varying its phase shift as a function of frequency. Generally, the filter is described by the frequency at which the phase shift crosses 90° (i.e., when the input and output signals go into quadrature – when there is a quarter wavelength of delay between them).

They are generally used to compensate for other undesired phase shifts that arise in the system, or for mixing with an unshifted version of the original to implement a notch comb filter.

They may also be used to convert a mixed phase filter into a minimum phase filter with an equivalent magnitude response or an unstable filter into a stable filter with an equivalent magnitude response.

Active Analog Implementation

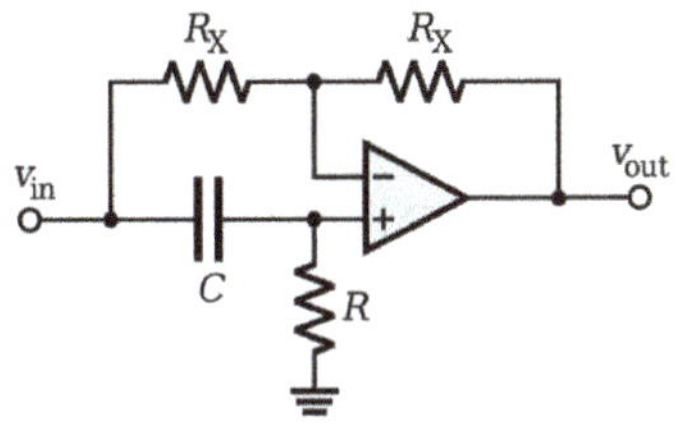

Figure 1: Schematic of an op amp all-pass filter

The operational amplifier circuit shown in Figure 1 implements an active all-pass filter with the transfer function

$$H(s) = \frac{sRC - 1}{sRC + 1},$$

which has one pole at -1/RC and one zero at 1/RC (i.e., they are *reflections* of each other across the imaginary axis of the complex plane). The magnitude and phase of H(iω) for some angular frequency ω are

$$|H(i\omega)| = 1 \quad \text{and} \quad \angle H(i\omega) = 180^\circ - 2\arctan(\omega RC).$$

As expected, the filter has unity-gain magnitude for all ω. The filter introduces a different delay at each frequency and reaches input-to-output *quadrature* at ω=1/RC (i.e., phase shift is 90 degrees).

This implementation uses a high-pass filter at the non-inverting input to generate the phase shift and negative feedback to compensate for the filter's attenuation.

- At high frequencies, the capacitor is a short circuit, thereby creating a unity-gain voltage buffer (i.e., no phase shift).
- At low frequencies and DC, the capacitor is an open circuit and the circuit is an inverting amplifier (i.e., +180 degree phase shift) with unity gain.
- At the corner frequency ω=1/RC of the high-pass filter (i.e., when input frequency is 1/(2πRC)), the circuit introduces a +90 degree shift (i.e., output is in quadrature with input; it is delayed by a quarter wavelength).

In fact, the phase shift of the all-pass filter is double the phase shift of the high-pass filter at its non-inverting input.

Interpretation as a Padé Approximation to a Pure Delay

The Laplace transform of a pure delay is given by

$$\exp\{-sT\},$$

where T is the delay (in seconds) and $s \in \mathbb{C}$ is complex frequency. This can be approximated using a Padé approximant, as follows:

$$\exp\{-sT\} = \frac{\exp\{-sT/2\}}{\exp\{sT/2\}} \approx \frac{1 - sT/2}{1 + sT/2},$$

where the last step was achieved via a first-order Taylor series expansion of the numerator and denominator. By setting $RC = T/2$ we recover $H(s)$ from above.

Implementation Using Low-pass Filter

A similar all-pass filter can be implemented by interchanging the position of the resistor and capacitor, which turns the high-pass filter into a low-pass filter. The result is a phase shifter with the same quadrature frequency but a −180 degree shift at high frequencies and no shift at low frequencies. In other words, the transfer function is negated, and so it has the same pole at −1/RC and reflected zero at 1/RC. Again, the phase shift of the all-pass filter is double the phase shift of the first-order filter at its non-inverting input.

Voltage Controlled Implementation

The resistor can be replaced with a FET in its *ohmic mode* to implement a voltage-controlled phase shifter; the voltage on the gate adjusts the phase shift. In electronic music, a phaser typically consists of two, four or six of these phase-shifting sections connected in tandem and summed with the original. A low-frequency oscillator (LFO) ramps the control voltage to produce the characteristic swooshing sound.

General Usage

These circuits are used as phase shifters and in systems of phase shaping and time delay. Filters such as the above can be cascaded with unstable or mixed-phase filters to create a stable or minimum-phase filter without changing the magnitude response of the system. For example, by proper choice of pole (and therefore zero), a pole of an unstable system that is in the right-hand plane can be canceled and reflected on the left-hand plane.

Passive Analog Implementation

The benefit to implementing all-pass filters with active components like operational amplifiers is that they do not require inductors, which are bulky and costly in integrated circuit designs. In other applications where inductors are readily available, all-pass filters can be implemented entirely without active components. There are a number of circuit topologies that can be used for this. The following are the most commonly used circuits.

Lattice Filter

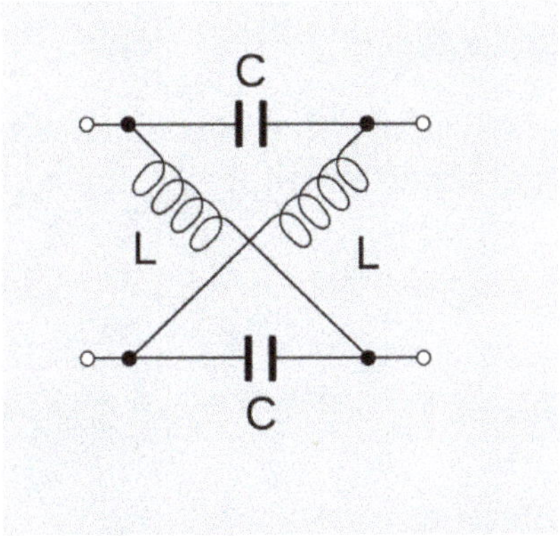

An all-pass filter using lattice topology

The lattice phase equaliser, or filter, is a filter composed of lattice, or X-sections. With single element branches it can produce a phase shift up to 180°, and with resonant branches it can produce phase shifts up to 360°. The filter is an example of a constant-resistance network (i.e., its image impedance is constant over all frequencies).

T-section Filter

The phase equaliser based on T topology is the unbalanced equivalent of the lattice filter and has the same phase response. While the circuit diagram may look like a low pass filter it is different in that the two inductor branches are mutually coupled. This results in transformer action between the two inductors and an all-pass response even at high frequency.

Bridged T-section Filter

The bridged T topology is used for delay equalisation, particularly the differential delay between two landlines being used for stereophonic sound broadcasts. This application requires that the filter has a linear phase response with frequency (i.e., constant group delay) over a wide bandwidth and is the reason for choosing this topology.

Digital Implementation

A Z-transform implementation of an all-pass filter with a complex pole at z_0 is

$$H(z) = \frac{z^{-1} - \overline{z_0}}{1 - z_0 z^{-1}}$$

which has a zero at $1/\overline{z_0}$, where $\overline{z}$ denotes the complex conjugate. The pole and zero sit at the same angle but have reciprocal magnitudes (i.e., they are *reflections* of each other across the boundary of the complex unit circle). The placement of this pole-zero pair for a given z_0 can be rotated in the complex plane by any angle and retain its all-pass magnitude characteristic. Complex pole-zero pairs in all-pass filters help control the frequency where phase shifts occur.

To create an all-pass implementation with real coefficients, the complex all-pass filter can be cascaded with an all-pass that substitutes $\overline{z_0}$ for z_0, leading to the Z-transform implementation

$$H(z) = \frac{z^{-1} - \overline{z_0}}{1 - z_0 z^{-1}} \times \frac{z^{-1} - z_0}{1 - \overline{z_0} z^{-1}} = \frac{z^{-2} - 2\Re(z_0) z^{-1} + |z_0|^2}{1 - 2\Re(z_0) z^{-1} + |z_0|^2 z^{-2}},$$

which is equivalent to the difference equation

$$y[k] - 2\Re(z_0) y[k-1] + |z_0|^2 y[k-2] = x[k-2] - 2\Re(z_0) x[k-1] + |z_0|^2 x[k],$$

where $y[k]$ is the output and $x[k]$ is the input at discrete time step k.

Filters such as the above can be cascaded with unstable or mixed-phase filters to create a stable or minimum-phase filter without changing the magnitude response of the system. For example, by proper choice of z_0, a pole of an unstable system that is outside of the unit circle can be canceled and reflected inside the unit circle.

Digital Filter

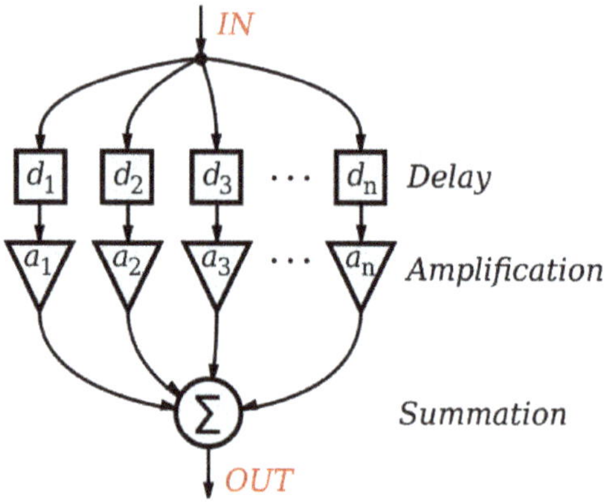

A general finite impulse response filter with n stages, each with an independent delay, d_i, and amplification gain, a_i.

In signal processing, a digital filter is a system that performs mathematical operations on a sampled, discrete-time signal to reduce or enhance certain aspects of that signal. This is in contrast to the other major type of electronic filter, the analog filter, which is an electronic circuit operating on continuous-time analog signals.

A digital filter system usually consists of an analog-to-digital converter to sample the input signal, followed by a microprocessor and some peripheral components such as memory to store data and filter coefficients etc. Finally a digital-to-analog converter to complete the output stage. Program Instructions (software) running on the microprocessor implement the digital filter by performing the necessary mathematical operations on the numbers received from the ADC. In some high performance applications, an FPGA or ASIC is used instead of a general purpose microprocessor, or a specialized DSP with specific paralleled architecture for expediting operations such as filtering.

Digital filters may be more expensive than an equivalent analog filter due to their increased complexity, but they make practical many designs that are impractical or impossible as analog filters. When used in the context of real-time analog systems, digital filters sometimes have problematic latency (the difference in time between the input and the response) due to the associated analog-to-digital and digital-to-analog conversions and anti-aliasing filters, or due to other delays in their implementation.

Digital filters are commonplace and an essential element of everyday electronics such as radios, cellphones, and AV receivers.

Characterization

A digital filter is characterized by its transfer function, or equivalently, its difference equation. Mathematical analysis of the transfer function can describe how it will respond to any input. As such, designing a filter consists of developing specifications appropriate to the problem (for example, a second-order low pass filter with a specific cutoff frequency), and then producing a transfer function which meets the specifications.

The transfer function for a linear, time-invariant, digital filter can be expressed as a transfer function in the Z-domain; if it is causal, then it has the form:

$$H(z) = \frac{B(z)}{A(z)} = \frac{b_0 + b_1 z^{-1} + b_2 z^{-2} + \cdots + b_N z^{-N}}{1 + a_1 z^{-1} + a_2 z^{-2} + \cdots + a_M z^{-M}}$$

where the order of the filter is the greater of N or M. See Z-transform's LCCD equation for further discussion of this transfer function.

This is the form for a recursive filter, which typically leads to an IIR infinite impulse response behaviour, but if the denominator is made equal to unity i.e. no feedback, then this becomes an FIR or finite impulse response filter.

Analysis Techniques

A variety of mathematical techniques may be employed to analyze the behaviour of a given digital filter. Many of these analysis techniques may also be employed in designs, and often form the basis of a filter specification.

Typically, one characterizes filters by calculating how they will respond to a simple input such as an impulse. One can then extend this information to compute the filter's response to more complex signals.

Impulse Response

The impulse response, often denoted $h[k]$ or h_k, is a measurement of how a filter will respond to the Kronecker delta function. For example, given a difference equation, one would set $x_0 = 1$ and $x_k = 0$ for $k \neq 0$ and evaluate. The impulse response is a characterization of the filter's behaviour. Digital filters are typically considered in two categories: infinite impulse response (IIR) and finite impulse response (FIR). In the case of linear time-invariant FIR filters, the impulse response is exactly equal to the sequence of filter coefficients:

$$y_n = \sum_{k=0}^{N} h_k x_{n-k}$$

IIR filters on the other hand are recursive, with the output depending on both current and previous inputs as well as previous outputs. The general form of an IIR filter is thus:

$$\sum_{m=0}^{M} a_m y_{n-m} = \sum_{k=0}^{N} b_k x_{n-k}$$

Plotting the impulse response will reveal how a filter will respond to a sudden, momentary disturbance.

Difference Equation

In discrete-time systems, the digital filter is often implemented by converting the transfer function to a linear constant-coefficient difference equation (LCCD) via the Z-transform. The discrete frequency-domain transfer function is written as the ratio of two polynomials. For example:

$$H(z) = \frac{(z+1)^2}{(z-\frac{1}{2})(z+\frac{3}{4})}$$

This is expanded:

$$H(z) = \frac{z^2 + 2z + 1}{z^2 + \frac{1}{4}z - \frac{3}{8}}$$

and to make the corresponding filter causal, the numerator and denominator are divided by the highest order of z:

$$H(z)=\frac{1+2z^{-1}+z^{-2}}{1+\frac{1}{4}z^{-1}-\frac{3}{8}z^{-2}}=\frac{Y(z)}{X(z)}$$

The coefficients of the denominator, a_k are the 'feed-backward' coefficients and the coefficients of the numerator are the 'feed-forward' coefficients, b_k. The resultant linear difference equation is:

$$y[n]=-\sum_{k=1}^{M}a_k y[n-k]+\sum_{k=0}^{N}b_k x[n-k]$$

or, for the example above:

$$\frac{Y(z)}{X(z)}=\frac{1+2z^{-1}+z^{-2}}{1+\frac{1}{4}z^{-1}-\frac{3}{8}z^{-2}}$$

rearranging terms:

$$\Rightarrow (1+\frac{1}{4}z^{-1}-\frac{3}{8}z^{-2})Y(z)=(1+2z^{-1}+z^{-2})X(z)$$

then by taking the inverse z-transform:

$$\Rightarrow y[n]+\frac{1}{4}y[n-1]-\frac{3}{8}y[n-2]=x[n]+2x[n-1]+x[n-2]$$

and finally, by solving for $y[n]$:

$$y[n]=-\frac{1}{4}y[n-1]+\frac{3}{8}y[n-2]+x[n]+2x[n-1]+x[n-2]$$

This equation shows how to compute the next output sample, $y[n]$, in terms of the past outputs, $y[n-p]$, the present input, $x[n]$, and the past inputs, $x[n-p]$.. Applying the filter to an input in this form is equivalent to a Direct Form I or II realization, depending on the exact order of evaluation.

Filter Design

The design of digital filters is a deceptively complex topic. Although filters are easily understood and calculated, the practical challenges of their design and implementation are significant and are the subject of much advanced research.

There are two categories of digital filter: the recursive filter and the nonrecursive filter. These are often referred to as infinite impulse response (IIR) filters and finite impulse response (FIR) filters, respectively.

Filter Realization

After a filter is designed, it must be *realized* by developing a signal flow diagram that describes the filter in terms of operations on sample sequences.

A given transfer function may be realized in many ways. Consider how a simple expression such as $ax+bx+c$ could be evaluated – one could also compute the equivalent $x(a+b)+c$. In the same way, all realizations may be seen as "factorizations" of the same transfer function, but different realizations will have different numerical properties. Specifically, some realizations are more efficient in terms of the number of operations or storage elements required for their implementation, and others provide advantages such as improved numerical stability and reduced round-off error. Some structures are better for fixed-point arithmetic and others may be better for floating-point arithmetic.

Direct Form I

A straightforward approach for IIR filter realization is direct form I, where the difference equation is evaluated directly. This form is practical for small filters, but may be inefficient and impractical (numerically unstable) for complex designs. In general, this form requires 2N delay elements (for both input and output signals) for a filter of order N.

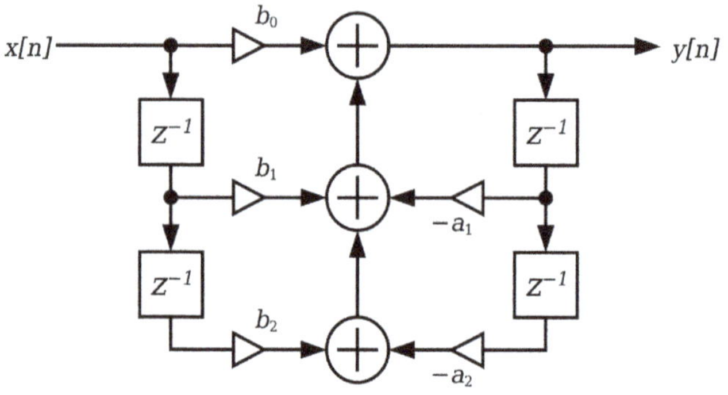

Direct Form II

The alternate direct form II only needs N delay units, where N is the order of the filter – potentially half as much as direct form I. This structure is obtained by reversing the order of the numerator and denominator sections of Direct Form I, since they are in fact two linear systems, and the commutativity property applies. Then, one will notice that there are two columns of delays (z^{-1}) that tap off the center net, and these can be combined since they are redundant, yielding the implementation as shown below.

The disadvantage is that direct form II increases the possibility of arithmetic overflow for filters of high Q or resonance. It has been shown that as Q increases, the round-off noise of both direct form topologies increases without bounds. This is because, conceptually, the signal is first passed through an all-pole filter (which normally boosts gain at the resonant frequencies) before the result of that is saturated, then passed through an all-zero filter (which often attenuates much of what the all-pole half amplifies).

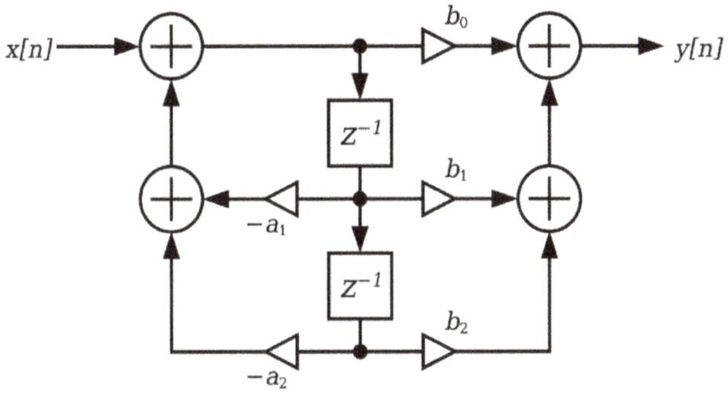

Cascaded Second-order Sections

A common strategy is to realize a higher-order (greater than 2) digital filter as a cascaded series of second-order "biquadratric" (or "biquad") sections. The advantage of this strategy is that the coefficient range is limited. Cascading direct form II sections results in N delay elements for filters of order N. Cascading direct form I sections results in $N + 2$ delay elements, since the delay elements of the input of any section (except the first section) are redundant with the delay elements of the output of the preceding section.

Other Forms

Other forms include:

- Direct form I and II transpose
- Series/cascade lower (typical second) order subsections
- Parallel lower (typical second) order subsections
    - Continued fraction expansion
- Lattice and ladder
    - One, two and three-multiply lattice forms
    - Three and four-multiply normalized ladder forms
    - ARMA structures
- State-space structures:

  - optimal (in the minimum noise sense): $(N+1)^2$ parameters
  - block-optimal and section-optimal: $4N-1$ parameters
  - input balanced with Givens rotation: $4N-1$ parameters
- Coupled forms: Gold Rader (normal), State Variable (Chamberlin), Kingsbury, Modified State Variable, Zölzer, Modified Zölzer
- Wave Digital Filters (WDF)
- Agarwal–Burrus (1AB and 2AB)
- Harris–Brooking
- ND-TDL
- Multifeedback
- Analog-inspired forms such as Sallen-key and state variable filters
- Systolic arrays

Comparison of Analog and Digital Filters

Digital filters are not subject to the component non-linearities that greatly complicate the design of analog filters. Analog filters consist of imperfect electronic components, whose values are specified to a limit tolerance (e.g. resistor values often have a tolerance of ±5%) and which may also change with temperature and drift with time. As the order of an analog filter increases, and thus its component count, the effect of variable component errors is greatly magnified. In digital filters, the coefficient values are stored in computer memory, making them far more stable and predictable.

Because the coefficients of digital filters are definite, they can be used to achieve much more complex and selective designs – specifically with digital filters, one can achieve a lower passband ripple, faster transition, and higher stopband attenuation than is practical with analog filters. Even if the design could be achieved using analog filters, the engineering cost of designing an equivalent digital filter would likely be much lower. Furthermore, one can readily modify the coefficients of a digital filter to make an adaptive filter or a user-controllable parametric filter. While these techniques are possible in an analog filter, they are again considerably more difficult.

Digital filters can be used in the design of finite impulse response filters. Analog filters do not have the same capability, because finite impulse response filters require delay elements.

Digital filters rely less on analog circuitry, potentially allowing for a better signal-to-noise ratio. A digital filter will introduce noise to a signal during analog low pass filtering, analog to digital conversion, digital to analog conversion and may introduce digital noise due to quantization. With analog filters, every component is a source of thermal noise (such as Johnson noise), so as the filter complexity grows, so does the noise.

However, digital filters do introduce a higher fundamental latency to the system. In an analog filter, latency is often negligible; strictly speaking it is the time for an electrical signal to propagate through the filter circuit. In digital systems, latency is introduced by delay elements in the digital signal path, and by analog-to-digital and digital-to-analog converters that enable the system to process analog signals.

In very simple cases, it is more cost effective to use an analog filter. Introducing a digital filter requires considerable overhead circuitry, as previously discussed, including two low pass analog filters.

Another argument for analog filters is low power consumption. Analog filters require substantially less power and are therefore the only solution when power requirements are tight.

When making an electrical circuit on a PCB it is generally easier to use a digital solution, because the processing units are highly optimized over the years. Making the same circuit with analog components would take up a lot more space when using discrete components. Two alternatives are FPAA's and ASIC's, but they are expensive for low quantities.

Types of Digital Filters

Many digital filters are based on the fast Fourier transform, a mathematical algorithm that quickly extracts the frequency spectrum of a signal, allowing the spectrum to be manipulated (such as to create band-pass filters) before converting the modified spectrum back into a time-series signal.

Another form of a digital filter is that of a state-space model. A well used state-space filter is the Kalman filter published by Rudolf Kalman in 1960.

Traditional linear filters are usually based on attenuation. Alternatively nonlinear filters can be designed, including energy transfer filters which allow the user to move energy in a designed way. So that unwanted noise or effects can be moved to new frequency bands either lower or higher in frequency, spread over a range of frequencies, split, or focused. Energy transfer filters complement traditional filter designs and introduce many more degrees of freedom in filter design. Digital energy transfer filters are relatively easy to design and to implement and exploit nonlinear dynamics.

Recursive Least Squares Filter

The Recursive least squares (RLS) is an adaptive filter which recursively finds the coefficients that minimize a weighted linear least squares cost function relating to the input signals. This is in contrast to other algorithms such as the least mean squares (LMS) that aim to reduce the mean square error. In the derivation of the RLS, the input signals are considered deterministic, while for the LMS and similar algorithm they are considered stochastic. Compared to most of its competitors, the RLS exhibits extremely fast convergence. However, this benefit comes at the cost of high computational complexity.

Motivation

RLS was discovered by Gauss but laid unused or ignored until 1950 when Plackett rediscovered the original work of Gauss from 1821. In general, the RLS can be used to solve any problem that can be solved by adaptive filters. For example, suppose that a signal d(n) is transmitted over an echoey, noisy channel that causes it to be received as

$$x(n)=\sum_{k=0}^{q} b_n(k)d(n-k)+v(n)$$

where $v(n)$ represents additive noise. We will attempt to recover the desired signal $d(n)$ by use of a $p+1$-tap FIR filter, $\mathbf{w}$:

$$\hat{d}(n)=\sum_{k=0}^{p} w_n(k)x(n-k)=\mathbf{w}_n^T\mathbf{x}_n$$

where $\mathbf{x}_n=[x(n) \quad x(n-1) \quad \ldots \quad x(n-p)]^T$ is the vector containing the $p+1$ most recent samples of $x(n)$. Our goal is to estimate the parameters of the filter $\mathbf{w}$, and at each time n we refer to the new least squares estimate by $\mathbf{w}_n$. As time evolves, we would like to avoid completely redoing the least squares algorithm to find the new estimate for $\mathbf{w}_{n+1}$, in terms of $\mathbf{w}_n$.

The benefit of the RLS algorithm is that there is no need to invert matrices, thereby saving computational power. Another advantage is that it provides intuition behind such results as the Kalman filter.

Discussion

The idea behind RLS filters is to minimize a cost function C by appropriately selecting the filter coefficients $\mathbf{w}_n$, updating the filter as new data arrives. The error signal $e(n)$ and desired signal $d(n)$ are defined in the negative feedback diagram below:

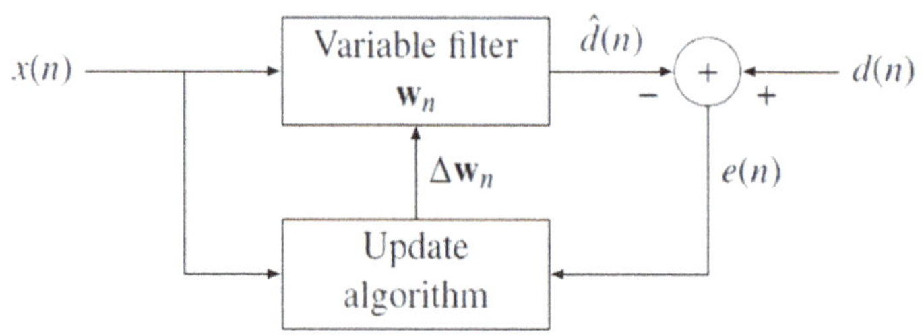

The error implicitly depends on the filter coefficients through the estimate $\hat{d}(n)$:

$$e(n) = d(n) - \hat{d}(n)$$

The weighted least squares error function C —the cost function we desire to minimize— being a function of e(n) is therefore also dependent on the filter coefficients:

$$C(\mathbf{w}_n) = \sum_{i=0}^{n} \lambda^{n-i} e^2(i)$$

where $0 < \lambda \le 1$ is the "forgetting factor" which gives exponentially less weight to older error samples.

The cost function is minimized by taking the partial derivatives for all entries k of the coefficient vector $\mathbf{w}_n$ and setting the results to zero

$$\frac{\partial C(\mathbf{w}_n)}{\partial w_n(k)} = \sum_{i=0}^{n} 2\lambda^{n-i} e(i) \frac{\partial e(i)}{\partial w_n(k)} = -\sum_{i=0}^{n} 2\lambda^{n-i} e(i) x(i-k) = 0 \qquad k = 0, 1, \cdots, p$$

Next, replace $e(n)$ with the definition of the error signal

$$\sum_{i=0}^{n} \lambda^{n-i} \left[d(i) - \sum_{l=0}^{p} w_n(l) x(i-l) \right] x(i-k) = 0 \qquad k = 0, 1, \cdots, p$$

Rearranging the equation yields

$$\sum_{l=0}^{p} w_n(l) \left[\sum_{i=0}^{n} \lambda^{n-i} x(i-l) x(i-k) \right] = \sum_{i=0}^{n} \lambda^{n-i} d(i) x(i-k) \qquad k = 0, 1, \cdots, p$$

This form can be expressed in terms of matrices

$$\mathbf{R}_x(n) \mathbf{w}_n = \mathbf{r}_{dx}(n)$$

where $\mathbf{R}_x(n)$ is the weighted sample covariance matrix for $x(n)$, and $\mathbf{r}_{dx}(n)$ is the equivalent estimate for the cross-covariance between $d(n)$ and $x(n)$. Based on this expression we find the coefficients which minimize the cost function as

$$\mathbf{w}_n = \mathbf{R}_x^{-1}(n) \mathbf{r}_{dx}(n)$$

This is the main result of the discussion.

Choosing λ

The smaller λ is, the smaller is the contribution of previous samples to the covariance matrix. This makes the filter *more* sensitive to recent samples, which means more fluctuations in the filter co-efficients. The $\lambda = 1$ case is referred to as the *growing window RLS algo-*

rithm. In practice, $\lambda = 1$ is usually chosen between 0.98 and 1. By using type-II maximum likelihood estimation the optimal λ can be estimated from a set of data.

Recursive Algorithm

The discussion resulted in a single equation to determine a coefficient vector which minimizes the cost function. In this section we want to derive a recursive solution of the form

$$\mathbf{w}_n = \mathbf{w}_{n-1} + \Delta\mathbf{w}_{n-1}$$

where $\Delta\mathbf{w}_{n-1}$ is a correction factor at time $n-1$. We start the derivation of the recursive algorithm by expressing the cross covariance $\mathbf{r}_{dx}(n)$ in terms of $\mathbf{r}_{dx}(n-1)$

$$\mathbf{r}_{dx}(n) = \sum_{i=0}^{n} \lambda^{n-i} d(i)\mathbf{x}(i)$$

$$= \sum_{i=0}^{n-1} \lambda^{n-i} d(i)\mathbf{x}(i) + \lambda^0 d(n)\mathbf{x}(n)$$

$$= \lambda\mathbf{r}_{dx}(n-1) + d(n)\mathbf{x}(n)$$

where $\mathbf{x}(i)$ is the $p+1$ dimensional data vector

$$\mathbf{x}(i) = [x(i), x(i-1), \ldots, x(i-p)]^T$$

Similarly we express $\mathbf{R}_x(n)$ in terms of $\mathbf{R}_x(n-1)$ by

$$\mathbf{R}_x(n) = \sum_{i=0}^{n} \lambda^{n-i} \mathbf{x}(i)\mathbf{x}^T(i)$$

$$= \lambda\mathbf{R}_x(n-1) + \mathbf{x}(n)\mathbf{x}^T(n)$$

In order to generate the coefficient vector we are interested in the inverse of the deterministic auto-covariance matrix. For that task the Woodbury matrix identity comes in handy. With

$A = \lambda\mathbf{R}_x(n-1)$ is $(p+1)$-by-$(p+1)$

$U = \mathbf{x}(n)$ is $(p+1)$-by-1

$$V = \mathbf{x}^T(n) \text{ is 1-by-}(p+1)$$

$$C = \mathbf{I}_1 \text{ is the 1-by-1 identity matrix}$$

The Woodbury matrix identity follows

$$\mathbf{R}_x^{-1}(n) = \left[\lambda \mathbf{R}_x(n-1) + \mathbf{x}(n)\mathbf{x}^T(n)\right]^{-1}$$

$$= \lambda^{-1}\mathbf{R}_x^{-1}(n-1)$$

$$-\lambda^{-1}\mathbf{R}_x^{-1}(n-1)\mathbf{x}(n)$$

$$\left\{1 + \mathbf{x}^T(n)\lambda^{-1}\mathbf{R}_x^{-1}(n-1)\mathbf{x}(n)\right\}^{-1} \mathbf{x}^T(n)\lambda^{-1}\mathbf{R}_x^{-1}(n-1)$$

To come in line with the standard literature, we define

$$\mathbf{P}(n) = \mathbf{R}_x^{-1}(n)$$

$$= \lambda^{-1}\mathbf{P}(n-1) - \mathbf{g}(n)\mathbf{x}^T(n)\lambda^{-1}\mathbf{P}(n-1)$$

where the *gain vecto* $g(n)$ is

$$\mathbf{g}(n) = \lambda^{-1}\mathbf{P}(n-1)\mathbf{x}(n)\left\{1 + \mathbf{x}^T(n)\lambda^{-1}\mathbf{P}(n-1)\mathbf{x}(n)\right\}^{-1}$$

$$= \mathbf{P}(n-1)\mathbf{x}(n)\left\{\lambda + \mathbf{x}^T(n)\mathbf{P}(n-1)\mathbf{x}(n)\right\}^{-1}$$

Before we move on, it is necessary to bring $\mathbf{g}(n)$ into another form

$$\mathbf{g}(n)\left\{1 + \mathbf{x}^T(n)\lambda^{-1}\mathbf{P}(n-1)\mathbf{x}(n)\right\} = \lambda^{-1}\mathbf{P}(n-1)\mathbf{x}(n)$$

$$\mathbf{g}(n) + \mathbf{g}(n)\mathbf{x}^T(n)\lambda^{-1}\mathbf{P}(n-1)\mathbf{x}(n) = \lambda^{-1}\mathbf{P}(n-1)\mathbf{x}(n)$$

Subtracting the second term on the left side yields

$$\mathbf{g}(n) = \lambda^{-1}\mathbf{P}(n-1)\mathbf{x}(n) - \mathbf{g}(n)\mathbf{x}^T(n)\lambda^{-1}\mathbf{P}(n-1)\mathbf{x}(n)$$

$$= \lambda^{-1}\left[\mathbf{P}(n-1) - \mathbf{g}(n)\mathbf{x}^T(n)\mathbf{P}(n-1)\right]\mathbf{x}(n)$$

With the recursive definition of $\mathbf{P}(n)$ the desired form follows

$$\mathbf{g}(n) = \mathbf{P}(n)\mathbf{x}(n)$$

Now we are ready to complete the recursion. As discussed

$$\mathbf{w}_n = \mathbf{P}(n)\mathbf{r}_{dx}(n)$$

$$= \lambda\mathbf{P}(n)\mathbf{r}_{dx}(n-1) + d(n)\mathbf{P}(n)\mathbf{x}(n)$$

The second step follows from the recursive definition of $\mathbf{r}_{dx}(n)$. Next we incorporate the recursive definition of $\mathbf{P}(n)$ together with the alternate form of $\mathbf{g}(n)$ and get

$$\mathbf{w}_n = \lambda\left[\lambda^{-1}\mathbf{P}(n-1) - \mathbf{g}(n)\mathbf{x}^T(n)\lambda^{-1}\mathbf{P}(n-1)\right]\mathbf{r}_{dx}(n-1) + d(n)\mathbf{g}(n)$$

$$= \mathbf{P}(n-1)\mathbf{r}_{dx}(n-1) - \mathbf{g}(n)\mathbf{x}^T(n)\mathbf{P}(n-1)\mathbf{r}_{dx}(n-1) + d(n)\mathbf{g}(n)$$

$$= \mathbf{P}(n-1)\mathbf{r}_{dx}(n-1) + \mathbf{g}(n)\left[d(n) - \mathbf{x}^T(n)\mathbf{P}(n-1)\mathbf{r}_{dx}(n-1)\right]$$

With $\mathbf{w}_{n-1} = \mathbf{P}(n-1)\mathbf{r}_{dx}(n-1)$ we arrive at the update equation

$$\mathbf{w}_n = \mathbf{w}_{n-1} + \mathbf{g}(n)\left[d(n) - \mathbf{x}^T(n)\mathbf{w}_{n-1}\right]$$

$$= \mathbf{w}_{n-1} + \mathbf{g}(n)\alpha(n)$$

where $\alpha(n) = d(n) - \mathbf{x}^T(n)\mathbf{w}_{n-1}$ is the *a priori* error. Compare this with the *a posteriori* error; the error calculated *after* the filter is updated:

$$e(n) = d(n) - \mathbf{x}^T(n)\mathbf{w}_n$$

That means we found the correction factor

$$\Delta\mathbf{w}_{n-1} = \mathbf{g}(n)\alpha(n)$$

This intuitively satisfying result indicates that the correction factor is directly proportional to both the error and the gain vector, which controls how much sensitivity is desired, through the weighting factor, λ.

RLS Algorithm Summary

The RLS algorithm for a *p*-th order RLS filter can be summarized as

Parameters: $p =$ filter order

$\lambda =$ forgetting factor

$\delta =$ value to initialize $\mathbf{P}(0)$

Initialization: $\mathbf{w}(n) = 0,$

$x(k) = 0, k = -p, \ldots, -1,$

$d(k) = 0, k = -p, \ldots, -1$

$\mathbf{P}(0) = \delta^{-1} I$ where I is the identity matrix of rank $p+1$

Computation: For $n = 1, 2, \ldots$

$$\mathbf{x}(n) = \begin{bmatrix} x(n) \\ x(n-1) \\ \vdots \\ x(n-p) \end{bmatrix}$$

$$\alpha(n) = d(n) - \mathbf{x}^T(n)\mathbf{w}(n-1)$$

$$\mathbf{g}(n) = \mathbf{P}(n-1)\mathbf{x}^*(n)\left\{\lambda + \mathbf{x}^T(n)\mathbf{P}(n-1)\mathbf{x}^*(n)\right\}^{-1}$$

$$\mathbf{P}(n) = \lambda^{-1}\mathbf{P}(n-1) - \mathbf{g}(n)\mathbf{x}^T(n)\lambda^{-1}\mathbf{P}(n-1)$$

$$\mathbf{w}(n) = \mathbf{w}(n-1) + \alpha(n)\mathbf{g}(n).$$

Note that the recursion for P follows an Algebraic Riccati equation and thus draws parallels to the Kalman filter.

Lattice Recursive Least Squares Filter (LRLS)

The Lattice Recursive Least Squares adaptive filter is related to the standard RLS except that it requires fewer arithmetic operations (order N). It offers additional advantages over conventional LMS algorithms such as faster convergence rates, modular structure, and insensitivity to variations in eigenvalue spread of the input correlation matrix. The LRLS algorithm described is based on *a posteriori* errors and includes the normalized form. The derivation is similar to the standard RLS algorithm and is based on the definition of $d(k)$. In the forward prediction case, we have $d(k) = x(k)$ with the input signal $x(k-1)$ as the most up to date sample. The backward prediction case is $d(k) = x(k-i-1)$, where i is the index of the sample in the past we want to predict, and the input signal $x(k)$ is the most recent sample.

Parameter Summary

$\kappa_f(k,i)$ is the forward reflection coefficient

$\kappa_b(k,i)$ is the backward reflection coefficient

$e_f(k,i)$ represents the instantaneous *a posteriori* forward prediction error

$e_b(k,i)$ represents the instantaneous *a posteriori* backward prediction error

$\xi^d_{b_{min}}(k,i)$ is the minimum least-squares backward prediction error

$\xi^d_{f_{min}}(k,i)$ is the minimum least-squares forward prediction error

$\gamma(k,i)$ is a conversion factor between *a priori* and *a posteriori* errors

$v_i(k)$ are the feedforward multiplier coefficients.

ϵ is a small positive constant that can be 0.01

LRLS Algorithm Summary

The algorithm for a LRLS filter can be summarized as

Initialization:

For i = 0,1,...,N

$$\delta(-1,i) = \delta_D(-1,i) = 0 \text{ (if x(k) = 0 for k < 0)}$$

$$\xi^d_{b_{min}}(-1,i) = \xi^d_{f_{min}}(-1,i) = \epsilon$$

$$\gamma(-1,i) = 1$$

$$e_b(-1,i) = 0$$

End

Computation:

For k ≥ 0

$$\gamma(k,0) = 1$$

$$e_b(k,0) = e_f(k,0) = x(k)$$

$$\xi^d_{b_{min}}(k,0) = \xi^d_{f_{min}}(k,0) = x^2(k) + \lambda \xi^d_{f_{min}}(k-1,0)$$

$$e(k,0) = d(k)$$

For i = 0,1,...,N

$$\delta(k,i) = \lambda \delta(k-1,i) + \frac{e_b(k-1,i)e_f(k,i)}{\gamma(k-1,i)}$$

$$\gamma(k,i+1) = \gamma(k,i) - \frac{e_b^2(k,i)}{\xi^d_{b_{min}}(k,i)}$$

$$\kappa_b(k,i) = \frac{\delta(k,i)}{\xi^d_{f_{min}}(k,i)}$$

$$\kappa_f(k,i) = \frac{\delta(k,i)}{\xi^d_{b_{min}}(k-1,i)}$$

$$e_b(k,i+1) = e_b(k-1,i) - \kappa_b(k,i)e_f(k,i)$$

$$e_f(k,i+1)=e_f(k,i)-\kappa_f(k,i)e_b(k-1,i)$$

$$\xi^d_{b_{min}}(k,i+1)=\xi^d_{b_{min}}(k-1,i)-\delta(k,i)\kappa_b(k,i)$$

$$\xi^d_{f_{min}}(k,i+1)=\xi^d_{f_{min}}(k,i)-\delta(k,i)\kappa_f(k,i)$$

Feedforward Filtering

$$\delta_D(k,i)=\lambda\delta_D(k-1,i)+\frac{e(k,i)e_b(k,i)}{\gamma(k,i)}$$

$$v_i(k)=\frac{\delta_D(k,i)}{\xi^d_{b_{min}}(k,i)}$$

$$e(k,i+1)=e(k,i)-v_i(k)e_b(k,i)$$

End

End

Normalized Lattice Recursive Least Squares Filter (NLRLS)

The normalized form of the LRLS has fewer recursions and variables. It can be calculated by applying a normalization to the internal variables of the algorithm which will keep their magnitude bounded by one. This is generally not used in real-time applications because of the number of division and square-root operations which comes with a high computational load.

Quadrature Mirror Filter

In digital signal processing, a quadrature mirror filter is a filter whose magnitude response is the mirror image around $\pi/2$ of that of another filter. Together these filters are known as the Quadrature Mirror Filter pair.

A filter $H_1(z)$ will be quadrature mirror filter of $H_0(z)$ if $H_1(z)=H_0(-z)$

The filter responses are symmetric about

$$\Omega=\pi/2$$

$$|H_1(e^{j\Omega})| = |H_0(e^{j(\pi-\Omega)})|$$

In audio/voice codecs, a quadrature mirror filter pair is often used to implement a filter bank that splits an input signal into two bands. The resulting high-pass and low-pass signals are often reduced by a factor of 2, giving a critically sampled two-channel representation of the original signal. The analysis filters are often related by the following formulae in addition to quadrate mirror property:

$|H_0(e^{j\Omega})|^2 + |H_1(e^{j\Omega})|^2 = 1$ where Ω is the frequency, and the sampling rate is normalized to 2π.

This is known as power complementary property. In other words, the power sum of the high-pass and low-pass filters is equal to 1.

Orthogonal wavelets -- the Haar wavelets and related Daubechies wavelets, Coiflets, and some developed by Mallat, are generated by scaling functions which, with the wavelet, satisfy a quadrature mirror filter relationship.

Further Description

The earliest wavelets were based on expanding a function in terms of rectangular steps, the Haar wavelets. This is usually a poor approximation, whereas Daubechies wavelets are among the simplest but most important families of wavelets. A linear filter that is zero for "smooth" signals, given a record of N points x_n is defined as:

$$y_n = \sum_{i=0}^{M-1} b_i x_{n-i}$$

It is desirable to have it vanish for a constant, so taking the order $m = 4$ for example:

$$b_0 \bullet 1 + b_1 \bullet 1 + b_2 \bullet 1 + b_3 \bullet 1 = 0$$

And to have it vanish for a linear ramp so that:

$$b_0 \bullet 0 + b_1 \bullet 1 + b_2 \bullet 2 + b_3 \bullet 3 = 0$$

A linear filter will vanish for any $x = \alpha n + \beta$, and this is all that can be done with a fourth order wavelet. Six terms will be needed to vanish a quadratic curve and so on given the other constraints to be included. Next an accompanying filter may be defined as:

$$z_n = \sum_{i=0}^{M-1} c_i x_{n-i}$$

This filter responds in an exactly opposite manner, being large for smooth signals and small for non-smooth signals. A linear filter is just a convolution of the signal with the filter's coefficients, so the series of the coefficients is the signal that the filter responds to maximally. Thus, the output of the second filter vanishes when the coefficients of the first one are input into it. The aim is to have:

$$\sum_{i=0}^{M-1} c_i b_i = 0$$

Where the associated time series flips the order of the coefficients because the linear filter is a convolution, and so both have the same index in this sum. A pair of filters with this property are defined as quadrature mirror filters. Even if the two resulting bands have been subsampled by a factor of 2, the relationship between the filters means that approximately perfect reconstruction is possible. That is, the two bands can then be upsampled, filtered again with the same filters and added together, to reproduce the original signal exactly (but with a small delay). (In practical implementations, numeric precision issues in floating-point arithmetic may affect the perfection of the reconstruction.)

Finite Impulse Response

In signal processing, a finite impulse response (FIR) filter is a filter whose impulse response (or response to any finite length input) is of *finite* duration, because it settles to zero in finite time. This is in contrast to infinite impulse response (IIR) filters, which may have internal feedback and may continue to respond indefinitely (usually decaying).

The impulse response (that is, the output in response to a Kronecker delta input) of an Nth-order discrete-time FIR filter lasts exactly $N + 1$ samples (from first nonzero element through last nonzero element) before it then settles to zero.

FIR filters can be discrete-time or continuous-time, and digital or analog.

Definition

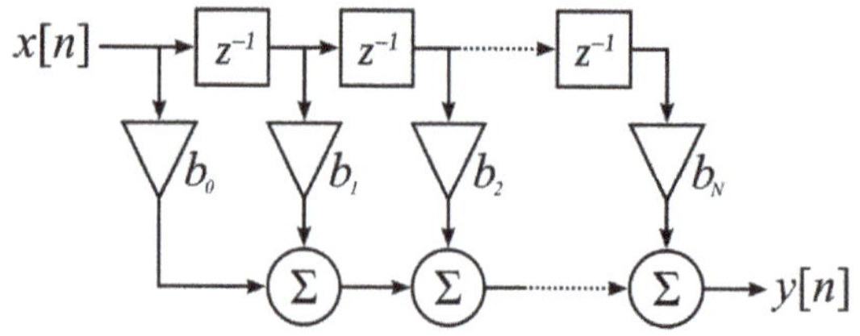

A direct form discrete-time FIR filter of order N. The top part is an N-stage delay line with $N + 1$ taps. Each unit delay is a z^{-1} operator in Z-transform notation.

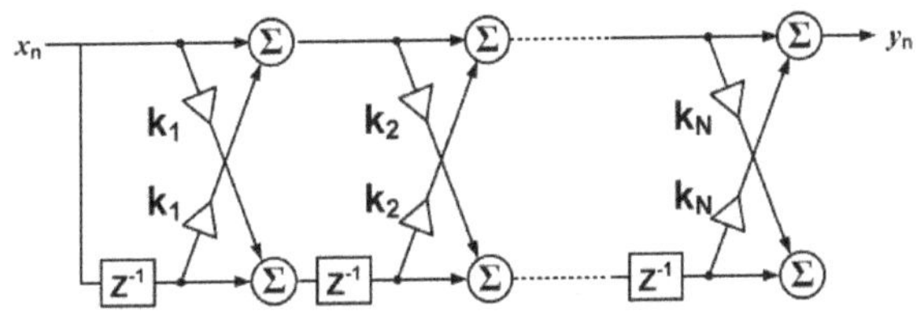

A lattice-form discrete-time FIR filter of order *N*. Each unit delay is a z^{-1} operator in Z-transform notation.

For a causal discrete-time FIR filter of order *N*, each value of the output sequence is a weighted sum of the most recent input values:

$$\begin{aligned} y[n] &= b_0 x[n] + b_1 x[n-1] + \cdots + b_N x[n-N] \\ &= \sum_{i=0}^{N} b_i \cdot x[n-i], \end{aligned}$$

where:

$x[n]$ is the input signal,

$y[n]$ is the output signal,

N is the filter order; an Nth-order filter has $(N+1)$ terms on the right-hand side

b_i is the value of the impulse response at the i'th instant for $0 \le i \le N$ of an Nth-order FIR filter. If the filter is a direct form FIR filter then b_i is also a co-efficient of the filter.

This computation is also known as discrete convolution.

The $x[n-i]$ in these terms are commonly referred to as *taps*, based on the structure of a tapped delay line that in many implementations or block diagrams provides the delayed inputs to the multiplication operations. One may speak of a *5th order/6-tap filter*, for instance.

The impulse response of the filter as defined is nonzero over a finite duration. Including zeros, the impulse response is the infinite sequence:

$$h[n] = \sum_{i=0}^{N} b_i \cdot \delta[n-i] = \begin{cases} b_n & 0 \le n \le N \\ 0 & \textbf{otherwise.} \end{cases}$$

If an FIR filter is non-causal, the range of nonzero values in its impulse response can start before $n = 0$, with the defining formula appropriately generalized.

Properties

An FIR filter has a number of useful properties which sometimes make it preferable to an infinite impulse response (IIR) filter. FIR filters:

- Require no feedback. This means that any rounding errors are not compounded by summed iterations. The same relative error occurs in each calculation. This also makes implementation simpler.
- Are inherently stable, since the output is a sum of a finite number of finite multiples of the input values, so can be no greater than $\sum |b_i|$ times the largest value appearing in the input.
- Can easily be designed to be linear phase by making the coefficient sequence symmetric. This property is sometimes desired for phase-sensitive applications, for example data communications, seismology, crossover filters, and mastering.

The main disadvantage of FIR filters is that considerably more computation power in a general purpose processor is required compared to an IIR filter with similar sharpness or selectivity, especially when low frequency (relative to the sample rate) cutoffs are needed. However many digital signal processors provide specialized hardware features to make FIR filters approximately as efficient as IIR for many applications.

Frequency Response

The filter's effect on the sequence x[n] is described in the frequency domain by the convolution theorem:

$$\underbrace{\mathcal{F}\{x * h\}}_{Y(\omega)} = \underbrace{\mathcal{F}\{x\}}_{X(\omega)} \cdot \underbrace{\mathcal{F}\{h\}}_{H(\omega)} \quad \text{and} \quad y[n] = x[n] * h[n] = \mathcal{F}^{-1}\{X(\omega) \cdot H(\omega)\},$$

where operators $\mathcal{F}$ and $\mathcal{F}^{-1}$ respectively denote the discrete-time Fourier transform (DTFT) and its inverse. Therefore, the complex-valued, multiplicative function $H(\omega)$ is the filter's frequency response. It is defined by a Fourier series:

$$H_{2\pi}(\omega) \stackrel{\text{def}}{=} \sum_{n=-\infty}^{\infty} h[n] \cdot \left(e^{i\omega}\right)^{-n} = \sum_{n=0}^{N} b_n \cdot \left(e^{i\omega}\right)^{-n},$$

where the added subscript denotes 2π-periodicity. Here ω represents frequency in normalized units (*radians/sample*). The substitution $\omega = 2\pi f$, favored by many filter design programs, changes the units of frequency (f) to *cycles/sample* and the periodicity to 1. When the x[n] sequence has a known sampling-rate, f_s *samples/second*, the substitution $\omega = 2\pi f / f_s$ changes the units of frequency (f) to *cycles/second* (hertz) and the periodicity to f_s. The value $\omega = \pi$ corresponds to a frequency of $f = \frac{f_s}{2}\,Hz$ $= \frac{1}{2}$ *cycles/sample*, which is the Nyquist frequency.

Transfer function

The frequency response, $H_{2\pi}(\omega)$, can also be written as $H(e^{i\omega})$, where function H is the Z-transform of the impulse response:

$$H(z) \stackrel{\text{def}}{=} \sum_{n=-\infty}^{\infty} h[n] \cdot z^{-n}.$$

z is a complex variable, and H(z) is a surface. One cycle of the periodic frequency response can be found in the region defined by $z = e^{i\omega}, -\pi \leq \omega \leq \pi,$ which is the unit circle of the z-plane. Filter transfer functions are often used to verify the stability of IIR designs. As we have already noted, FIR designs are inherently stable.

Filter Design

An FIR filter is designed by finding the coefficients and filter order that meet certain specifications, which can be in the time-domain (e.g. a matched filter) and/or the frequency domain (most common). Matched filters perform a cross-correlation between the input signal and a known pulse-shape. The FIR convolution is a cross-correlation between the input signal and a time-reversed copy of the impulse-response. Therefore, the matched-filter's impulse response is "designed" by sampling the known pulse-shape and using those samples in reverse order as the coefficients of the filter.

When a particular frequency response is desired, several different design methods are common:

1. Window design method
2. Frequency Sampling method
3. Weighted least squares design
4. Parks-McClellan method (also known as the Equiripple, Optimal, or Minimax method). The Remez exchange algorithm is commonly used to find an optimal equiripple set of coefficients. Here the user specifies a desired frequency response, a weighting function for errors from this response, and a filter order N. The algorithm then finds the set of $(N+1)$ coefficients that minimize the maximum deviation from the ideal. Intuitively, this finds the filter that is as close as you can get to the desired response given that you can use only $(N+1)$

coefficients. This method is particularly easy in practice since at least one text includes a program that takes the desired filter and *N*, and returns the optimum coefficients.

5. Equiripple FIR filters can be designed using the FFT algorithms as well. The algorithm is iterative in nature. You simply compute the DFT of an initial filter design that you have using the FFT algorithm (if you don't have an initial estimate you can start with h[n]=delta[n]). In the Fourier domain or FFT domain you correct the frequency response according to your desired specs and compute the inverse FFT. In time-domain you retain only N of the coefficients (force the other coefficients to zero). Compute the FFT once again. Correct the frequency response according to specs.

Software packages like MATLAB, GNU Octave, Scilab, and SciPy provide convenient ways to apply these different methods.

Window Design Method

In the window design method, one first designs an ideal IIR filter and then truncates the infinite impulse response by multiplying it with a finite length window function. The result is a finite impulse response filter whose frequency response is modified from that of the IIR filter. Multiplying the infinite impulse by the window function in the time domain results in the frequency response of the IIR being convolved with the Fourier transform (or DTFT) of the window function. If the window's main lobe is narrow, the composite frequency response remains close to that of the ideal IIR filter.

The ideal response is usually rectangular, and the corresponding IIR is a sinc function. The result of the frequency domain convolution is that the edges of the rectangle are tapered, and ripples appear in the passband and stopband. Working backward, one can specify the slope (or width) of the tapered region (*transition band*) and the height of the ripples, and thereby derive the frequency domain parameters of an appropriate window function. Continuing backward to an impulse response can be done by iterating a filter design program to find the minimum filter order. Another method is to restrict the solution set to the parametric family of Kaiser windows, which provides closed form relationships between the time-domain and frequency domain parameters. In general, that method will not achieve the minimum possible filter order, but it is particularly convenient for automated applications that require dynamic, on-the-fly, filter design.

The window design method is also advantageous for creating efficient half-band filters, because the corresponding sinc function is zero at every other sample point (except the center one). The product with the window function does not alter the zeros, so almost half of the coefficients of the final impulse response are zero. An appropriate implementation of the FIR calculations can exploit that property to double the filter's efficiency.

Moving Average Example

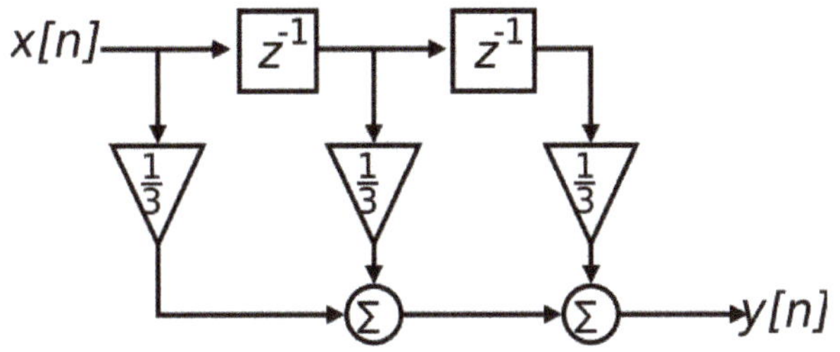

Fig. (a) Block diagram of a simple FIR filter (2nd-order/3-tap filter in this case, implementing a moving average)

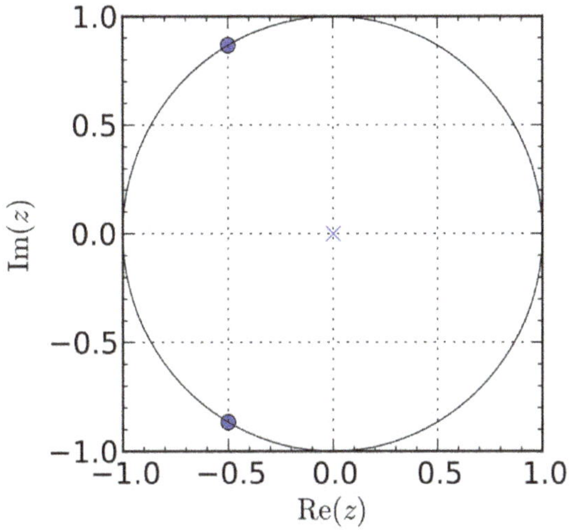

Fig. (b) Pole–zero diagram of a second-order FIR filter

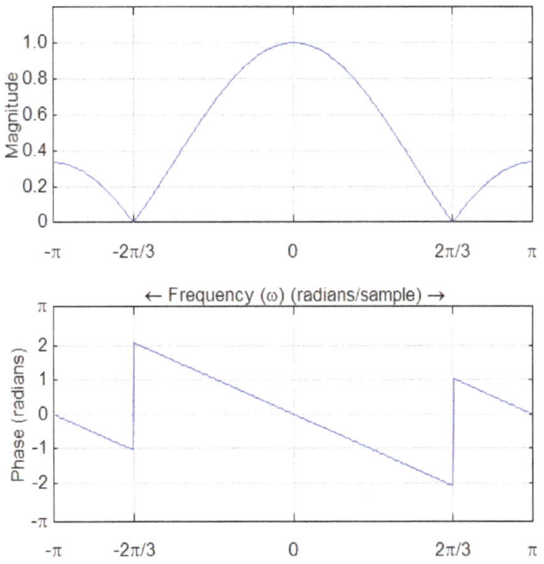

Fig. (c) Magnitude and phase responses

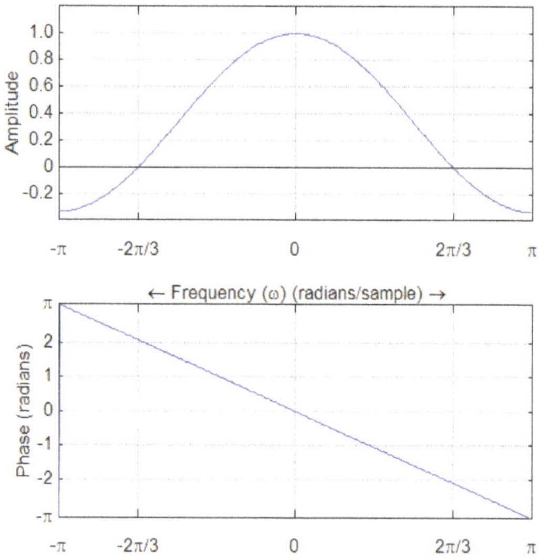

Fig. (d) Amplitude and phase responses

A moving average filter is a very simple FIR filter. It is sometimes called a boxcar filter, especially when followed by decimation. The filter coefficients, $b_0,\ldots,b_N$, are found via the following equation:

$$b_i = \frac{1}{N+1}$$

To provide a more specific example, we select the filter order:

$$N = 2$$

The impulse response of the resulting filter is:

$$h[n] = \frac{1}{3}\delta[n] + \frac{1}{3}\delta[n-1] + \frac{1}{3}\delta[n-2]$$

The Fig. (a) on the right shows the block diagram of a 2nd-order moving-average filter discussed below. The transfer function is:

$$H(z) = \frac{1}{3} + \frac{1}{3}z^{-1} + \frac{1}{3}z^{-2} = \frac{1}{3}\frac{z^2+z+1}{z^2}.$$

Fig. (b) on the right shows the corresponding pole–zero diagram. Zero frequency (DC) corresponds to (1,0), positive frequencies advancing counterclockwise around the circle to the Nyquist frequency at (-1,0). Two poles are located at the origin, and two zeros are located at $z_1 = -\frac{1}{2} + j\frac{\sqrt{3}}{2}$, $z_2 = -\frac{1}{2} - j\frac{\sqrt{3}}{2}$.

The frequency response, in terms of normalized frequency ω, is:

$$H\left(e^{j\omega}\right)=\frac{1}{3}+\frac{1}{3}e^{-j\omega}+\frac{1}{3}e^{-j2\omega}.$$

Fig. (c) on the right shows the magnitude and phase components of $H\left(e^{j\omega}\right)$. But plots like these can also be generated by doing a discrete Fourier transform (DFT) of the impulse response. And because of symmetry, filter design or viewing software often displays only the [0,π] region. The magnitude plot indicates that the moving-average filter passes low frequencies with a gain near 1 and attenuates high frequencies, and is thus a crude low-pass filter. The phase plot is linear except for discontinuities at the two frequencies where the magnitude goes to zero. The size of the discontinuities is π, representing a sign reversal. They do not affect the property of linear phase. That fact is illustrated in Fig. (d).

Infinite Impulse Response

Infinite impulse response (IIR) is a property applying to many linear time-invariant systems. Common examples of linear time-invariant systems are most electronic and digital filters. Systems with this property are known as *IIR systems* or *IIR filters*, and are distinguished by having an impulse response which does not become exactly zero past a certain point, but continues indefinitely. This is in contrast to a finite impulse response in which the impulse response $h(t)$ *does* become exactly zero at times $t > T$ for some finite T, thus being of finite duration.

In practice, the impulse response, even of IIR systems, usually approaches zero and can be neglected past a certain point. However the physical systems which give rise to IIR or FIR (*finite impulse response*) responses are dissimilar, and therein lies the importance of the distinction. For instance, analog electronic filters composed of resistors, capacitors, and/or inductors (and perhaps linear amplifiers) are generally IIR filters. On the other hand, discrete-time filters (usually digital filters) based on a tapped delay line *employing no feedback* are necessarily FIR filters. The capacitors (or inductors) in the analog filter have a "memory" and their internal state never completely relaxes following an impulse. But in the latter case, after an impulse has reached the end of the tapped delay line, the system has no further memory of that impulse and has returned to its initial state; its impulse response beyond that point is exactly zero.

Implementation and Design

Although almost all analog electronic filters are IIR, digital filters may be either IIR or FIR. The presence of feedback in the topology of a discrete-time filter (such as the block diagram shown below) generally creates an IIR response. The z domain transfer function of an IIR filter contains a non-trivial denominator, describing those feedback terms. The transfer function of an FIR filter, on the other hand, has only a numerator as expressed in the general form derived below. All of the a_i coefficients with $i > 0$ (feedback terms) are zero and the filter has no finite poles.

The transfer functions pertaining to IIR analog electronic filters have been extensively studied and optimized for their amplitude and phase characteristics. These continuous-time filter functions are described in the Laplace domain. Desired solutions can be transferred to the case of discrete-time filters whose transfer functions are expressed in the z domain, through the use of certain mathematical techniques such as the bilinear transform, impulse invariance, or pole–zero matching method. Thus digital IIR filters can be based on well-known solutions for analog filters such as the Chebyshev filter, Butterworth filter, and Elliptic filter, inheriting the characteristics of those solutions.

Transfer Function Derivation

Digital filters are often described and implemented in terms of the difference equation that defines how the output signal is related to the input signal:

$$\begin{aligned} y[n] &= \frac{1}{a_0}(b_0 x[n] + b_1 x[n-1] + \cdots + b_P x[n-P] \\ &\quad - a_1 y[n-1] - a_2 y[n-2] - \cdots - a_Q y[n-Q]) \end{aligned}$$

where:

- P is the feedforward filter order
- b_i are the feedforward filter coefficients
- Q is the feedback filter order
- a_i are the feedback filter coefficients
- $x[n]$ is the input signal
- $y[n]$ is the output signal.

A more condensed form of the difference equation is:

$$y[n] = \frac{1}{a_0}\left(\sum_{i=0}^{P} b_i x[n-i] - \sum_{j=1}^{Q} a_j y[n-j]\right)$$

which, when rearranged, becomes:

$$\sum_{j=0}^{Q} a_j y[n-j] = \sum_{i=0}^{P} b_i x[n-i]$$

To find the transfer function of the filter, we first take the Z-transform of each side of the above equation, where we use the time-shift property to obtain:

$$\sum_{j=0}^{Q} a_j z^{-j} Y(z) \quad , \quad = \sum_{i=0}^{P} b_i z^{-i} X(z)$$

We define the transfer function to be:

$$H(z) = \frac{Y(z)}{X(z)} \quad , \quad = \frac{\sum_{i=0}^{P} b_i z^{-i}}{\sum_{j=0}^{Q} a_j z^{-j}}$$

Considering that in most IIR filter designs coefficient a_0 is 1, the IIR filter transfer function takes the more traditional form:

$$H(z) = \frac{\sum_{i=0}^{P} b_i z^{-i}}{1 + \sum_{j=1}^{Q} a_j z^{-j}}$$

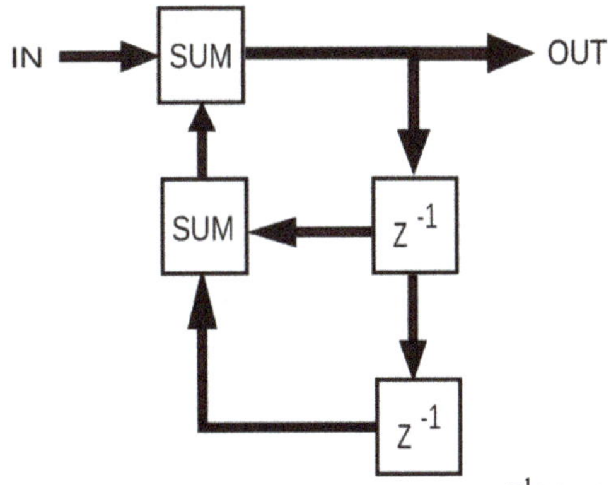

An example of a block diagram of an IIR filter. The z^{-1} block is a unit delay.

Stability

The transfer function allows one to judge whether or not a system is bounded-input, bounded-output (BIBO) stable. To be specific, the BIBO stability criterion requires that the ROC of the system includes the unit circle. For example, for a causal system, all poles of the transfer function have to have an absolute value smaller than one. In other words, all poles must be located within a unit circle in the z -plane.

The poles are defined as the values of $H(z)$ which make the denominator of $H(z)$ equal to 0:

$$0 = \sum_{j=0}^{Q} a_j z^{-j}$$

Clearly, if $a_j \neq 0$ then the poles are not located at the origin of the z -plane. This is in contrast to the FIR filter where all poles are located at the origin, and is therefore always stable.

IIR filters are sometimes preferred over FIR filters because an IIR filter can achieve a much sharper transition region roll-off than an FIR filter of the same order.

Example

Let the transfer function $H(z)$ of a discrete-time filter be given by:

$$H(z) = \frac{B(z)}{A(z)} = \frac{1}{1 - az^{-1}}$$

governed by the parameter a, a real number with $0 < |a| < 1$. $H(z)$ is stable and causal with a pole at a. The time-domain impulse response can be shown to be given by:

$$h(n) = a^n u(n)$$

where $u(n)$ is the unit step function. It can be seen that $h(n)$ is non-zero for all $n \geq 0$, thus an impulse response which continues infinitely.

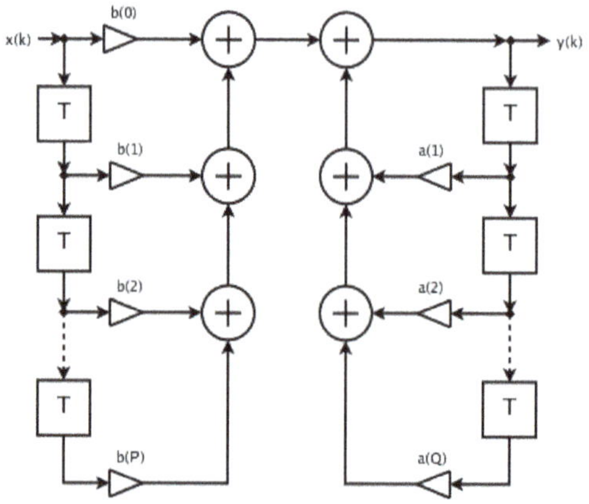

IIR filter example

Advantages and Disadvantages

The main advantage digital IIR filters have over FIR filters is their efficiency in implementation, in order to meet a specification in terms of passband, stopband, ripple, and/or roll-off. Such a set of specifications can be accomplished with a lower order (Q in the above formulae) IIR filter than would be required for an FIR filter meeting the same requirements. If implemented in a signal processor, this implies a correspondingly fewer number of calculations per time step; the computational savings is often of a rather large factor.

On the other hand, FIR filters can be easier to design, for instance, to match a particular frequency response requirement. This is particularly true when the requirement is not one of the usual cases (high-pass, low-pass, notch, etc.) which have been studied and optimized for analog filters. Also FIR filters can be easily made to be linear phase (constant group delay vs frequency)—a property that is not easily met using IIR filters and then only as an approximation (for instance with the Bessel filter). Another issue regarding digital IIR filters is the potential for limit cycle behavior when idle, due to the feedback system in conjunction with quantization.

References

- Miroslav D. Lutovac, Dejan V. Tošić, Brian Lawrence Evans, Filter Design for Signal Processing Using MATLAB and Mathematica, Miroslav Lutovac, 2001 ISBN 0201361302.
- B. A. Shenoi, Introduction to Digital Signal Processing and Filter Design, John Wiley & Sons, 2005 ISBN 0471656380.
- L. D. Paarmann, Design and Analysis of Analog Filters: A Signal Processing Perspective, Springer, 2001 ISBN 0792373731.
- Rabiner, Lawrence R., and Gold, Bernard, 1975: Theory and Application of Digital Signal Processing (Englewood Cliffs, New Jersey: Prentice-Hall, Inc.) ISBN 0-13-914101-4
- A. Antoniou (2006). Digital Signal Processing: Signals, Systems, and Filters. McGraw-Hill, New York, NY. doi:10.1036/0071454241. ISBN 0-07-145424-1.
- S.K. Mitra (1998). Digital Signal Processing: A Computer-Based Approach. McGraw-Hill, New York, NY. ISBN 0-07-286546-6.
- A.V. Oppenheim; R.W. Schafer; J.R. Buck (1999). Discrete-Time Signal Processing. Prentice-Hall, Upper Saddle River, NJ. ISBN 0-13-754920-2.
- Widrow, Bernard; Stearns, Samuel D. (1985). Adaptive Signal Processing. Englewood Cliffs, NJ: Prentice Hall. ISBN 0-13-004029-0.
- Hayes, Monson H. (1996). "9.4: Recursive Least Squares". Statistical Digital Signal Processing and Modeling. Wiley. p. 541. ISBN 0-471-59431-8.
- Weifeng Liu, Jose Principe and Simon Haykin, Kernel Adaptive Filtering: A Comprehensive Introduction, John Wiley, 2010, ISBN 0-470-44753-2

8

Devices Related to Digital Signal Processing

Devices related to digital signal processing are analog-to-digital converter, digital-to-analog converter, time-to-digital converter and reconstruction filter. Analog-to-digital converters are systems that help in transforming an analog signal to a digital signal whereas a digital-to-analog convertor is a device that converts a digital signal into an analog signal. This chapter helps discusses in detail the devices that are related to digital signal processing.

Analog-to-digital Converter

4-channel stereo multiplexed analog-to-digital converter WM8775SEDS made by Wolfson Microelectronics placed on an X-Fi Fatality Pro sound card.

In electronics, an analog-to-digital converter (ADC, A/D, A–D, or A-to-D) is a system that converts an analog signal, such as a sound picked up by a microphone or light entering a digital camera, into a digital signal. An ADC may also provide an isolated measurement such as an electronic device that converts an input analog voltage or current to a digital number proportional to the magnitude of the voltage or current.

Typically the digital output is a two's complement binary number that is proportional to the input, but there are other possibilities.

There are several ADC architectures. Due to the complexity and the need for precisely matched components, all but the most specialized ADCs are implemented as integrated circuits (ICs).

A digital-to-analog converter (DAC) performs the reverse function; it converts a digital signal into an analog signal.

Explanation

The conversion involves quantization of the input, so it necessarily introduces a small amount of error. Furthermore, instead of continuously performing the conversion, an ADC does the conversion periodically, sampling the input. The result is a sequence of digital values that have been converted from a continuous-time and continuous-amplitude analog signal to a discrete-time and discrete-amplitude digital signal.

An ADC is defined by its bandwidth and its signal-to-noise ratio. The bandwidth of an ADC is characterized primarily by its sampling rate. The dynamic range of an ADC is influenced by many factors, including the resolution, linearity and accuracy (how well the quantization levels match the true analog signal), aliasing and jitter. The dynamic range of an ADC is often summarized in terms of its effective number of bits (ENOB), the number of bits of each measure it returns that are on average not noise. An ideal ADC has an ENOB equal to its resolution. ADCs are chosen to match the bandwidth and required signal-to-noise ratio of the signal to be quantized. If an ADC operates at a sampling rate greater than twice the bandwidth of the signal, then perfect reconstruction is possible given an ideal ADC and neglecting quantization error. The presence of quantization error limits the dynamic range of even an ideal ADC. However, if the dynamic range of the ADC exceeds that of the input signal, its effects may be neglected resulting in an essentially perfect digital representation of the input signal.

Resolution

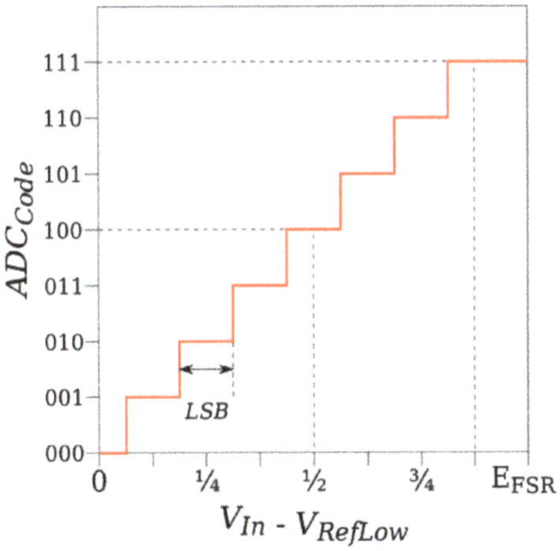

Fig. 1. An 8-level ADC coding scheme.

The resolution of the converter indicates the number of discrete values it can produce over the range of analog values. The resolution determines the magnitude of the quantization error and therefore determines the maximum possible average signal to noise ratio for an ideal ADC without the use of oversampling. The values are usually stored electronically in binary form, so the resolution is usually expressed in bits. In consequence, the number of discrete values available, or "levels", is assumed to be a power of two. For example, an ADC with a resolution of 8 bits can encode an analog input to one in 256 different levels, since $2^8 = 256$. The values can represent the ranges from 0 to 255 (i.e. unsigned integer) or from −128 to 127 (i.e. signed integer), depending on the application.

Resolution can also be defined electrically, and expressed in volts. The minimum change in voltage required to guarantee a change in the output code level is called the least significant bit (LSB) voltage. The resolution Q of the ADC is equal to the LSB voltage. The voltage resolution of an ADC is equal to its overall voltage measurement range divided by the number of intervals:

$$Q = \frac{E_{\mathrm{FSR}}}{2^M - 1},$$

where M is the ADC's resolution in bits and E_{FSR} is the full scale voltage range (also called 'span'). E_{FSR} is given by

$$E_{\mathrm{FSR}} = V_{\mathrm{RefHi}} - V_{\mathrm{RefLow}},$$

where V_{RefHi} and V_{RefLow} are the upper and lower extremes, respectively, of the voltages that can be coded.

Normally, the number of voltage intervals is given by

$$N = 2^M - 1,$$

where M is the ADC's resolution in bits.

That is, one voltage interval is assigned in between two consecutive code levels.

Example:

- Coding scheme as in figure 1 (assume input signal x(t) = Acos(t), A = 5V)
- Full scale measurement range = -5 to 5 volts
- ADC resolution is 8 bits: 2^8 = 256 quantization levels (codes)
- ADC voltage resolution, Q = (10 V – 0 V) / (256 - 1) = 10 V / 255 ≈ 0.039 V ≈ 39 mV.

In practice, the useful resolution of a converter is limited by the best signal-to-noise ratio (SNR) that can be achieved for a digitized signal. An ADC can resolve a signal to only a certain number of bits of resolution, called the effective number of bits (ENOB). One effective bit of resolution changes the signal-to-noise ratio of the digitized signal by 6 dB, if the resolution is limited by the ADC. If a preamplifier has been used prior to A/D conversion, the noise introduced by the amplifier can be an important contributing factor towards the overall SNR.

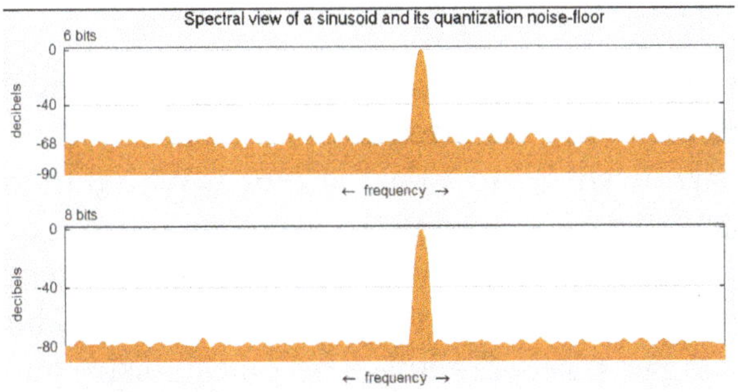

Comparison of quantizing a sinusoid to 64 levels (6 bits) and 256 levels (8 bits). The additive noise created by 6-bit quantization is 12 dB greater than the noise created by 8-bit quantization. When the spectral distribution is flat, as in this example, the 12 dB difference manifests as a measurable difference in the noise floors.

Quantization Error

Quantization error is the noise introduced by quantization in an ideal ADC. It is a rounding error between the analog input voltage to the ADC and the output digitized value. The noise is non-linear and signal-dependent. In an ideal analog-to-digital converter, where the quantization error is uniformly distributed between −1/2 LSB and +1/2 LSB, and the signal has a uniform distribution covering all quantization levels, the Signal-to-quantization-noise ratio (SQNR) can be calculated from

$$\mathrm{SQNR} = 20\log_{10}(2^Q) \approx 6.02 \cdot Q \text{ dB}$$

Where Q is the number of quantization bits. For example, a 16-bit ADC has a maximum signal-to-noise ratio of 6.02 × 16 = 96.3 dB, and therefore the quantization error is 96.3 dB below the maximum level. Quantization error is distributed from DC to the Nyquist frequency, consequently if part of the ADC's bandwidth is not used (as in oversampling), some of the quantization error will fall out of band, effectively improving the SQNR. In an oversampled system, noise shaping can be used to further increase SQNR by forcing more quantization error out of the band.

Dither

In ADCs, performance can usually be improved using dither. This is a very small amount of random noise (white noise), which is added to the input before conversion. Its effect is to cause the state of the LSB to randomly oscillate between 0 and 1 in the presence of very low levels of input, rather than sticking at a fixed value. Rather than the signal simply getting cut off altogether at this low level (which is only being quantized to a resolution of 1 bit), it extends the effective range of signals that the ADC can convert, at the expense of a slight increase in noise – effectively the quantization error is diffused across a series of noise values which is far less objectionable than a hard cutoff. The result is an accurate representation of the signal over time. A suitable filter at the output of the system can thus recover this small signal variation.

An audio signal of very low level (with respect to the bit depth of the ADC) sampled without dither sounds extremely distorted and unpleasant. Without dither the low level may cause the least significant bit to "stick" at 0 or 1. With dithering, the true level of the audio may be calculated by averaging the actual quantized sample with a series of other samples [the dither] that are recorded over time. A virtually identical process, also called dither or dithering, is often used when quantizing photographic images to a fewer number of bits per pixel—the image becomes noisier but to the eye looks far more realistic than the quantized image, which otherwise becomes banded. This analogous process may help to visualize the effect of dither on an analogue audio signal that is converted to digital. Dithering is also used in integrating systems such as electricity meters. Since the values are added together, the dithering produces results that are more exact than the LSB of the analog-to-digital converter. Note that dither can only increase the resolution of a sampler, it cannot improve the linearity, and thus accuracy does not necessarily improve.

Accuracy

An ADC has several sources of errors. Quantization error and (assuming the ADC is intended to be linear) non-linearity are intrinsic to any analog-to-digital conversion. These errors are measured in a unit called the least significant bit (LSB). In the above example of an eight-bit ADC, an error of one LSB is 1/256 of the full signal range, or about 0.4%.

Non-linearity

All ADCs suffer from non-linearity errors caused by their physical imperfections, causing their output to deviate from a linear function (or some other function, in the case of a deliberately non-linear ADC) of their input. These errors can sometimes be mitigated by calibration, or prevented by testing. Important parameters for linearity are integral non-linearity (INL) and differential non-linearity (DNL). These non-linearities reduce the dynamic range of the signals that can be digitized by the ADC, also reducing the effective resolution of the ADC.

Jitter

When digitizing a sine wave $x(t) = A\sin(2\pi f_0 t)$, the use of a non-ideal sampling clock will result in some uncertainty in when samples are recorded. Provided that the actual sampling time *uncertainty* due to the *clock jitter* is $\text{Ä}t$, the error caused by this phenomenon can be estimated as $E_{ap} \leq |x'(t)\Delta t| \leq 2A\pi f_0 \Delta t.$. This will result in additional recorded noise that will reduce the effective number of bits (ENOB) below that predicted by quantization error alone. The error is zero for DC, small at low frequencies, but significant when high frequencies have high amplitudes. This effect can be ignored if it is drowned out by the *quantizing error*. Jitter requirements can be calculated using the following formula: $\Delta t < \frac{1}{2^q \pi f_0}$, where q is the number of ADC bits.

Output size (bits)	Signal Frequency						
	1 Hz	1 kHz	10 kHz	1 MHz	10 MHz	100 MHz	1 GHz
8	1,243 µs	1.24 µs	124 ns	1.24 ns	124 ps	12.4 ps	1.24 ps
10	311 µs	311 ns	31.1 ns	311 ps	31.1 ps	3.11 ps	0.31 ps
12	77.7 µs	77.7 ns	7.77 ns	77.7 ps	7.77 ps	0.78 ps	0.08 ps
14	19.4 µs	19.4 ns	1.94 ns	19.4 ps	1.94 ps	0.19 ps	0.02 ps
16	4.86 µs	4.86 ns	486 ps	4.86 ps	0.49 ps	0.05 ps	–
18	1.21 µs	1.21 ns	121 ps	1.21 ps	0.12 ps	–	–
20	304 ns	304 ps	30.4 ps	0.30 ps	0.03 ps	–	–

Clock jitter is caused by phase noise. The resolution of ADCs with a digitization bandwidth between 1 MHz and 1 GHz is limited by jitter. When sampling audio signals at 44.1 kHz, the anti-aliasing filter should have eliminated all frequencies above 22 kHz. The input frequency (in this case, < 22 kHz), not the ADC clock frequency, is the determining factor with respect to jitter performance.

Sampling Rate

The analog signal is continuous in time and it is necessary to convert this to a flow of digital values. It is therefore required to define the rate at which new digital values are sampled from the analog signal. The rate of new values is called the *sampling rate* or *sampling frequency* of the converter. A continuously varying bandlimited signal can be sampled (that is, the signal values at intervals of time T, the sampling time, are measured and stored) and then the original signal can be *exactly* reproduced from the discrete-time values by an interpolation formula. The accuracy is limited by quantization error. However, this faithful reproduction is only possible if the sampling rate is higher than twice the highest frequency of the signal. This is essentially what is embodied in the Shannon-Nyquist sampling theorem. Since a practical ADC cannot make an instantaneous conversion, the input value must necessarily be held constant during the time that the converter performs a conversion (called the *conversion time*). An input circuit called a sample and hold performs this task—in most cases by using a capacitor to store the analog voltage at the input, and using an electronic switch or gate to disconnect the capacitor from the input. Many ADC integrated circuits include the sample and hold subsystem internally.

Aliasing

An ADC works by sampling the value of the input at discrete intervals in time. Provided that the input is sampled above the Nyquist rate, defined as twice the highest frequency of interest, then all frequencies in the signal can be reconstructed. If frequencies above half the Nyquist rate are sampled, they are incorrectly detected as lower frequencies, a process referred to as aliasing. Aliasing occurs because instantaneously sampling a function at two or fewer times per cycle results in missed cycles, and therefore the appearance of an incorrectly lower frequency. For example, a 2 kHz sine wave being sampled at 1.5 kHz would be reconstructed as a 500 Hz sine wave.

To avoid aliasing, the input to an ADC must be low-pass filtered to remove frequencies above half the sampling rate. This filter is called an *anti-aliasing filter*, and is essential for a practical ADC system that is applied to analog signals with higher frequency content. In applications where protection against aliasing is essential, oversampling may be used to greatly reduce or even eliminate it. Although aliasing in most systems is unwanted, it should also be noted that it can be exploited to provide simultaneous down-mixing of a band-limited high frequency signal (see undersampling and frequency mixer). The alias is effectively the lower heterodyne of the signal frequency and sampling frequency.

Oversampling

Signals are often sampled at the minimum rate required, for economy, with the result that the quantization noise introduced is white noise spread over the whole pass band of the converter. If a signal is sampled at a rate much higher than the Nyquist rate and then digitally filtered to limit it to the signal bandwidth there are the following advantages:

- digital filters can have better properties (sharper rolloff, phase) than analogue filters, so a sharper anti-aliasing filter can be realised and then the signal can be downsampled giving a better result
- a 20-bit ADC can be made to act as a 24-bit ADC with 256× oversampling
- the signal-to-noise ratio due to quantization noise will be higher than if the whole available band had been used. With this technique, it is possible to obtain an effective resolution larger than that provided by the converter alone
- The improvement in SNR is 3 dB (equivalent to 0.5 bits) per octave of oversampling which is not sufficient for many applications. Therefore, oversampling is usually coupled with noise shaping. With noise shaping, the improvement is 6L+3 dB per octave where L is the order of loop filter used for noise shaping. e.g. – a 2nd order loop filter will provide an improvement of 15 dB/octave.

Oversampling is typically used in audio frequency ADCs where the required sampling rate (typically 44.1 or 48 kHz) is very low compared to the clock speed of typical transistor circuits (>1 MHz). In this case, by using the extra bandwidth to distribute quantization error onto out of band frequencies, the accuracy of the ADC can be greatly increased at no cost. Furthermore, as any aliased signals are also typically out of band, aliasing can often be completely eliminated using very low cost filters.

Relative Speed and Precision

The speed of an ADC varies by type. The Wilkinson ADC is limited by the clock rate which is processable by current digital circuits. Currently, frequencies up to 300 MHz are possible. For a successive-approximation ADC, the conversion time scales with the logarithm of the resolution, e.g. the number of bits. Thus for high resolution, it is possible that the successive-approximation ADC is faster than the Wilkinson. However, the time consuming steps in the Wilkinson are digital, while those in the successive-approximation are analog. Since analog is inherently slower than digital, as the resolution increases, the time required also increases. Thus there are competing processes at work. Flash ADCs are certainly the fastest type of the three. The conversion is basically performed in a single parallel step. For an 8-bit unit, conversion takes place in a few tens of nanoseconds.

There is, as expected, somewhat of a tradeoff between speed and precision. Flash ADCs have drifts and uncertainties associated with the comparator levels. This results in poor linearity. For successive-approximation ADCs, poor linearity is also present, but less so than for flash ADCs. Here, non-linearity arises from accumulating errors from the subtraction processes. Wilkinson ADCs have the highest linearity of the three. These have the best differential non-linearity. The other types require channel smoothing to achieve the level of the Wilkinson.

Sliding Scale Principle

The sliding scale or randomizing method can be employed to greatly improve the linearity of any type of ADC, but especially flash and successive approximation types. For any ADC the mapping from input voltage to digital output value is not exactly a floor or ceiling function as it should be. Under normal conditions, a pulse of a particular amplitude is always converted to a digital value. The problem lies in that the ranges of analog values for the digitized values are not all of the same width, and the differential linearity decreases proportionally with the divergence from the average width. The sliding scale principle uses an averaging effect to overcome this phenomenon. A random, but known analog voltage is added to the sampled input voltage. It is then converted to digital form, and the equivalent digital amount is subtracted, thus restoring it to its original value. The advantage is that the conversion has taken place at a random point. The statistical distribution of the final levels is decided by a weighted average over a region of the range of the ADC. This in turn desensitizes it to the width of any specific level.

Types

These are the most common ways of implementing an electronic ADC:

Direct-conversion

A direct-conversion ADC or flash ADC has a bank of comparators sampling the input signal in parallel, each firing for their decoded voltage range. The comparator bank feeds a logic circuit that generates a code for each voltage range. Direct conversion is very fast, capable of gigahertz sampling rates, but usually has only 8 bits of resolution or fewer, since the number of comparators needed, $2^N - 1$, doubles with each additional bit, requiring a large, expensive circuit. ADCs of this type have a large die size, a high input capacitance, high power dissipation, and are prone to produce glitches at the output (by outputting an out-of-sequence code). Scaling to newer submicrometre technologies does not help as the device mismatch is the dominant design limitation. They are often used for video, wideband communications or other fast signals in optical storage.

Successive Approximation

A successive-approximation ADC uses a comparator to successively narrow a range that contains the input voltage. At each successive step, the converter compares the input voltage to the output of an internal digital to analog converter which might represent the midpoint of a selected voltage range. At each step in this process, the approximation is stored in a successive approximation register (SAR). For example, consider an input voltage of 6.3 V and the initial range is 0 to 16 V. For the first step, the input 6.3 V is compared to 8 V (the midpoint of the 0–16 V range). The comparator reports that the input voltage is less than 8 V, so the SAR is updated to narrow the range to 0–8 V. For the second step, the input voltage is compared to 4 V (midpoint of 0–8). The comparator reports the input voltage is above 4 V, so the SAR is updated to reflect the input voltage is in the range 4–8 V. For the third step, the input voltage is compared with 6 V (halfway between 4 V and 8 V); the comparator reports the input voltage is greater than 6 volts, and search range becomes 6–8 V. The steps are continued until the desired resolution is reached.

Ramp-compare

A ramp-compare ADC produces a saw-tooth signal that ramps up or down then quickly returns to zero. When the ramp starts, a timer starts counting. When the ramp voltage matches the input, a comparator fires, and the timer's value is recorded. Timed ramp converters require the least number of transistors. The ramp time is sensitive to temperature because the circuit generating the ramp is often a simple oscillator. There are two solutions: use a clocked counter driving a DAC and then use the comparator to preserve the counter's value, or calibrate the timed ramp. A special advantage of the ramp-compare system is that comparing a second signal just requires another comparator, and another register to store the voltage value. A very simple (non-linear) ramp-converter can be implemented with a microcontroller and one resistor and capacitor. Vice versa, a filled capacitor can be taken from an integrator, time-to-amplitude converter, phase detector, sample and hold circuit, or peak and hold circuit and discharged. This has the advantage that a slow comparator cannot be disturbed by fast input changes.

Wilkinson

The Wilkinson ADC was designed by D. H. Wilkinson in 1950. The Wilkinson ADC is based on the comparison of an input voltage with that produced by a charging capacitor. The capacitor is allowed to charge until its voltage is equal to the amplitude of the input pulse (a comparator determines when this condition has been reached). Then, the capacitor is allowed to discharge linearly, which produces a ramp voltage. At the point when the capacitor begins to discharge, a gate pulse is initiated. The gate pulse remains on until the capacitor is completely discharged. Thus the duration of the gate pulse is directly proportional to the amplitude of the input pulse. This gate pulse operates a linear gate which receives pulses from a high-frequency oscillator clock. While the gate is open, a discrete number of clock pulses pass through the linear gate and are counted by the address register. The time the linear gate is open is proportional to the amplitude of the input pulse, thus the number of clock pulses recorded in the address register is proportional also. Alternatively, the charging of the capacitor could be monitored, rather than the discharge.

Integrating

An integrating ADC (also dual-slope or multi-slope ADC) applies the unknown input voltage to the input of an integrator and allows the voltage to ramp for a fixed time period (the run-up period). Then a known reference voltage of opposite polarity is applied to the integrator and is allowed to ramp until the integrator output returns to zero (the run-down period). The input voltage is computed as a function of the reference voltage, the constant run-up time period, and the measured run-down time period. The run-down time measurement is usually made in units of the converter's clock, so longer integration times allow for higher resolutions. Likewise, the speed of the converter can be improved by sacrificing resolution. Converters of this type (or variations on the concept) are used in most digital voltmeters for their linearity and flexibility.

Delta-encoded

A delta-encoded ADC or counter-ramp has an up-down counter that feeds a digital to analog converter (DAC). The input signal and the DAC both go to a comparator. The comparator controls the counter. The circuit uses negative feedback from the comparator to adjust the counter until the DAC's output is close enough to the input signal.

The number is read from the counter. Delta converters have very wide ranges and high resolution, but the conversion time is dependent on the input signal level, though it will always have a guaranteed worst-case. Delta converters are often very good choices to read real-world signals. Most signals from physical systems do not change abruptly. Some converters combine the delta and successive approximation approaches; this works especially well when high frequencies are known to be small in magnitude.

Pipeline

A pipeline ADC (also called subranging quantizer) uses two or more steps of subranging. First, a coarse conversion is done. In a second step, the difference to the input signal is determined with a digital to analog converter (DAC). This difference is then converted finer, and the results are combined in a last step. This can be considered a refinement of the successive-approximation ADC wherein the feedback reference signal consists of the interim conversion of a whole range of bits (for example, four bits) rather than just the next-most-significant bit. By combining the merits of the successive approximation and flash ADCs this type is fast, has a high resolution, and only requires a small die size.

Sigma-delta

A sigma-delta ADC (also known as a delta-sigma ADC) oversamples the desired signal by a large factor and filters the desired signal band. Generally, a smaller number of bits than required are converted using a Flash ADC after the filter. The resulting signal, along with the error generated by the discrete levels of the Flash, is fed back and subtracted from the input to the filter. This negative feedback has the effect of noise shaping the error due to the Flash so that it does not appear in the desired signal frequencies. A digital filter (decimation filter) follows the ADC which reduces the sampling rate, filters off unwanted noise signal and increases the resolution of the output (sigma-delta modulation, also called delta-sigma modulation).

Time-interleaved

A time-interleaved ADC uses M parallel ADCs where each ADC samples data every M:th cycle of the effective sample clock. The result is that the sample rate is increased M times compared to what each individual ADC can manage. In practice, the individual differences between the M ADCs degrade the overall performance reducing the SFDR. However, technologies exist to correct for these time-interleaving mismatch errors.

Intermediate FM Stage

An ADC with intermediate FM stage first uses a voltage-to-frequency converter to convert the desired signal into an oscillating signal with a frequency proportional to the voltage of the desired signal, and then uses a frequency counter to convert that frequency into a digital count proportional to the desired signal voltage. Longer integration times allow for higher resolutions. Likewise, the speed of the converter can be improved by sacrificing resolution. The two parts of the ADC may be widely separated, with the frequency signal passed through an opto-isolator or transmitted wirelessly. Some such ADCs use sine wave or square wave frequency modulation; others use pulse-frequency modulation. Such ADCs were once the most popular way to show a digital display of the status of a remote analog sensor.

Other Types

There can be other ADCs that use a combination of electronics and other technologies. A time-stretch analog-to-digital converter (TS-ADC) digitizes a very wide bandwidth analog signal, that cannot be digitized by a conventional electronic ADC, by time-stretching the signal prior to digitization. It commonly uses a photonic preprocessor frontend to time-stretch the signal, which effectively slows the signal down in time and compresses its bandwidth. As a result, an electronic backend ADC, that would have been too slow to capture the original signal, can now capture this slowed down signal. For continuous capture of the signal, the frontend also divides the signal into multiple segments in addition to time-stretching. Each segment is individually digitized by a separate electronic ADC. Finally, a digital signal processor rearranges the samples and removes any distortions added by the frontend to yield the binary data that is the digital representation of the original analog signal.

Commercial

Commercial ADCs are usually implemented as integrated circuits. Most converters sample with 6 to 24 bits of resolution, and produce fewer than 1 megasample per second. Thermal noise generated by passive components such as resistors masks the measurement when higher resolution is desired. For audio applications and in room temperatures, such noise is usually a little less than 1 μV (microvolt) of white noise. If the MSB corresponds to a standard 2 V of output signal, this translates to a noise-limited performance that is less than 20~21 bits, and obviates the need for any dithering. As of February 2002, Mega- and giga-sample per second converters are available. Mega-sample converters are required in digital video cameras, video capture cards, and TV tuner cards to convert full-speed analog video to digital video files. Commercial converters usually have ±0.5 to ±1.5 LSB error in their output.

In many cases, the most expensive part of an integrated circuit is the pins, because they make the package larger, and each pin has to be connected to the integrated circuit's silicon. To save pins, it is common for slow ADCs to send their data one bit at a time over a serial interface to the computer, with the next bit coming out when a clock signal changes state, say from 0 to 5 V. This saves quite a few pins on the ADC package, and in many cases, does not make the overall design any more complex (even microprocessors which use memory-mapped I/O only need a few bits of a port to implement a serial bus to an ADC). Commercial ADCs often have several inputs that feed the same converter, usually through an analog multiplexer. Different models of ADC may include sample and hold circuits, instrumentation amplifiers or differential inputs, where the quantity measured is the difference between two voltages.

Applications

Music Recording

Analog-to-digital converters are integral to 2000s era music reproduction technology and digital audio workstation-based sound recording. People often produce music on computers using an analog recording and therefore need analog-to-digital converters to create the pulse-code modulation (PCM) data streams that go onto compact discs and digital music files. The current crop of analog-to-digital converters utilized in music can sample at rates up to 192 kilohertz. Considerable literature exists on these matters, but commercial considerations often play a significant role. Most high-profile recording studios record in 24-bit/192-176.4 kHz pulse-code modulation (PCM) or in Direct Stream Digital (DSD) formats, and then downsample or decimate the signal for Red-Book CD production (44.1 kHz) or to 48 kHz for commonly used radio and television broadcast applications.

Digital Signal Processing

People must use ADCs to process, store, or transport virtually any analog signal in digital form. TV tuner cards, for example, use fast video analog-to-digital converters. Slow on-chip 8, 10, 12, or 16 bit analog-to-digital converters are common in microcontrollers. Digital storage oscilloscopes need very fast analog-to-digital converters, also crucial for software defined radio and their new applications.

Scientific Instruments

Digital imaging systems commonly use analog-to-digital converters in digitizing pixels. Some radar systems commonly use analog-to-digital converters to convert signal strength to digital values for subsequent signal processing. Many other in situ and remote sensing systems commonly use analogous technology. The number of binary bits in the resulting digitized numeric values reflects the resolution, the number of unique discrete levels of quantization (signal processing). The correspondence between the analog signal and the digital signal depends on the quantization error. The quantization process must occur at an adequate speed, a constraint that may limit the resolution of the digital signal. Many sensors in scientific instruments produce an analog signal; temperature, pressure, pH, light intensity etc. All these signals can be amplified and fed to an ADC to produce a digital number proportional to the input signal.

Rotary Encoder

Some non-electronic or only partially electronic devices, such as rotary encoders, can also be considered ADCs. Typically the digital output of an ADC will be a two's complement binary number that is proportional to the input. An encoder might output a Gray code.

Electrical Symbol

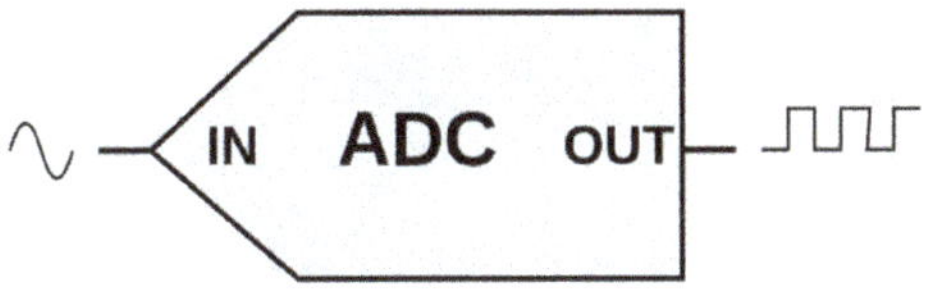

Testing

Testing an Analog to Digital Converter requires an analog input source and hardware to send control signals and capture digital data output. Some ADCs also require an accurate source of reference signal.

The key parameters to test a SAR ADC are:

1. DC offset error
2. DC gain error
3. Signal to noise ratio (SNR)
4. Total harmonic distortion (THD)
5. Integral non linearity (INL)
6. Differential non linearity (DNL)
7. Spurious free dynamic range
8. Power dissipation

Digital-to-analog Converter

In electronics, a digital-to-analog converter (DAC, D/A, D–A, D2A, or D-to-A) is a device that converts a digital signal into an analog signal. An analog-to-digital converter (ADC) performs the reverse function.

There are several DAC architectures; the suitability of a DAC for a particular application is determined by six main parameters: physical size, power consumption, resolution, maximum sampling frequency, accuracy and cost. Due to the complexity and the need for precisely matched components, all but the most specialized DACs are implemented as integrated circuits (ICs). Digital-to-analog conversion can degrade a signal, so a DAC should be specified that has insignificant errors in terms of the application.

8-channel Cirrus Logic CS4382 digital-to-analog converter as used in a sound card.

DACs are commonly used in music players to convert digital data streams into analog audio signals. They are also used in televisions and mobile phones to convert digital video data into analog video signals which connect to the screen drivers to display monochrome or color images. These two applications use DACs at opposite ends of the speed/resolution trade-off. The audio DAC is a low speed high resolution type while the video DAC is a high speed low to medium resolution type. Discrete DACs would typically be extremely high speed low resolution power hungry types, as used in military radar systems. Very high speed test equipment, especially sampling oscilloscopes, may also use discrete DACs.

Overview

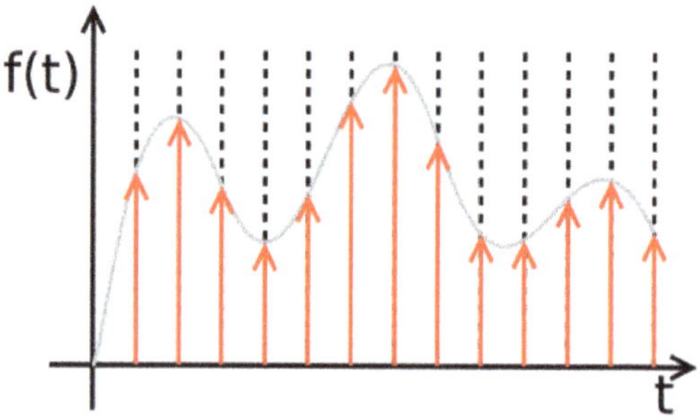

Ideally sampled signal.

A DAC converts an abstract finite-precision number (usually a fixed-point binary number) into a physical quantity (e.g., a voltage or a pressure). In particular, DACs are often used to convert finite-precision time series data to a continually varying physical signal.

An *ideal* DAC converts the abstract numbers into a conceptual sequence of impulses that are then processed by a reconstruction filter using some form of interpolation to fill in data between the impulses. A conventional *practical* DAC converts the numbers into a piecewise constant function made up of a sequence of rectangular functions that is modeled with the zero-order hold. Other DAC methods such as those based on delta-sigma modulation) produce a pulse-density modulated output that can be similarly filtered to produce a smoothly varying signal.

As per the Nyquist–Shannon sampling theorem, a DAC can reconstruct the original signal from the sampled data provided that its bandwidth meets certain requirements (e.g., a baseband signal with bandwidth less than the Nyquist frequency). Digital sampling introduces quantization error that manifests as low-level noise added to the reconstructed signal.

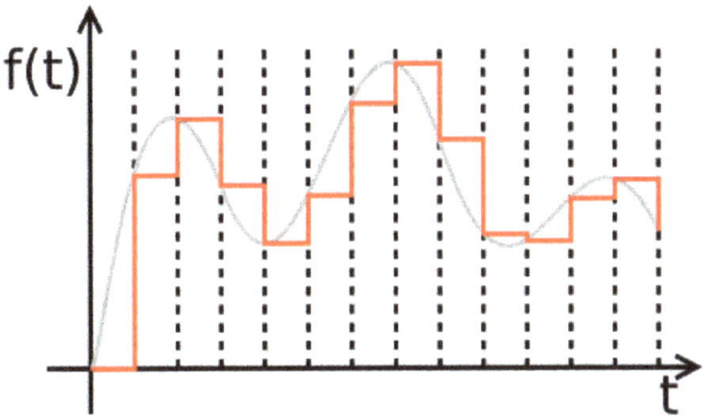

Piecewise constant output of a conventional DAC lacking a reconstruction filter. In a practical DAC, a filter or the finite bandwidth of the device smooths out the step response into a continuous curve.

Instead of impulses, a conventional *practical* DAC updates the analog voltage at uniform sampling intervals, which is then interpolated via a reconstruction filter to continuously varied levels.

These numbers are written to the DAC, typically with a clock signal that causes each number to be latched in sequence, at which time the DAC output voltage changes rapidly from the previous value to the value represented by the currently latched number. The effect of this is that the output voltage is *held* in time at the current value until the next input number is latched, resulting in a piecewise constant or staircase-shaped output. This is equivalent to a zero-order hold operation and has an effect on the frequency response of the reconstructed signal.

The fact that DACs output a sequence of piecewise constant values (known as zero-order hold in sample data textbooks) or rectangular pulses causes multiple harmonics above the Nyquist frequency. Usually, these are removed with a low pass filter acting as a reconstruction filter in applications that require it.

Other DAC methods (e.g., methods based on delta-sigma modulation) produce a pulse-density modulated signal that can then be filtered in a similar way to produce a smoothly varying signal.

Applications

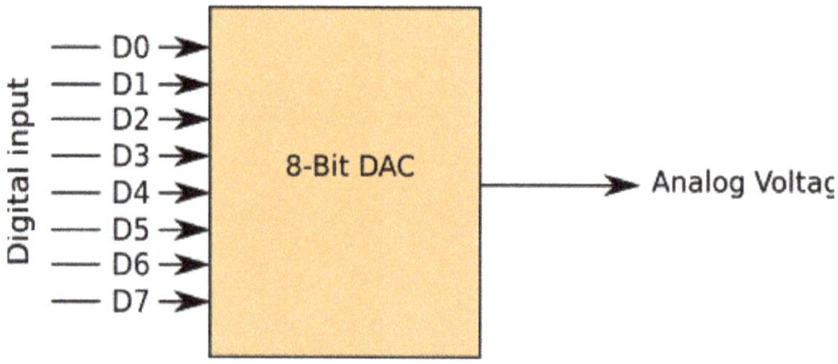

A simplified functional diagram of an 8-bit DAC

DACs and ADCs are part of an enabling technology that has contributed greatly to the digital revolution. To illustrate, consider a typical long-distance telephone call. The caller's voice is converted into an analog electrical signal by a microphone, then the analog signal is converted to a digital stream by an ADC. The digital stream is then divided into network packets where it may be sent along with other digital data, not necessarily audio. The packets are then received at the destination, but each packet may take a completely different route and may not even arrive at the destination in the correct time order. The digital voice data is then extracted from the packets and assembled into a digital data stream. A DAC converts this back into an analog electrical signal, which drives an audio amplifier, which in turn drives a loudspeaker, which finally produces sound.

Audio

Most modern audio signals are stored in digital form (for example MP3s and CDs) and in order to be heard through speakers they must be converted into an analog signal. DACs are therefore found in CD players, digital music players, and PC sound cards.

Specialist standalone DACs can also be found in high-end hi-fi systems. These normally take the digital output of a compatible CD player or dedicated transport (which is basically a CD player with no internal DAC) and convert the signal into an analog line-level output that can then be fed into an amplifier to drive speakers.

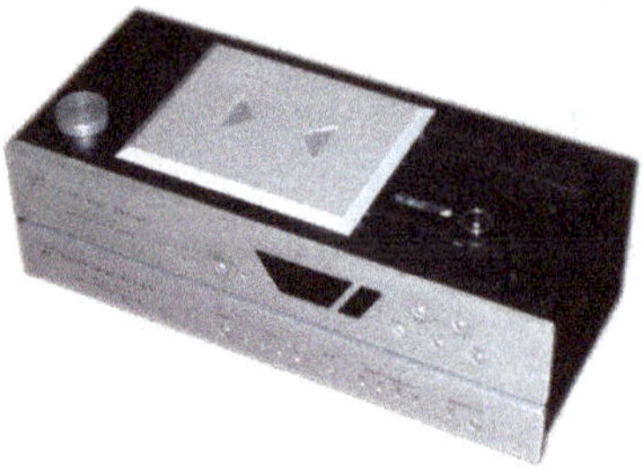

Top-loading CD player and external digital-to-analog converter.

Similar digital-to-analog converters can be found in digital speakers such as USB speakers, and in sound cards.

In VoIP (Voice over IP) applications, the source must first be digitized for transmission, so it undergoes conversion via an analog-to-digital converter, and is then reconstructed into analog using a DAC on the receiving party's end.

Video

Video sampling tends to work on a completely different scale altogether thanks to the highly nonlinear response both of cathode ray tubes (for which the vast majority of digital video foundation work was targeted) and the human eye, using a "gamma curve" to provide an appearance of evenly distributed brightness steps across the display's full dynamic range - hence the need to use RAMDACs in computer video applications with deep enough colour resolution to make engineering a hardcoded value into the DAC for each output level of each channel impractical (e.g. an Atari ST or Sega Genesis would require 24 such values; a 24-bit video card would need 768...). Given this inherent distortion, it is not unusual for a television or video projector to truthfully claim a linear contrast ratio (difference between darkest and brightest output levels) of 1000:1 or greater, equivalent to 10 bits of audio precision even though it may only accept signals with 8-bit precision and use an LCD panel that only represents 6 or 7 bits per channel.

Video signals from a digital source, such as a computer, must be converted to analog form if they are to be displayed on an analog monitor. As of 2007, analog inputs were more commonly used than digital, but this changed as flat panel displays with DVI and/or HDMI connections became more widespread. A video DAC is, however, incorporated in any digital video player with analog outputs. The DAC is usually integrated with some memory (RAM), which contains conversion tables for gamma correction, contrast and brightness, to make a device called a RAMDAC.

A device that is distantly related to the DAC is the digitally controlled potentiometer, used to control an analog signal digitally.

Mechanical

An unusual application of digital-to-analog conversion was the whiffletree electromechanical digital-to-analog converter linkage in the IBM Selectric typewriter.

A one-bit mechanical actuator assumes two positions: one when on, another when off. The motion of several one-bit actuators can be combined and weighted with a whiffletree mechanism to produce finer steps. The IBM Selectric typewriter uses such as system. When a typewriter key is pressed, it moves a metal bar (interposer) down that has several lugs. The lugs are the information bits. When a key is pressed, its interposer is moved by the motor. If a lug is present at one position, it will move the corresponding selector bail (bar); if the lug is not present, the selector bail stays where it is. The discrete motions of the bails are combined by a whiffle tree, and the output controls the rotation and tilt of the Selectric's typeball.

Types

The most common types of electronic DACs are:

- The pulse-width modulator, the simplest DAC type. A stable current or voltage is switched into a low-pass analog filter with a duration determined by the digital input code. This technique is often used for electric motor speed control, but has many other applications as well.
- Oversampling DACs or interpolating DACs such as the delta-sigma DAC, use a pulse density conversion technique. The oversampling technique allows for the use of a lower resolution DAC internally. A simple 1-bit DAC is often chosen because the oversampled result is inherently linear. The DAC is driven with a pulse-density modulated signal, created with the use of a low-pass filter, step nonlinearity (the actual 1-bit DAC), and negative feedback loop, in a technique called delta-sigma modulation. This results in an effective high-pass filter acting on the quantization (signal processing) noise, thus steering this noise out of the low frequencies of interest into the megahertz frequencies of little interest, which is called noise shaping. The quantization noise at these high frequencies is removed or greatly attenuated by use of an analog low-pass filter at the output (sometimes a simple RC low-pass circuit is sufficient). Most very high resolution DACs (greater than 16 bits) are of this type due to its high linearity and low cost. Higher oversampling rates can relax the specifications of the output low-pass filter and enable further suppression of quantization noise. Speeds of greater than 100 thousand samples per second (for example, 192 kHz) and resolutions of 24 bits are attainable with delta-sigma DACs. A short comparison with pulse-width modulation shows that a 1-bit DAC with a simple first-order integrator would have to run at 3 THz (which is physically unrealizable) to achieve 24 meaningful bits of resolution, requiring a higher-order low-pass filter in the noise-shaping loop. A single integrator is a low-pass filter with a frequency response inversely proportional to frequency and using one such integrator in the noise-shaping loop is a first order delta-sigma modulator. Multiple higher order topologies (such as MASH) are used to achieve higher degrees of noise-shaping with a stable topology.
- The binary-weighted DAC, which contains individual electrical components for each bit of the DAC connected to a summing point. These precise voltages or currents sum to the correct output value. This is one of the fastest conversion methods but suffers from poor accuracy because of the high precision required for each individual voltage or current. Such high-precision components are expensive, so this type of converter is usually limited to 8-bit resolution or less.

- Switched resistor DAC contains a parallel resistor network. Individual resistors are enabled or bypassed in the network based on the digital input.
- Switched current source DAC, from which different current sources are selected based on the digital input.
- Switched capacitor DAC contains a parallel capacitor network. Individual capacitors are connected or disconnected with switches based on the input.

- The R-2R ladder DAC which is a binary-weighted DAC that uses a repeating cascaded structure of resistor values R and 2R. This improves the precision due to the relative ease of producing equal valued-matched resistors (or current sources).
- The Successive-Approximation or Cyclic DAC, which successively constructs the output during each cycle. Individual bits of the digital input are processed each cycle until the entire input is accounted for.
- The thermometer-coded DAC, which contains an equal resistor or current-source segment for each possible value of DAC output. An 8-bit thermometer DAC would have 255 segments, and a 16-bit thermometer DAC would have 65,535 segments. This is perhaps the fastest and highest precision DAC architecture but at the expense of high cost. Conversion speeds of >1 billion samples per second have been reached with this type of DAC.
- Hybrid DACs, which use a combination of the above techniques in a single converter. Most DAC integrated circuits are of this type due to the difficulty of getting low cost, high speed and high precision in one device.
    - The segmented DAC, which combines the thermometer-coded principle for the most significant bits and the binary-weighted principle for the least significant bits. In this way, a compromise is obtained between precision (by the use of the thermometer-coded principle) and number of resistors or current sources (by the use of the binary-weighted principle). The full binary-weighted design means 0% segmentation, the full thermometer-coded design means 100% segmentation.
- Most DACs, shown earlier in this list, rely on a constant reference voltage to create their output value. Alternatively, a *multiplying DAC* takes a variable input voltage for their conversion. This puts additional design constraints on the bandwidth of the conversion circuit.

Performance

DACs are very important to system performance. The most important characteristics of these devices are:

Resolution

The number of possible output levels the DAC is designed to reproduce. This is usually stated as the number of bits it uses, which is the base two logarithm of the number of levels. For instance a 1 bit DAC is designed to reproduce 2 (2^1) levels while an 8 bit DAC is designed for 256 (2^8) levels. Resolution is related to the effective number of bits which is a measurement of the actual resolution attained by the DAC. Resolution determines color depth in video applications and audio bit depth in audio applications.

Maximum Sampling Rate

A measurement of the maximum speed at which the DACs circuitry can operate and still produce the correct output. As stated above, the Nyquist–Shannon sampling theorem defines a relationship between this and the bandwidth of the sampled signal.

Monotonicity

The ability of a DAC's analog output to move only in the direction that the digital input moves (i.e., if the input increases, the output doesn't dip before asserting the correct output.) This characteristic is very important for DACs used as a low frequency signal source or as a digitally programmable trim element.

Total Harmonic Distortion and Noise (THD+N)

A measurement of the distortion and noise introduced to the signal by the DAC. It is expressed as a percentage of the total power of unwanted harmonic distortion and noise that accompany the desired signal. This is a very important DAC characteristic for dynamic and small signal DAC applications.

Dynamic Range

A measurement of the difference between the largest and smallest signals the DAC can reproduce expressed in decibels. This is usually related to resolution and noise floor.

Other measurements, such as phase distortion and jitter, can also be very important for some applications, some of which (e.g. wireless data transmission, composite video) may even *rely* on accurate production of phase-adjusted signals.

Linear PCM audio sampling usually works on the basis of each bit of resolution being equivalent to 6 decibels of amplitude (a 2x increase in volume or precision).

Non-linear PCM encodings (A-law / μ-law, ADPCM, NICAM) attempt to improve their effective dynamic ranges by a variety of methods - logarithmic step sizes between the output signal strengths represented by each data bit (trading greater quantisation distortion of loud signals for better performance of quiet signals)

Figures of Merit

- Static performance:
    - Differential nonlinearity (DNL) shows how much two adjacent code analog values deviate from the ideal 1 LSB step.
    - Integral nonlinearity (INL) shows how much the DAC transfer characteristic deviates from an ideal one. That is, the ideal characteristic is usually a straight line; INL shows how much the actual voltage at a given code value differs from that line, in LSBs (1 LSB steps).
    - Gain
    - Offset
    - Noise is ultimately limited by the thermal noise generated by passive components such as resistors. For audio applications and in room temperatures, such noise is usually a little less than 1 μV (microvolt) of white noise. This limits performance to less than 20~21 bits even in 24-bit DACs.
- Frequency domain performance
    - Spurious-free dynamic range (SFDR) indicates in dB the ratio between the powers of the converted main signal and the greatest undesired spur.
    - Signal-to-noise and distortion ratio (SNDR) indicates in dB the ratio between the powers of the converted main signal and the sum of the noise and the generated harmonic spurs
    - i-th harmonic distortion (HDi) indicates the power of the i-th harmonic of the converted main signal
    - Total harmonic distortion (THD) is the sum of the powers of all HDi
    - If the maximum DNL error is less than 1 LSB, then the D/A converter is guaranteed to be monotonic. However, many monotonic converters may have a maximum DNL greater than 1 LSB.
- Time domain performance:
    - Glitch impulse area (glitch energy)
    - Response uncertainty
    - Time nonlinearity (TNL)

Time-to-digital Converter

In electronic instrumentation and signal processing, a time to digital converter (abbreviated TDC) is a device for recognizing events and providing a digital representation of the time they occurred. For example, a TDC might output the time of arrival for each incoming pulse. Some applications wish to measure the time interval between two events rather than some notion of an absolute time.

In electronics time-to-digital converters (TDCs) or time digitizers are devices commonly used to measure a time interval and convert it into digital (binary) output. In some cases interpolating TDCs are also called time counters (TCs).

TDCs are used in many different applications, where the time interval between two signal pulses (start and stop pulse) should be determined. Measurement is started and stopped, when either the rising or the falling edge of a signal pulse crosses a set threshold. These requirements are fulfilled in many physical experiments, like time-of-flight and lifetime measurements in atomic and high energy physics, experiments that involve laser ranging and electronic research involving the testing of integrated circuits and high-speed data transfer.

Application

TDCs are used in applications where measurement events happen infrequently, such as high energy physics experiments, where the sheer number of data channels in most detectors ensures that each channel will be excited only infrequently by particles such as electrons, photons, and ions.

Coarse Measurement

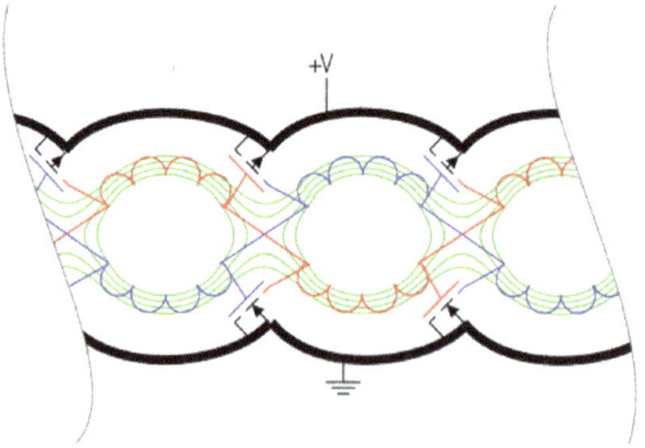

A CMOS (rotary) traveling wave oscillator or delay line or distributed amplifier runs at a flip-flop compatible frequency, but has sharper edges and sub-edge resolution

If the required time resolution is not high, then counters can be used to make the conversion.

Basic Counter

In its simplest implementation, a TDC is simply a high-frequency counter that increments every clock cycle. The current contents of the counter represents the current time. When an event occurs, the counter's value is captured in an output register.

In that approach, the measurement is an integer number of clock cycles, so the measurement is quantized to a clock period. To get finer resolution, a faster clock is needed. The accuracy of the measurement depends upon the stability of the clock frequency.

Typically a TDC uses a crystal oscillator reference frequency for good long term stability. High stability crystal oscillators are usually relative low frequency such as 10 MHz (or 100 ns resolution). To get better resolution, a phase-locked loop frequency multiplier can be used to generate a faster clock. One might, for example, multiply the crystal reference oscillator by 100 to get a clock rate of 1 GHz (1 ns resolution).

Counter Technology

High clock rates impose additional design constraints on the counter: if the clock period is short, it is difficult to update the count. Binary counters, for example, need a fast carry architecture because they essentially add one to the previous counter value. A solution is using a hybrid counter architecture. A Johnson counter, for example, is a fast non-binary counter. It can be used to count very quickly the low order count; a more conventional binary counter can be used to accumulate the high order count. The fast counter is sometime called a prescaler.

The speed of counters fabricated in CMOS-technology is limited by the capacitance between the gate and the channel and by the resistance of the channel and the signal traces. The product of both is the cut-off-frequency. Modern chip technology allows multiple metal layers and therefore coils with a large number of windings to be inserted into the chip. This allows designers to peak the device for a specific frequency, which may lie above the cut-off-frequency of the original transistor.

A peaked variant of the Johnson counter is the traveling-wave counter which also achieves sub-cycle resolution. Other methods to achieve sub-cycle resolution include analog-to-digital converters and vernier Johnson counters.

Measuring a Time Interval

In most situations, the user does not want to just capture an arbitrary time that an event occurs, but wants to measure a time interval, the time between a start event and a stop event.

That can be done by measuring an arbitrary time both the start and stop events and subtracting. The measurement can be off by two counts.

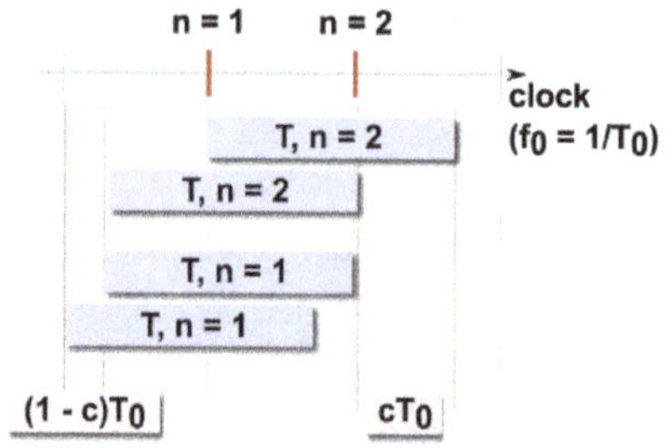

sketch of the coarse counting method in TDCs: showing measurements of T in various relations to the clock pulses

The subtraction can be avoided if the counter is held at zero until the start event, counts during the interval, and then stops counting after the stop event.

Coarse counters base on a reference clock with signals generated at a stable frequency f_0. When the start signal is detected the counter starts counting clock signals and terminates counting after the stop signal is detected. The time interval T between start and stop is then

$$T = n \cdot T_0$$

with n, the number of counts and $T_0 = 1/f_0$, the period of the reference clock.

Statistical Counter

Since start, stop and clock signal are asynchronous, there is a uniform probability distribution of the start and stop signal-times between two subsequent clock pulses. This detuning of the start and stop signal from the clock pulses is called quantization error.

For a series of measurements on the same constant and asynchronous time interval one measures two different numbers of counted clock pulses n_1 and n_2. These occur with probabilities

$$p(n_1) = 1 - c$$

$$q(n_2) = c$$

with $c = Frc(T/T_0)$ the fractional part of T/T_0. The value for the time interval is then obtained by

$$T = (p \cdot n_1 + q \cdot n_2) \cdot T_0$$

Measuring a time interval using a coarse counter with the averaging method described above is relatively time consuming because of the many repetitions that are needed to determine the probabilities p and q. In comparison to the other methods described later on, a coarse counter has a very limited resolution (1ns in case of a 1 GHz reference clock), but satisfies with its theoretically unlimited measuring range.

Fine Measurement

In contrast to the coarse counter in the previous section, fine measurement methods with much better accuracy but far smaller measuring range are presented here. Analogue methods like time interval stretching or double conversion as well as digital methods like tapped delay lines and the Vernier method are under examination. Though the analogue methods still obtain better accuracies, digital time interval measurement is often preferred due to its flexibility in integrated circuit technology and its robustness against external perturbations like temperature changes.

The counter implementation's accuracy is limited by the clock frequency. If time is measured by whole counts, then the resolution is limited to the clock period. For example, a 10 MHz clock has a resolution of 100 ns. To get resolution finer than a clock period, there are time interpolation circuits. These circuits measure the fraction of a clock period: that is, the time between a clock event and the event being measured. The interpolation circuits often require a significant amount of time to perform their function; consequently, the TDC needs a quiet interval before the next measurement.

Ramp Interpolator

When counting is not feasible because the clock rate would be too high, analog methods can be used. Analog methods are often used to measure intervals that are between 10 and 200 ns. These methods often use a capacitor that is charged during the interval being measured. Initially, the capacitor is discharged to zero volts. When the start event occurs, the capacitor is charged with a constant current I_1; the constant current causes the voltage v on the capacitor to increase linearly with time. The rising voltage is called the fast ramp. When the stop event occurs, the charging current is stopped. The voltage on the capacitor v is directly proportional to the time interval T and can be measured with an analog-to-digital converter (ADC). The resolution of such a system is in the range of 1 to 10 ps.

Although a separate ADC can be used, the ADC step is often integrated into the interpolator. A second constant current I_2 is used to discharge the capacitor at a constant but much slower rate (the slow ramp). The slow ramp might be 1/1000 of the fast ramp. This discharge effectively "stretches" the time interval; it will take 1000 times as long for the capacitor to discharge to zero volts. The stretched interval can be measured with a counter. The measurement is similar to a dual-slope analog converter.

The dual-slope conversion can take a long time: a thousand or so clock ticks in the scheme described above. That limits how often a measurement can be made (dead time). Resolution of 1 ps with a 100 MHz (10 ns) clock requires a stretch ratio of 10,000 and implies a conversion time of 150 μs. To decrease the conversion time, the interpolator circuit can be used twice in a residual interpolator technique. The fast ramp is used initially as above to determine the time. The slow ramp is only at 1/100. The slow ramp will cross zero at some time during the clock period. When the ramp crosses zero, the fast ramp is turned on again to measure the crossing time ($t_{residual}$). Consequently, the time can be determined to 1 part in 10,000.

Interpolators are often used with a stable system clock. The start event is asynchronous, but the stop event is a following clock. For convenience, imagine that the fast ramp rises exactly 1 volt during a 100 ns clock period. Assume the start event occurs at 67.3 ns after a clock pulse; the fast ramp integrator is triggered and starts rising. The asynchronous start event is also routed through a synchronizer that takes at least two clock pulses. By the next clock pulse, the ramp has risen to .327 V. By the second clock pulse, the ramp has risen to 1.327 V and the synchronizer reports the start event has been seen. The fast ramp is stopped and the slow ramp starts. The synchronizer output can be used to capture system time from a counter. After 1327 clocks, the slow ramp returns to its starting point, and interpolator knows that the event occurred 132.7 ns before the synchronizer reported.

The interpolator is actually more involved because there are synchronizer issues and current switching is not instantaneous. Also, the interpolator must calibrate the height of the ramp to a clock period.

Vernier

Vernier Interpolator

The vernier method is more involved. The method involves a triggerable oscillator and a coincidence circuit. At the event, the integer clock count is stored and the oscillator is started. The triggered oscillator has a slightly different frequency than the clock oscillator. For sake of argument, say the triggered oscillator has a period that is 1 ns faster than the clock. If the event happened 67 ns after the last clock, then the triggered oscillator transition will slide by −1 ns after each subsequent clock pulse. The triggered oscillator will be at 66 ns after the next clock, at 65 ns after the second clock, and so forth. A coincidence detector looks for when the triggered oscillator and the clock transition at the same time, and that indicates the fraction time that needs to be added.

The interpolator design is more involved. The triggerable clock must be calibrated to clock. It must also start quickly and cleanly.

Vernier Method

The Vernier method is a digital version of the time stretching method. Two only slightly detuned oscillators (with frequencies f_1 and f_2) start their signals with the arrival of the start and the stop signal. As soon as the leading edges of the oscillator signals coincide the measurement ends and the number of periods of the oscillators (n_1 and n_2 respectively) lead to the original time interval T:

$$T = \frac{n_1 - 1}{f_1} - \frac{n_2 - 1}{f_2}$$

Since highly reliable oscillators with stable and accurate frequency are still quite a challenge one also realizes the vernier method via two tapped delay lines using two slightly different cell delay times τ. This setting is called differential delay line or vernier delay line.

In the example presented here the first delay line affiliated with the start signal contains cells of D-flip-flops with delay τ_L which are initially set to transparent. During the transition of the start signal through one of those cells, the signal is delayed by τ_L and the state of the flip-flop is sampled as transparent. The second delay line belonging to the stop signal is composed of a series of non-inverting buffers with delay $\tau_B < \tau_L$. Propagating through its channel the stop signal latches the flip-flops of the start signal's delay line. As soon as the stop signal passes the start signal, the latter is stopped and all leftover flip-flops are sampled opaque. Analogous to the above case of the oscillators the wanted time interval T is then

$$T = n \cdot (\tau_1 - \tau_2)$$

with n the number of cells marked as transparent.

Tapped Delay Line

Digital Image Processing

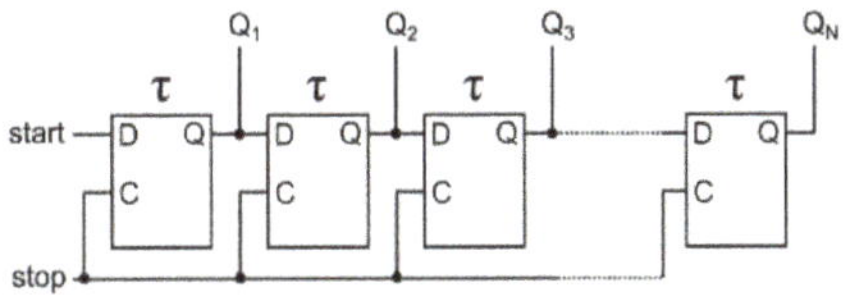

circuit diagram of a tapped delay line

In general a tapped delay line contains a number of cells with well defined delay times. Propagating through this line the start signal is delayed. The state of the line is sampled at the time of the arrival of the stop signal. This can be realized for example with a line of D-flip-flop cells with a delay time τ. The start signal propagates through this line of transparent flip-flops and is delayed by a certain number of them. The output of each flip-flop is sampled on the fly. The stop signal latches all flip-flops while propagating through its channel undelayed and the start signal cannot propagate further. Now the time interval between start and stop signal is proportional to the number of flip-flops that were sampled as transparent.

Hybrid Measurement

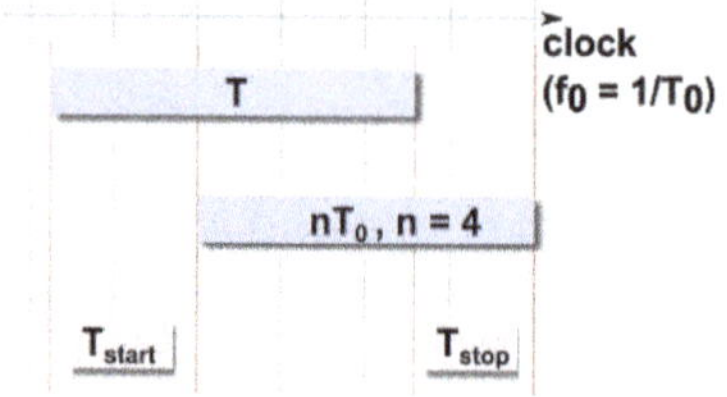

Sketch of the Nutt interpolation method

Counters can measure long intervals but have limited resolution. Interpolators have high resolution but they cannot measure long intervals. A hybrid approach can achieve both long intervals and high resolution. The long interval can be measured with a counter. The counter information is supplemented with two time interpolators: one interpolator measures the (short) interval between the start event and a following clock event, and the second interpolator measure the interval between the stop event and a following clock event. The basic idea has some complications: the start and stop events are asynchronous, and one or both might happen close to a clock pulse. The counter and interpolators must agree on matching the start and end clock events. To accomplish that goal, synchronizers are used.

The common hybrid approach is the Nutt method. In this example the fine measurement circuit measures the time between start and stop pulse and the respective second nearest clock pulse of the coarse counter (T_{start}, T_{stop}), detected by the synchronizer. Thus the wanted time interval is

$$T = nT_0 + T_{start} - T_{stop}$$

with n the number of counter clock pulses and T_0 the period of the coarse counter.

History

Time measurement has played a crucial role in the understanding of nature from the earliest times. Starting with sun, sand or water driven clocks we are able to use clocks today, based on the most precise caesium resonators.

The first direct predecessor of a TDC was invented in the year 1942 by Bruno Rossi for the measurement of muon lifetimes. It was designed as a time-to-amplitude-converter, constantly charging a capacitor during the measured time interval. The corresponding voltage is directly proportional to the time interval under examination.

While the basic concepts (like Vernier methods (Pierre Vernier 1584-1638) and time stretching) of dividing time into measurable intervals are still up-to-date, the implementation changed a lot during the past 50 years. Starting with vacuum tubes and ferrite pot-core transformers those ideas are implemented in complementary metal-oxide-semiconductor (CMOS) design today.

Errors

Some information from

Regarding even the fine measuring methods presented, there are still errors one may wish remove or at least to consider. Non-linearities of the time-to-digital conversion for example can be identified by taking a large number of measurements of a poissonian distributed source (statistical code density test). Small deviations from the uniform distribution reveal the non-linearities. Inconveniently the statistical code density method is quite sensitive to external temperature changes. Thus stabilizing delay or phase-locked loop (DLL or PLL) circuits are recommended.

In a similar way, offset errors (non-zero readouts at $T = 0$) can be removed.

For long time intervals, the error due to instabilities in the reference clock (jitter) plays a major role. Thus clocks of superior quality are needed for such TDCs.

Furthermore, external noise sources can be eliminated in postprocessing by robust estimation methods.

Configurations

TDCs are currently built as stand-alone measuring devices in physical experiments or as system components like PCI cards. They can be made up of either discrete or integrated circuits.

Circuit design changes with the purpose of the TDC, which can either be a very good solution for single-shot TDCs with long dead times or some trade-off between dead-time and resolution for multi-shot TDCs.

Delay Generator

The time-to-digital converter measures the time between a start event and a stop event. There is also a digital-to-time converter or delay generator. The delay generator converts a number to a time delay. When the delay generator gets a start pulse at its input, then it outputs a stop pulse after the specified delay. The architectures for TDC and delay generators are similar. Both use counters for long, stable, delays. Both must consider the problem of clock quantization errors.

For example, the Tektronix 7D11 Digital Delay uses a counter architecture. A digital delay may be set from 100 ns to 1 s in 100 ns increments. An analog circuit provides an additional fine delay of 0 to 100 ns. A 5 MHz reference clock drives a phase-locked loop to produce a stable 500 MHz clock. It is this fast clock that is gated by the (fine-delayed) start event and determines the main quantization error. The fast clock is divided down to 10 MHz and fed to main counter. The instrument quantization error depends primarily on the 500 MHz clock (2 ns steps), but other errors also enter; the instrument is specified to have 2.2 ns of jitter. The recycle time is 575 ns.

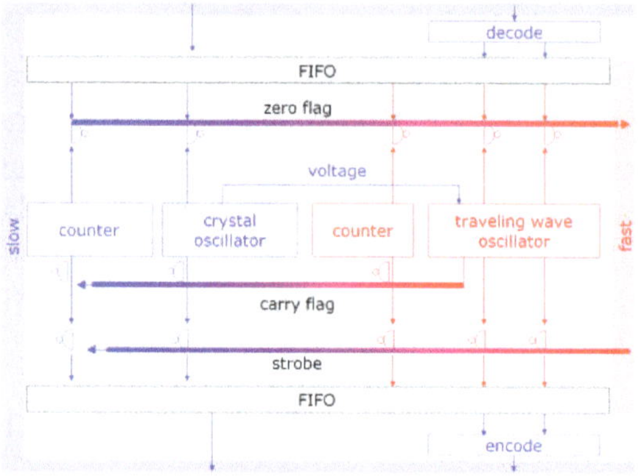

Similarity between a TDC (bottom) and a Delay Generator (top, but needs bottom for trigger). The strobe is gated by the oscillator to avoid a race with the carry bit

Just as a TDC may use interpolation to get finer than one clock period resolution, a delay generator may use similar techniques. The Hewlett-Packard 5359A High Resolution Time Synthesizer provides delays of 0 to 160 ms, has an accuracy of 1 ns, and achieves a typical jitter of 100 ps. The design uses a triggered phase-locked oscillator that runs at 200 MHz. Interpolation is done with a ramp, an 8-bit digital-to-analog converter, and a comparator. The resolution is about 45 ps.

When the start pulse is received, then counts down and outputs a stop pulse. For low jitter the synchronous counter has to feed a zero flag from the most significant bit down to the least significant bit and then combine it with the output from the Johnson counter.

A digital-to-analog converter (DAC) could be used to achieve sub-cycle resolution, but it is easier to either use vernier Johnson counters or traveling-wave Johnson counters.

The delay generator can be used for pulse width modulation, e.g. to drive a MOSFET to load a Pockels cell within 8 ns with a specific charge.

The output of a delay generator can gate a digital-to-analog converter and so pulses of a variable height can be generated. This allows matching to low levels needed by analog electronics, higher levels for ECL and even higher levels for TTL. If a series of DACs is gated in sequence, variable pulse shapes can be generated to account for any transfer function.

Reconstruction Filter

In a mixed-signal system (analog and digital), a reconstruction filter is used to construct a smooth analog signal from a digital input, as in the case of a digital to analog converter (DAC) or other sampled data output device.

Sampled Data Reconstruction Filters

The sampling theorem describes why the input of an ADC requires a low-pass analog electronic filter, called the anti-aliasing filter: the sampled *input* signal must be bandlimited to prevent aliasing (here meaning waves of higher frequency being *recorded* as a lower frequency).

For the same reason, the output of a DAC requires a low-pass analog filter, called a reconstruction filter - because the *output* signal must be bandlimited, to prevent imaging (meaning Fourier coefficients being reconstructed as spurious high-frequency 'mirrors'). This is an implementation of the Whittaker–Shannon interpolation formula.

Ideally, both filters should be brickwall filters, constant phase delay in the pass-band with constant flat frequency response, and zero response from the Nyquist frequency. This can be achieved by a filter with a 'sinc' impulse response.

Implementation

While in theory a DAC outputs a series of discrete Dirac impulses, in practice, a real DAC outputs pulses with finite bandwidth and width. Both idealized Dirac pulses, zero-order held steps and other output pulses, if unfiltered, would contain spurious high-frequency content when compared to the original signal. Thus, the reconstruction filter smooths the waveform to remove image frequencies (copies) above the Nyquist limit. In doing so, it reconstructs the continuous time signal (whether originally sampled, or modelled by digital logic) corresponding to the digital time sequence.

Practical filters have non-flat frequency or phase response in the pass band and incomplete suppression of the signal elsewhere. The ideal sinc waveform has an infinite response to a signal, in both the positive and negative time directions, which is impossible to perform in real time – as it would require infinite delay. Consequently, real reconstruction filters typically either allow some energy above the Nyquist rate, attenuate some in-band frequencies, or both. For this reason, oversampling may be used to ensure that frequencies of interest are accurately reproduced without excess energy being emitted out of band.

In systems that have both, the anti-aliasing filter and a reconstruction filter may be of identical design. For example, both the input and the output for audio equipment may be sampled at 44.1 kHz. In this case, both audio filters block as much as possible above 22 kHz and pass as much as possible below 20 kHz.

Alternatively, a system may have no reconstruction filter and simply tolerate some energy being wasted reproducing higher frequency images of the primary signal spectrum.

Image Processing

In image processing, digital reconstruction filters are used both to recreate images from samples as in medical imaging and for resampling. A number of comparisons have been made, by various criteria; one observation is that reconstruction can be improved if the *derivative* of the signal is also known, in addition to the amplitude, and conversely that also performing derivative reconstruction can improve signal reconstruction methods.

Resampling may be referred to as decimation or interpolation, accordingly as the sampling rate decreases or increases – as in sampling and reconstruction generally, the same criteria generally apply in both cases, and thus the same filter can be used.

For resampling, in principle the analog image is reconstructed, then sampled, and this is necessary for general changes in resolution. For integer ratios of sampling rate, one may simplify by sampling the impulse response of the continuous reconstruction filter to produce a discrete resampling filter, then using the discrete resampling filter to directly resample the image. For decimation by an integer amount, only a single sampled filter is necessary; for interpolation by an integer amount, different samplings are needed for different phases – for instance, if one is upsampling by a factor of 4, then one sampled filter is used for the half-way point, while a different sampled filter is used for the point 1/4 of the way from one point to another.

A subtlety in image processing is that (linear) signal processing assumes linear luminance – that doubling a pixel value doubles the luminance of the output. However, images are frequently gamma encoded, notably in the sRGB color space, so luminance is not linear. Thus to apply a linear filter, one must first gamma decode the values – and if resampling, one must gamma decode, resample, then gamma encode.

Common Filters

The most common day-to-day filters are:

- nearest-neighbor interpolation, with kernel the box filter – for downsampling, this corresponding to averaging;
- bilinear interpolation, with kernel the tent filter;
- bicubic interpolation, with kernel a cubic spline – this latter has a free parameter, with each value of the parameter yielding a different interpolation filter.

These are in increasing order of stopband suppression (anti-aliasing), and decreasing speed

For reconstruction purposes, a variety of kernels are used, many of which can be interpreted as approximating the sinc function, either by windowing or by giving a spline approximation, either by cubics or higher order splines. In the case of windowed sinc filters, the frequency response of the reconstruction filter can be understood in terms of the frequency response of the window, as the frequency response of a windowed filter is the convolution of the original response (for sinc, a brick-wall) with the frequency response of the window. Among these, the Lanczos window and Kaiser window are frequently praised.

Another class of reconstruction filters include the Gaussian for various widths, or cardinal B-splines of higher order – the box filter and tent filter being the 0th and 1st order cardinal B-splines. These filters fail to be interpolating filters, since their impulse response do not vanish at all non-zero original sample points – for 1:1 resampling, they are not the identity, but rather blur. On the other hand, being nonnegative, they do not introduce any overshoot or ringing artifacts, and by being wider in the time domain they can be narrower in the frequency domain (by the Fourier uncertainty principle), though at the cost of blurring, which is reflected in passband roll-off ("scalloping").

In photography, a great variety of interpolation filters exist, some proprietary, for which opinions are mixed. Evaluation is often subjective, with reactions being varied, and some arguing that at realistic resampling ratios, there is little difference between them, as compared with bicubic, though for higher resampling ratios behavior is more varied.

Wavelet Reconstruction Filters

Reconstruction filters are also used when "reconstructing" a waveform or an image from a collection of wavelet coefficients. In medical imaging, a common technique is to use a number of 2D X-ray photos or MRI scans to "reconstruct" a 3D image.

References

- Knoll, Glenn F. (1989). Radiation Detection and Measurement (2nd ed.). New York: John Wiley & Sons. ISBN 0471815047.
- Nicholson, P. W. (1974). Nuclear Electronics. New York: John Wiley & Sons. pp. 315–316. ISBN 0471636975.

9

Allied Fields of Digital Signal Processing

This text provides a plethora of the allied fields of digital signal processing for better comprehension. Audio signal processing, digital image processing and statistical signal processing are discussed in this chapter. Audio signal processing is the modification of audio signals with the help of audio effect whereas digital image processing is the use of computer algorithms to implement image processing on digital images. This section is a compilation of various fields of digital signal processing of the broader subject matter.

Audio Signal Processing

Audio signal processing or audio processing is the intentional alteration of audio signals often through an audio effect or effects unit. As audio signals may be electronically represented in either digital or analog format, signal processing may occur in either domain. Analog processors operate directly on the electrical signal, while digital processors operate mathematically on the digital representation of that signal.

History

Audio signals are electronic representations of sound waves—longitudinal waves which travel through air, consisting of compressions and rarefactions. The energy contained in audio signals is typically measured in decibels. Audio processing was necessary for early radio broadcasting, as there were many problems with studio to transmitter links.

Analog Signals

"Analog" indicates something that is mathematically represented by a set of continuous values; for example, the analog clock uses constantly moving hands on a physical clock face, where moving the hands directly alters the information that clock is providing. Thus, an analog signal is one represented by a continuous stream of data, in this case along an electrical circuit in the form of voltage, current or charge changes *(compare with digital signals below)*. Analog signal processing (ASP) then involves physically altering the continuous signal by changing the voltage or current or charge via various electrical means.

Historically, before the advent of widespread digital technology, ASP was the only method by which to manipulate a signal. Since that time, as computers and software became more advanced, digital signal processing has become the method of choice.

Digital Signals

A digital representation expresses the pressure wave-form as a sequence of symbols, usually binary numbers. This permits signal processing using digital circuits such as microprocessors and computers. Although such a conversion can be prone to loss, most modern audio systems use this approach as the techniques of digital signal processing are much more powerful and efficient than analog domain signal processing.

Application Areas

Processing methods and application areas include storage, level compression, data compression, transmission, enhancement (e.g., equalization, filtering, noise cancellation, echo or reverb removal or addition, etc.)

Audio Broadcasting

Traditionally the most important audio processing (in audio broadcasting) takes place just before the transmitter. Studio audio processing is limited in the modern era due to digital audio systems (mixers, routers) being pervasive in the studio.

In audio broadcasting, the audio processor must

- prevent or minimize overmodulation,
- compensate for non-linear transmitters (a potential issue with medium wave and shortwave broadcasting) and
- adjust overall loudness to desired level

Techniques

Audio unprocessed by reverb and delay is metaphorically referred to as “dry”, while processed audio is referred to as “wet”.

- *echo* - to simulate the effect of reverberation in a large hall or cavern, one or several delayed signals are added to the original signal. To be perceived as echo, the delay has to be of order 35 milliseconds or above. Short of actually playing a sound in the desired environment, the effect of echo can be implemented using either digital or analog methods. Analog echo effects are implemented using tape delays and/or spring reverbs. When large numbers of delayed signals are mixed over several seconds, the resulting sound has the effect of being presented in a large room, and it is more commonly called reverberation or reverb for short.

- *flanger* - to create an unusual sound, a delayed signal is added to the original signal with a continuously variable delay (usually smaller than 10 ms). This effect is now done electronically using DSP, but originally the effect was created by playing the same recording on two synchronized tape players, and then mixing the signals together. As long as the machines were synchronized, the mix would sound more-or-less normal, but if the operator placed his finger on the flange of one of the players (hence "flanger"), that machine would slow down and its signal would fall out-of-phase with its partner, producing a phasing effect. Once the operator took his finger off, the player would speed up until its tachometer was back in phase with the master, and as this happened, the phasing effect would appear to slide up the frequency spectrum. This phasing up-and-down the register can be performed rhythmically.
- *phaser* - another way of creating an unusual sound; the signal is split, a portion is filtered with an all-pass filter to produce a phase-shift, and then the unfiltered and filtered signals are mixed. The phaser effect was originally a simpler implementation of the flanger effect since delays were difficult to implement with analog equipment. Phasers are often used to give a "synthesized" or electronic effect to natural sounds, such as human speech. The voice of C-3PO from Star Wars was created by taking the actor's voice and treating it with a phaser.
- *chorus* - a delayed signal is added to the original signal with a constant delay. The delay has to be short in order not to be perceived as echo, but above 5 ms to be audible. If the delay is too short, it will destructively interfere with the un-delayed signal and create a flanging effect. Often, the delayed signals will be slightly pitch shifted to more realistically convey the effect of multiple voices.
- *equalization* - different frequency bands are attenuated or boosted to produce desired spectral characteristics. Moderate use of equalization (often abbreviated as "EQ") can be used to "fine-tune" the tone quality of a recording; extreme use of equalization, such as heavily cutting a certain frequency can create more unusual effects.
- *filtering* - Equalization is a form of filtering. In the general sense, frequency ranges can be emphasized or attenuated using low-pass, high-pass, band-pass or band-stop filters. Band-pass filtering of voice can simulate the effect of a telephone because telephones use band-pass filters.
- *overdrive* effects such as the use of a fuzz box can be used to produce distorted sounds, such as for imitating robotic voices or to simulate distorted radiotelephone traffic (e.g., the radio chatter between starfighter pilots in the science fiction film *Star Wars*). The most basic overdrive effect involves *clipping* the signal when its absolute value exceeds a certain threshold.
- *pitch shift* - this effect shifts a signal up or down in pitch. For example, a signal may be shifted an octave up or down. This is usually applied to the entire sig-

nal, and not to each note separately. Blending the original signal with shifted duplicate(s) can create harmonies from one voice. Another application of pitch shifting is pitch correction. Here a musical signal is tuned to the correct pitch using digital signal processing techniques. This effect is ubiquitous in karaoke machines and is often used to assist pop singers who sing out of tune. It is also used intentionally for aesthetic effect in such pop songs as Cher's *Believe* and Madonna's *Die Another Day*.

- *time stretching* - the complement of pitch shift, that is, the process of changing the speed of an audio signal without affecting its pitch.
- *resonators* - emphasize harmonic frequency content on specified frequencies. These may be created from parametric EQs or from delay-based comb-filters.
- *robotic voice effects* are used to make an actor's voice sound like a synthesized human voice.
- *synthesizer* - generate artificially almost any sound by either imitating natural sounds or creating completely new sounds.
- *modulation* - to change the frequency or amplitude of a carrier signal in relation to a predefined signal. Ring modulation, also known as amplitude modulation, is an effect made famous by Doctor Who's Daleks and commonly used throughout sci-fi.
- *compression* - the reduction of the dynamic range of a sound to avoid unintentional fluctuation in the dynamics. Level compression is not to be confused with audio data compression, where the amount of data is reduced without affecting the amplitude of the sound it represents.
- *3D audio effects* - place sounds outside the stereo basis
- *reverse echo* - a swelling effect created by reversing an audio signal and recording echo and/or delay while the signal runs in reverse. When played back forward the last echos are heard before the effected sound creating a rush like swell preceding and during playback. Jimmy Page of Led Zeppelin used this effect in the bridge of "Whole Lotta Love".
- *active noise control*- a method for reducing unwanted sound
- *wave field synthesis* - a spatial audio rendering technique for the creation of virtual acoustic environments

Digital Image Processing

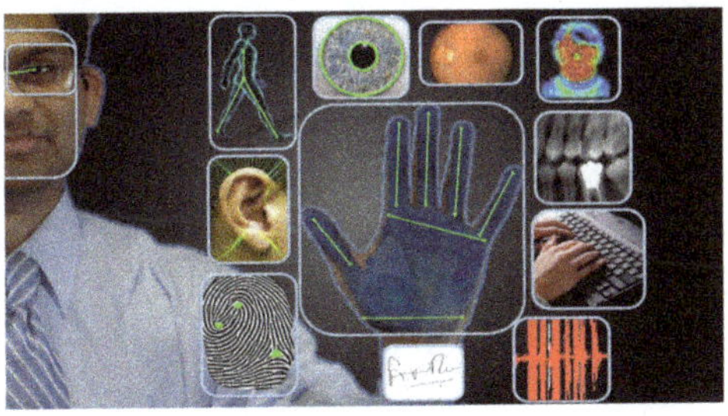

Digital image processing is the use of computer algorithms to perform image processing on digital images. As a subcategory or field of digital signal processing, digital image processing has many advantages over analog image processing. It allows a much wider range of algorithms to be applied to the input data and can avoid problems such as the build-up of noise and signal distortion during processing. Since images are defined over two dimensions (perhaps more) digital image processing may be modeled in the form of multidimensional systems.

History

Many of the techniques of digital image processing, or digital picture processing as it often was called, were developed in the 1960s at the Jet Propulsion Laboratory, Massachusetts Institute of Technology, Bell Laboratories, University of Maryland, and a few other research facilities, with application to satellite imagery, wire-photo standards conversion, medical imaging, videophone, character recognition, and photograph enhancement. The cost of processing was fairly high, however, with the computing equipment of that era. That changed in the 1970s, when digital image processing proliferated as cheaper computers and dedicated hardware became available. Images then could be processed in real time, for some dedicated problems such as television standards conversion. As general-purpose computers became faster, they started to take over the role of dedicated hardware for all but the most specialized and computer-intensive operations.

With the fast computers and signal processors available in the 2000s, digital image processing has become the most common form of image processing and generally, is used because it is not only the most versatile method, but also the cheapest.

Digital image processing technology for medical applications was inducted into the Space Foundation Space Technology Hall of Fame in 1994.

In 2002 Raanan Fattel, introduced Gradient domain image processing, a new way to process images in which the differences between pixels are manipulated rather than the pixel values themselves.

Tasks

Digital image processing allows the use of much more complex algorithms, and hence, can offer both more sophisticated performance at simple tasks, and the implementation of methods which would be impossible by analog means.

In particular, digital image processing is the only practical technology for:

- Classification
- Feature extraction
- Multi-scale signal analysis

- Pattern recognition
- Projection

Some techniques which are used in digital image processing include:

- Anisotropic diffusion
- Hidden Markov models
- Image editing
- Image restoration
- Independent component analysis
- Linear filtering
- Neural networks
- Partial differential equations
- Pixelation
- Principal components analysis
- Self-organizing maps
- Wavelets
- Applications
- Digital imaging

Digital Camera Images

Digital cameras generally include specialized digital image processing hardware – either dedicated chips or added circuitry on other chips – to convert the raw data from their image sensor into a color-corrected image in a standard image file format

Film

Westworld (1973) was the first feature film to use the digital image processing to pixellate photography to simulate an android's point of view.

Statistical Signal Processing

Statistical signal processing is an area of Applied Mathematics and Signal Processing that treats signals as stochastic processes, dealing with their statistical properties (e.g., mean, covariance, etc.). Because of its very broad range of application Statistical signal processing is taught at the graduate level in either Electrical Engineering, Applied Mathematics, Pure Mathematics/Statistics, or even Biomedical Engineering and Physics departments around the world, although important applications exist in almost all scientific fields.

Basic Signal Model

In many applications, a signal is modeled as functions consisting of both a deterministic and a stochastic component. A simple example and also a common model of many statistical systems is a signal $y(t)$ that consists of a deterministic part $x(t)$ added to noise which can be modeled in many situations as white Gaussian noise $w(t)$:

$$y(t) = x(t) + w(t)$$

where

$$w(t) \sim \mathcal{N}(0, \sigma^2)$$

White noise simply means that the noise process is completely uncorrelated. As a result, its autocorrelation function is an impulse:

$$R_{ww}(\tau) = \sigma^2 \delta(\tau)$$

where

$\delta(\tau)$ is the Dirac delta function.

Given information about a statistical system and the random variable from which it is derived, we can increase our knowledge of the output signal; conversely, given the statistical properties of the output signal, we can infer the properties of the underlying random variable. These statistical techniques are developed in the fields of estimation theory, detection theory, and numerous related fields that rely on statistical information to maximize their efficiency.

Example

The Computation of Average Transients (CAT) is used routinely in FT-NMR spectroscopy (nuclear magnetic resonance) to improve the signal-noise ratio of nmr spectra. The signal is measured repeatedly n times and then averaged.

$$\bar{y} = \frac{1}{n} \sum_i y(t)_i = x(t) + \frac{1}{n} \sum_i w(t)_i$$

Assuming that the noise is white and that its variance is constant in time it follows by error propagation that

$$\sigma(\bar{y}) = \frac{1}{\sqrt{n}}\sigma$$

Thus, if 10,000 measurements are averaged the signal to noise ratio is increased by a factor of 100, enabling the measurement of ^{13}C NMR spectra at natural abundance (1.1%) of ^{13}C.

References

- Atti, Andreas Spanias, Ted Painter, Venkatraman (2006). Audio signal processing and coding ([Online-Ausg.] ed.). Hoboken, NJ: John Wiley & Sons. p. 464. ISBN 0-471-79147-4.
- "Space Technology Hall of Fame:Inducted Technologies/1994". Space Foundation. 1994. Retrieved 7 January 2010.
- A Brief, Early History of Computer Graphics in Film, Larry Yaeger, 16 August 2002 (last update), retrieved 24 March 2010

Permissions

All chapters in this book are published with permission under the Creative Commons Attribution Share Alike License or equivalent. Every chapter published in this book has been scrutinized by our experts. Their significance has been extensively debated. The topics covered herein carry significant information for a comprehensive understanding. They may even be implemented as practical applications or may be referred to as a beginning point for further studies.

We would like to thank the editorial team for lending their expertise to make the book truly unique. They have played a crucial role in the development of this book. Without their invaluable contributions this book wouldn't have been possible. They have made vital efforts to compile up to date information on the varied aspects of this subject to make this book a valuable addition to the collection of many professionals and students.

This book was conceptualized with the vision of imparting up-to-date and integrated information in this field. To ensure the same, a matchless editorial board was set up. Every individual on the board went through rigorous rounds of assessment to prove their worth. After which they invested a large part of their time researching and compiling the most relevant data for our readers.

The editorial board has been involved in producing this book since its inception. They have spent rigorous hours researching and exploring the diverse topics which have resulted in the successful publishing of this book. They have passed on their knowledge of decades through this book. To expedite this challenging task, the publisher supported the team at every step. A small team of assistant editors was also appointed to further simplify the editing procedure and attain best results for the readers.

Apart from the editorial board, the designing team has also invested a significant amount of their time in understanding the subject and creating the most relevant covers. They scrutinized every image to scout for the most suitable representation of the subject and create an appropriate cover for the book.

The publishing team has been an ardent support to the editorial, designing and production team. Their endless efforts to recruit the best for this project, has resulted in the accomplishment of this book. They are a veteran in the field of academics and their pool of knowledge is as vast as their experience in printing. Their expertise and guidance has proved useful at every step. Their uncompromising quality standards have made this book an exceptional effort. Their encouragement from time to time has been an inspiration for everyone.

The publisher and the editorial board hope that this book will prove to be a valuable piece of knowledge for students, practitioners and scholars across the globe.

Index

www.ingramcontent.com/pod-product-compliance
Lightning Source LLC
Chambersburg PA
CBHW061930190326
41458CB00009B/2710
9781635490879